String Theory

2nd edition

by Andrew Zimmerman Jones and Alessandro Sfondrini

String Theory For Dummies®, 2nd edition

Published by
Wiley Publishing, Inc.
111 River St.
Hoboken, NJ 07030-5774
www.wiley.com

Published by Wiley Publishing, Inc., Indianapolis, Indiana

Published simultaneously in Canada

For general information on our other products and services, please contact our Customer Care Department within the U.S. at 877-762-2974, outside the U.S. at 317-572-3993, or fax 317-572-4002.

For technical support, please visit https://hub.wiley.com/community/support/dummies.

Wiley also publishes its books in a variety of electronic formats. Some content that appears in print may not be available in electronic books.

Library of Congress Control Number: 2022939030

ISBN: 978-1-119-88897-0 (pbk); ISBN: 978-1-119-88898-7 (ebk); ISBN: 978-1-119-88899-4 (ebk)

SKY10061362_112923

Contents at a Glance

Table of Contents

Introduction

Why are scientists so excited about string theory? Because string theory is the most likely candidate for a successful theory of quantum gravity — a theory that scientists hope will unite two major physical laws of the universe into one. Right now, these laws (quantum physics and general relativity) describe two totally different types of behavior in totally different ways, and in the realm where neither theory works completely, we really don't know what's going on!

Understanding the implications of string theory means understanding profound aspects of our reality at the most fundamental levels. Is there only one law of nature or infinitely many? Why does our universe follow the laws it does? Is time travel possible? How many dimensions does our universe possess? Physicists are passionately seeking answers to these questions.

Indeed, string theory is a fascinating topic, a scientific revolution that promises to transform our understanding of the universe. As you'll see, these types of revolutions have happened before, and this book helps you understand how physics has developed in the past, as well as how it may develop in the future.

This book contains some ideas that will probably, in the coming years, turn out to be completely false. (We can guarantee this was true about the first edition, and we have no reason to expect it won't also be true of this second edition.) It contains other ideas that may ultimately prove to be fundamental laws of our universe, perhaps forming the foundation for entirely new fields of science and technology. No one knows what the future holds for string theory.

About This Book

In this book, we aim to give a clear understanding of the ever-evolving scientific subfield known as string theory. The media is abuzz with talk about this "theory of everything," and when you're done with this book, you should know what they're talking about (probably better than they do, most of the time).

In writing this book, we've attempted to serve several masters. First and foremost among them has been scientific accuracy, followed closely by entertainment value. Along the way, we've also done our best to use language that you can understand no matter your scientific background, and we've certainly tried to keep any mathematics to a minimum.

We set out to achieve the following goals with this book:

- Provide the information needed to understand string theory (including established physics concepts that predate string theory).
- Establish the successes of string theory so far.
- Lay out the avenues of study that are attempting to gain more evidence for string theory.
- Explore the bizarre (and speculative) implications of string theory.
- Present the critical viewpoints in opposition to string theory, as well as some alternatives that may bear fruit if string theory proves to be false.
- Have some fun along the way.
- Avoid mathematics at all costs. (You're welcome!)

We hope you, good reader, find that we've been successful at meeting these goals.

And while time may flow in only one direction (Or does it? We explore this in Chapter 17), your reading of this book may not. String theory is a complex scientific topic that includes a lot of interconnected concepts, so jumping between concepts isn't quite as easy as it may be in some other *For Dummies* reference books. We've tried to help you out by including quick reminders and providing cross-references to other chapters where necessary. So feel free to wander the pages to your heart's content, knowing that if you get lost, you can work your way back to the information you need.

Foolish Assumptions

About the only assumption we've made in writing this book is that you're reading it because you want to know something about string theory. We've even tried not to assume that you *enjoy* reading physics books. (We do, but we try not to project our own strangeness onto others.)

We have assumed that you have a passing acquaintance with basic physics concepts — maybe you took a physics class in high school or have watched some of the scientific programs about gravity, light waves, black holes, or other physics-related topics on cable channels or your local PBS station. You don't need a degree in physics to follow the explanations in this book, although without a degree in physics you may be amazed that anyone can make sense of any theory so disconnected from our everyday experience. (Even with a physics degree, it can boggle the mind.)

As is customary in string theory books for the general public, the mathematics has been avoided. You need a graduate degree in mathematics or physics to follow the mathematical equations at the heart of string theory, and we've assumed that you don't have either one. Don't worry — while a complete understanding of string theory is rooted firmly in the advanced mathematical concepts of geometry and quantum field theory, we've used a combination of text and figures to explain the fascinating ideas behind string theory.

Icons Used in This Book

Throughout the book, you'll find icons in the margins that are designed to help you navigate the text. Here's what these icons mean:

REMEMBER

Although everything in this book is important, some information is more important than other information. This icon points out information that will definitely be useful later in the book.

TIP

In science, theories are often explained with analogies, thought experiments, or other helpful examples that present complex mathematical concepts in a way that is more intuitively understandable. This icon indicates that one of these examples or hints is being offered.

TECHNICAL STUFF

Sometimes we go into detail that you don't need to know to follow the basic discussion and that's a bit more technical (or mathematical) than you may be interested in. This icon points out that information, which you can skip without losing the thread of the discussion.

Beyond the Book

In addition to what you're reading right now, this book also comes with a free access-anywhere Cheat Sheet. To get it, simply go to www.dummies.com and look for String Theory for Dummies Cheat Sheet in the Search box.

If you want to learn more about some of the ideas that laid the basis or string theory, you can also check out *Einstein for Dummies* by Carlos I. Calle.

Where to Go from Here

The *For Dummies* books are organized in such a way that you can surf through any of the chapters and find useful information without having to start at Chapter 1. We (naturally) encourage you to read the whole book, but this structure makes it very easy to start with the topics that interest you the most.

If you have no idea what string theory is, then we recommend looking at Chapter 1 as a starting point, then moving through Chapters 2–3 for a basic overview of what we're talking about. Chapter 4 focuses on laying some foundational ideas about how theoretical science advances. If your physics is rusty, pay close attention to Chapters 5–9, which cover the history and current status of the major physics concepts that pop up over and over again.

If you're familiar with string theory but want some more details, jump straight to Chapters 10 and 11, where we explain how string theory came about and reached its current status. Chapters 12 and 13 go a bit deeper into the specifics, including the recent insights from the holographic principle. Chapter 14 offers some ways of testing the theory, while Chapters 15–17 take concepts from string theory and apply them to some fascinating topics in theoretical physics.

Some of you, however, may want to figure out what all the recent fuss is with people arguing across the blogosphere about string theory. For that, we recommend jumping straight to Chapter 18, which addresses some of the major criticisms of string theory. Chapters 19 and 20 focus heavily on other theories that may either help expand or replace string theory, so they're a good place to go from there.

1
Introducing String Theory

IN THIS PART . . .

Understand the basics of string theory.

Grasp the fundamentals of quantum gravity.

Explore the accomplishments and failures of string theory.

IN THIS CHAPTER

» Knowing that string theory is based on vibrating strings of energy

» Understanding the key elements of string theory

» Hoping to explain the entire universe with string theory

» Studying string theory could be the driving scientific goal of the 21st century

Chapter 1
So What Is String Theory Anyway?

String theory is a work in progress, so trying to pin down exactly what string theory is, or what its fundamental elements are, can be kind of tricky. Regardless, that's exactly what we try to do in this chapter.

In this chapter, you gain a basic understanding of string theory. We outline the key elements of string theory, which provide the foundation for most of this book. We also discuss the possibility that string theory is the starting point for a "theory of everything," which would define all of our universe's physical laws in one simple (or not so simple) mathematical formula. Finally, we look at the reasons why you should care about string theory.

String Theory: Seeing What Vibrating Strings Can Tell Us about the Universe

String theory is a physics theory that models the fundamental particles and interactions in the universe by representing everything in terms of vibrating filaments of energy, called strings. Like all modern physical theories, this image is actually expressed in a precise mathematical language that eventually results in quantitative as well as qualitative predictions.

In this theory, *strings* of energy represent the most fundamental aspect of nature. String theory also predicts other fundamental objects, called *branes,* which emerge as a natural generalization of the strings. All the matter in our universe consists of the vibrations of these strings (and branes). One important result of string theory is that gravity is a natural consequence of the theory, which is why scientists believe that string theory may hold the answer to possibly uniting gravity with the other forces that affect matter.

TIP

We want to reiterate something important: String theory is a *mathematical* theory. It's based on mathematical equations that can be interpreted in certain ways. If you've never studied physics before, this may seem odd, but *all* physical theories are expressed in the language of mathematics. In this book, we avoid the mathematics and try to get to the heart of what the theory is telling us about the physical universe.

REMEMBER

At present, no one knows exactly what the "final" version of string theory, which will precisely reproduce the universe as we know it, should look like. Scientists have some vague notions about the general elements that will exist within the theory, but no one has come up with the final list of equations that represents all of string theory in our universe, and experiments haven't yet been able to confirm it (though they haven't successfully refuted it, either). Physicists have created simplified versions of a stringy universe, but none quite describes our universe . . . yet.

Using tiny and huge concepts to create a theory of everything

String theory is a type of high-energy theoretical physics, practiced largely by particle physicists. It's an evolution of *quantum field theory* (see the sidebar "What is quantum field theory?"), which is the current framework that describes the particles and forces in our universe (except gravity). String theory famously predicts that the universe should have more spatial dimensions than the three we observe. It also shows that, in principle, the extra dimensions within the theory can be wrapped up into a very small size (a process called *compactification*) in a way that reproduces fundamental particles like the photon or the electron. This is the power of string theory — using the fundamental strings, and the way extra dimensions are compactified, to provide a unified description of all the particles and forces known to modern physics.

Among the forces that need to be described is, of course, gravity. Superficially, gravity has been the simplest force for humans to grasp since the time of Galileo. There is, however, more than meets the eye, as Einstein discovered: Gravity is a theory of the geometry of space and time. For this reason, it's notoriously hard to marry the ideas of quantum physics with gravity. String theory does incorporate both gravity and quantum physics in a natural way. You can even say that *string theory is a theory of quantum gravity* because it's impossible to construct any string theory without gravity.

Still, not every aspect of gravity is understood from string theory. Importantly, the established theory of gravity, general relativity, has a fluid, dynamic space-time, and one aspect of string theory that's still being worked on is getting that type of space-time to emerge out of the theory.

WHAT IS QUANTUM FIELD THEORY?

Physicists use *fields* to describe the things that don't just have a particular position but exist at every point in space. For example, you can think about the temperature in a room as a field — it may be different near an open window than near a hot stove, and you could imagine measuring the temperature at every single point in the room. A *field theory,* then, is a set of rules that tell you how some field will behave, such as how the temperature in the room changes over time.

In Chapters 7 and 8, you find out about one of the most important achievements of the 20th century: the development of *quantum theory*. This refers to principles that lead to seemingly bizarre physical phenomena that nonetheless appear to occur in the subatomic world.

When you combine these two concepts, you get *quantum field theory:* a field theory that obeys the principles of quantum theory. All modern particle physics is described by quantum field theories.

The major achievements of string theory are concepts you can't see, unless you know how to interpret the physics equations. String theory deals with rather extreme amounts of energy; that's why it's hard to test its predictions directly with experiments. Yet it has revealed profound mathematical relationships within the equations, which leads physicists to believe that they must be true. We discuss these properties and relationships — known by jargon that describes various symmetries and dualities, the cancellation of anomalies, and the explanation of black hole entropy — in Chapters 10 and 11.

In recent years, there has been much public debate over string theory, waged within newsrooms and across the internet. We address these issues in Part 5, but they come down to fundamental questions about how science should be pursued. String theorists believe that their methods are sound, while the critics believe they're questionable because they stray too far from contact with experimentation — the true core of physics. Time, and experimental evidence, will tell which side has made the better argument.

A quick look at where string theory has been

String theory was originally developed in 1968 as an attempt to explain the behavior of *hadrons* (such as protons and neutrons, the particles that make up an atomic nucleus) inside particle accelerators. Physicists later realized this theory could also be used to explain some aspects of gravity. For more than a decade, string theory was abandoned by most physicists, mainly because it required a large number of extra, unseen dimensions. It rose to prominence again in the mid-1980s, when physicists were able to prove it was a mathematically consistent theory.

In the mid-1990s, string theory was updated to become a more complex theory, called *M-theory,* which contains more objects than just strings. These new objects were called *branes,* and they could have anywhere from zero to nine dimensions. The earlier string theories (which now also include branes) were seen as approximations of the more complete M-theory.

REMEMBER

Technically, the modern M-theory is more than the traditional string theory, but the name "string theory" is still often used for M-theory and its various offspring theories. (Even the original superstring theories have been shown to include branes.) Our convention in this book is to refer to theories that contain branes, which are variants of M-theory and the original string theories, using the term "string theory."

Introducing the Key Elements of String Theory

Five key ideas are at the heart of string theory and come up again and again. It's best for you to become familiar with these key concepts right off the bat.

- String theory predicts that all objects in our universe are composed of vibrating filaments (and membranes) of energy.
- String theory attempts to reconcile general relativity (gravity) with quantum physics.
- String theory provides a way of unifying all the fundamental forces of the universe.
- String theory predicts a new connection (called *supersymmetry*) between two fundamentally different types of particles, bosons and fermions.
- String theory predicts a number of extra (usually unobservable) dimensions to the universe.

We introduce you to the very basics of these ideas in the following sections.

Strings and branes

When the theory was originally developed in the 1970s, the filaments of energy in string theory were considered to be one-dimensional objects: strings. (*One-dimensional* indicates that a string has only one dimension, length, as opposed to, say, a square, which has both length and height dimensions.)

These strings came in two forms: closed strings and open strings. An open string has ends that don't touch each other, while a closed string is a loop with no open end. It was eventually found that these early strings, called Type I strings, could go through five basic types of interactions, as Figure 1-1 shows.

TIP

The interactions are based on a string's ability to have its ends join and split apart. Because the ends of open strings can join together to form closed strings, you can't construct a string theory without closed strings. This is a manifestation of the *dualities* of string theory, which you will encounter in Chapter 11 and that resulted in the proposal of M-theory.

REMEMBER

This proved to be important because closed strings have properties that make physicists believe they might describe gravity! In other words, physicists began to realize that instead of just being a theory of matter particles, string theory may be able to explain gravity and the behavior of particles.

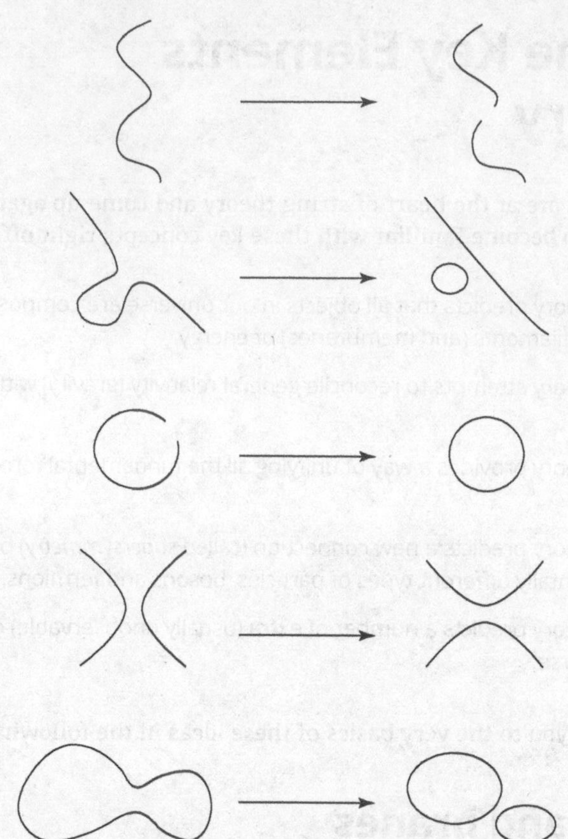

FIGURE 1-1: Type I strings can go through five fundamental interactions, based on different ways of joining and splitting.

Over the years, it was discovered that the theory required objects other than strings. These objects can be seen as sheets, or *branes*. Strings can attach at one or both ends to these branes. Figure 1-2 shows a 2-dimensional brane (called a 2-brane). (See Chapter 11 for more about branes.)

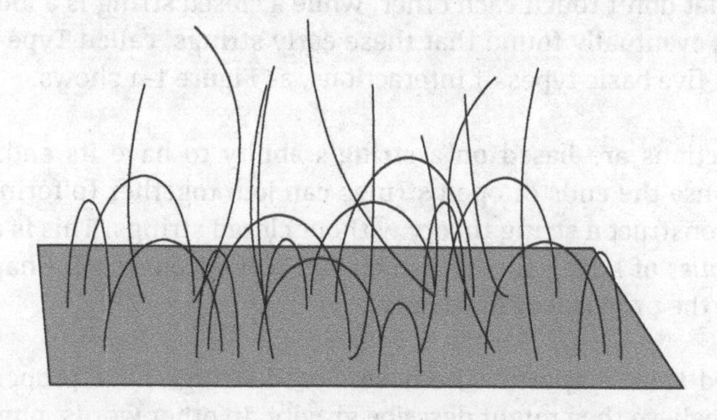

FIGURE 1-2: In string theory, strings attach themselves to branes.

Quantum gravity

Modern physics has two basic scientific laws: quantum physics and general relativity. These two scientific laws represent radically different fields of study. *Quantum physics* studies the very smallest objects in nature, while *relativity* tends to study nature on the scale of planets, galaxies, and the universe as a whole. (Obviously, gravity affects small particles too, and relativity accounts for this as well, but the effect is usually tiny.) Theories that attempt to unify quantum physics and relativity are theories of *quantum gravity*, and the most promising of all such theories today is string theory.

The closed strings of string theory (see the preceding section) correspond to the behavior expected for gravity. Specifically, they have properties that match the long-sought-after *graviton*, a particle that would carry the force of gravity between objects.

Quantum gravity is the subject of Chapter 2, where we cover this idea in much greater depth.

Unification of forces

Hand in hand with the question of quantum gravity, string theory attempts to unify the four forces in the universe — electromagnetic force, the strong nuclear force, the weak nuclear force, and gravity — together into one unified theory. In our universe, these fundamental forces appear as four different phenomena, but string theorists believe that in the early universe (when there were incredibly high energy levels), these forces are all described by different types of strings interacting with each other.

That such a unification may be possible isn't entirely surprising to physicists because they discovered 50 years ago that two of the forces are actually one and the same: The electromagnetic force and the weak force can be combined in the "electroweak" force. (If you've never heard of some of these forces, don't worry! We discuss them individually in greater detail in Chapter 2 and throughout Part 2.)

Supersymmetry

All particles in the universe can be divided into two types: bosons and fermions. (These types of particles are explained in more detail in Chapter 8.) String theory predicts that a type of connection, called *supersymmetry*, exists between these two particle types. Under supersymmetry, a fermion must exist for every boson and a boson for every fermion. Unfortunately, experiments have not yet detected these extra particles. (The latest particle that physicists have found is the Higgs boson, which is not one of the supersymmetric partners.)

Supersymmetry is a specific mathematical relationship between certain elements of physics equations. It was discovered outside string theory, although its incorporation into string theory transformed the theory into supersymmetric string theory (or superstring theory) in the mid-1970s. (See Chapter 10 for more specifics about supersymmetry.)

One benefit of supersymmetry is that it balances out string theory's equations by allowing certain terms to cancel out. Without supersymmetry, the equations result in physical inconsistencies, such as infinite values and imaginary energy levels.

Because scientists haven't observed the particles predicted by supersymmetry, this is still a theoretical assumption. Many physicists believe that the reason no one has observed the particles is because it takes a lot of energy to generate them. (Energy is related to mass by Einstein's famous $E = mc^2$ equation, so it takes energy to create a particle.) They may have existed in the early universe, but as the universe cooled off and energy spread out after the big bang, these particles would have collapsed into the lower-energy states that we observe today. (We may not think of our current universe as particularly low energy, but compared to the intense heat of the first few moments after the big bang, it certainly is.)

TIP

In other words, the strings vibrating as higher-energy particles lost energy and transformed from one type of particle (one type of vibration) into another, lower-energy type of vibration.

Scientists hope that astronomical observations or experiments with particle accelerators will uncover some of these higher-energy supersymmetric particles, providing support for this prediction of string theory.

Extra dimensions

Another mathematical result of string theory is that the theory makes sense only in a world with more than three space dimensions! (Our universe has three dimensions of space: left/right, up/down, and front/back.) Two possible explanations currently exist for the location of the extra dimensions.

- The extra space dimensions (generally six of them) are curled up (*compactified,* in string theory terminology) to incredibly small sizes, so we never perceive them.
- We are stuck on a 3-dimensional brane, and the extra dimensions extend off of it and are inaccessible to us.

A major area of research among string theorists is on mathematical models of how these extra dimensions could be related to our own. Some of the recent results have predicted that scientists may soon be able to detect the extra dimensions (if they exist) in upcoming experiments because they may be larger than previously expected. (See Chapter 15 for more about extra dimensions.)

Understanding the Aim of String Theory

To many physicists, the goal of string theory is to be a "theory of everything" — that is, to be the single physical theory that, at the most fundamental level, describes all of physical reality. If successful, string theory could explain many of the fundamental questions about our universe.

To others, the goal is more modest: String theory is a working theory of quantum gravity, and arguably the only one we truly understand. Studying string theory can produce important insights into the nature of quantum gravity, one of the key open questions in physics.

Quantizing gravity

The major accomplishment of string theory is providing a quantum theory of gravity. The current theory of gravity, general relativity, doesn't allow for the results of quantum physics. Because quantum physics places limitations on the behavior of small objects, it creates major inconsistencies when we're trying to examine the universe at extremely small scales. (See Chapter 7 for more on quantum physics.)

Therefore, the fact that string theory manages to marry general relativity and quantum physics is by itself remarkable. Not only that, but it has also led to spectacular advances in our understanding of quantum gravity, including the holographic principle, which you find in Chapter 13.

Unifying forces

Currently, four fundamental forces (more precisely called "interactions" by physicists) are known to physics: gravity, electromagnetic force, weak nuclear force, and strong nuclear force. String theory creates a framework in which all four of these interactions were once a part of the same unified force of the universe.

Under this theory, as the early universe cooled off after the big bang, this unified force began to break apart into the different forces we experience today. Experiments at high energies may someday allow us to detect the unification of these forces, although such experiments are well outside our current realm of technology.

Explaining matter and mass

One of the major goals of current string theory research is to construct a solution of string theory that contains the particles that actually exist in our universe.

String theory started out as a theory to explain particles, such as hadrons, as the different higher vibrational modes of a string. In most current formulations of string theory, the matter observed in our universe comes from the lowest-energy vibrations of strings and branes. (The higher-energy vibrations represent more energetic particles that don't currently exist in our universe if not for a very short time.)

The mass of these fundamental particles comes from the ways that these strings and branes are wrapped in the extra dimensions that are compactified within the theory, in ways that are rather messy and detailed.

For example, consider a simplified case where the extra dimensions are curled up in the shape of a donut (called a *torus* by mathematicians and physicists), as in Figure 1-3.

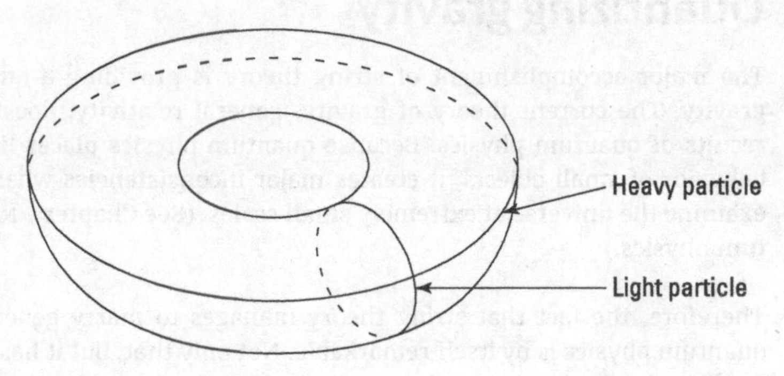

FIGURE 1-3: Strings wrap around extra dimensions to create particles with different masses.

A string has two ways to wrap once around this shape:

- A short loop around the tube, through the middle of the donut
- A long loop wrapping around the entire length of the donut (like a string wraps around a yo-yo)

TIP

The short loop would be a lighter particle, while the long loop is a heavier particle. As you wrap strings around the torus-shaped compactified dimensions, you get new particles with different masses.

One of the major reasons string theory has caught on is that this idea — that length translates into mass — is so straightforward and elegant. The compactified dimensions in string theory are much more elaborate than a simple torus, but they work the same way in principle.

It's even possible (though harder to visualize) for a string to wrap in both directions simultaneously — which would, again, give us yet another particle with yet another mass. Branes can also wrap around extra dimensions, creating even more possibilities.

Defining space and time

In many versions of string theory, the extra dimensions of space are compactified into a very tiny size, so they're unobservable to our current technology. Trying to look at space smaller than this compactified size would provide results that don't match our understanding of space-time. (As you see in Chapter 2, the behavior of space-time at these small scales is one of the reasons for a search for quantum gravity.) One of string theory's major obstacles is attempting to figure out how space-time can emerge from the theory.

As a rule, though, string theory is built upon Einstein's notion of space-time (see Chapter 6). Einstein's theory has three space dimensions and one time dimension. String theory predicts a few more space dimensions but doesn't change the fundamental rules of the game all that much, at least at low energies.

At present, it's unclear whether string theory can make sense of the fundamental nature of space and time any more than Einstein did. In string theory, it's almost as if the space and time dimensions of the universe are a backdrop to the interactions of strings, with no real meaning on their own.

Some proposals for how to address this have been developed, mainly focusing on space-time as an emergent phenomenon — that is, the space-time comes out of the sum total of all the string interactions in a way that hasn't yet been completely worked out within the theory.

However, these approaches don't meet some physicists' bar for compelling scientific evidence, leading to criticism of the theory. String theory's biggest competitor, loop quantum gravity, uses the quantization of space and time as the starting point of its own theory, as Chapter 19 explains. Some believe that this will ultimately be another approach to the same basic theory.

Appreciating the Theory's Amazing (and Controversial) Implications

Although string theory is fascinating in its own right, what may prove to be even more intriguing are the possibilities that result from it. These topics are explored in greater depth throughout the book and are the focus of Parts 3 and 4.

Landscape of possible theories

One of the most unexpected and disturbing discoveries of string theory is that instead of one single model of the universe, it turns out that there may be a huge number of possible models (or, more precisely, possible solutions to the theory) — maybe as many as 10^{500} different solutions! (That's a 1 followed by 500 zeroes!) While this huge number has prompted a crisis among some string theorists, others have embraced it as a virtue, claiming it means that string theory is very rich.

In order to wrap their minds around so many possible models, some string theorists have turned toward the *anthropic principle*, which tries to explain properties of our universe as a result of our presence in it. Others instead argue that the vast number of possible string models is to be expected — or even that *it's a feature, not a bug*, so to say. For them, it's just a matter of nailing down the particular string solution that does describe our universe.

With such a large number of theories available, the anthropic principle allows a physicist to use the fact that we're here to choose among only those theories that have physical parameters that allow us to be here. In other words, our very presence dictates the choice of physical law — or is it merely that our presence is an observable piece of data, like the speed of light?

REMEMBER

The use of the anthropic principle is one of the most controversial aspects of modern string theory. Even some of the strongest string theory supporters have expressed concern over its application because of the sordid (and somewhat unscientific) applications it has been used for in the past and their feeling that all that's needed is an observation of our universe, without anything anthropic applied at all.

As anthropic-principle skeptics are quick to point out, physicists adopt the anthropic principle only when they have no other options, and they abandon it if something better comes along. It remains to be seen if string theorists will find another way to maneuver through the string theory landscape. (Chapter 12 has more details about the anthropic principle.)

The universe as a hologram

In the mid-1990s, two physicists came up with an idea called the *holographic principle*. In this theory, if you have a volume of space, you can take all the information contained in that space and show that it corresponds to information "written" on the surface of the space. As odd as it seems, this holographic principle may be key in resolving a major mystery of black holes that has existed for more than 30 years!

Many physicists believe that the holographic principle will be one of the fundamental physical principles that will allow insights into a greater understanding of string theory. (Check out Chapter 13 for more on the holographic principle.)

Why Is String Theory So Important?

String theory yields many fascinating subjects for thought, but you may be wondering about the practical importance of it. For one thing, string theory is the next step in our growing understanding of the universe. If that's not practical enough, then there's this consideration: Your tax money goes to fund scientific research, and (a tiny fraction of) the people trying to get that money want to use it to study string theory (or its alternatives).

A completely honest string theorist would be forced to say that there are probably no practical applications for string theory, at least in the foreseeable future. This admission doesn't look that great on either the cover of a book or at the top of a webpage, so it gets spiced up with talk about discovering parallel universes, extra time dimensions, and new fundamental symmetries of nature. They might exist, but the theory's predictions make it seem that they're unlikely to ever be particularly useful, so far as we know.

Better understanding the nature of the universe is a good goal in its own right — as old as humanity, some might say — but when you're looking at funding multibillion-dollar particle accelerators or research satellite programs, you might want something tangible for your money. Unfortunately, there's no reason to think that string theory is going to give you anything practical.

Does this mean that exploring string theory isn't important? No, and it's our hope that reading Part 2 of this book will help illuminate the key at the heart of the search for string theory, or any new scientific truth.

REMEMBER

No one knows where a scientific theory will lead until the theory is developed and tested.

In 1905, when Albert Einstein first presented his famous equation $E = mc^2$, he thought it was an intriguing relationship, but he had no idea that it would result in something as potent as the atomic bomb. He had no way of knowing that the corrections to time calculations demanded by special relativity and general relativity would someday be required to get the worldwide global positioning system (GPS) to operate accurately (more on GPS in Chapter 6).

Quantum physics, which on the surface is about as theoretical a study as they come, is the basis for the laser and the transistor, two pieces of technology that are at the heart of modern computers and communication systems.

Even though we don't know what a purely theoretical concept like string theory may lead to, history has shown that it will almost certainly lead somewhere profound.

For an example of the unexpected nature of scientific progress, consider the discovery and study of electricity, which was originally seen as a mere parlor trick. To be sure, you could predict some technologies from the discovery of electricity, such as the lightbulb. But some of the most profound discoveries are things that may never have been predicted — radio and television, the computer, the internet, the cell phone, and so on.

The impact of science extends into culture as well. Another by-product of electricity is rock and roll music, which was created with the advent of electric guitars and other electric musical instruments.

If electricity can lead to rock and roll and the internet, then imagine what sort of unpredicted (and potentially unpredictable) cultural and technological advances string theory could lead to!

IN THIS CHAPTER

» Squaring off: Gravity and quantum physics just don't get along

» Seeing four types of particle interactions

» Hoping to tie all of physics into one equation with quantum gravity

Chapter 2
The Physics Road Dead-Ends at Quantum Gravity

Physicists like to group concepts together into neat little boxes with labels, but sometimes the theories they try to put together just don't want to get along. Right now, nature's fundamental physical laws can fit into one of two boxes: general relativity or quantum physics. But concepts from one box just don't work together well with concepts from the other box.

Any theory that can get these two physics concepts to work together would be called a *theory of quantum gravity.* String theory is currently the most likely candidate for a successful theory of quantum gravity.

In this chapter, we explain why scientists want (and need) a theory of quantum gravity. We begin by giving an overview of the scientific understanding of gravity, which is defined by Einstein's theory of general relativity, and our understanding of matter and the other forces of nature, in terms of quantum mechanics. With these fundamental tools in place, we then explain the ways in which these two theories clash with each other, which provides the basis for quantum gravity. Finally, we outline various attempts to unify these theories and the forces of physics into one coherent system, and the failures they've run into.

Understanding Two Schools of Thought on Gravity

Physicists are searching for a theory of quantum gravity because the current laws governing gravity don't work in all situations. Specifically, the theory of gravity seems to "break down" (that is, the equations become physically meaningless) in certain circumstances that we describe later in the chapter. To understand what this means, you must first understand a bit about what physicists know about gravity.

Gravity is an attractive force that binds objects together, seemingly across any amount of distance. The formulation of the classical theory of gravity by Sir Isaac Newton was one of the greatest achievements of physics. Two centuries later, the reinvention of gravity by Albert Einstein placed him in the pantheon of indisputably great scientific thinkers.

Unless you're a physicist, you probably take gravity for granted. It's an amazing force, able to hold the heavens together while being overcome by a 3-year-old on a swing — but not for long. At the scale of an atom, gravity is irrelevant compared to the electromagnetic force. In fact, a simple magnet can overcome the entire force of the planet Earth to pick up metallic objects, from paper clips to automobiles.

Newton's law of gravity: Gravity as force

Sir Isaac Newton developed his theory of gravity in the late 1600s. This amazing theory involved bringing together an understanding of astronomy and the principles of motion (known as *mechanics* or *kinematics*) into one comprehensive framework that also required the invention of a new form of mathematics: calculus. In Newton's gravitational theory, objects are drawn together by a physical force that spans vast distances of space.

The key is that gravity binds all objects together (much like the Force in *Star Wars*). The apple falling from a tree and the moon's motion around Earth are two manifestations of the exact same fundamental force.

The relationship that Newton discovered was a mathematical relationship (he did, after all, have to invent calculus to get it all to work out), just like relativity, quantum mechanics, and string theory.

A MATTER OF MASS

When we say that the force between objects is proportional to the mass of the two objects, you may think this means that heavier things fall faster than lighter things. For example, wouldn't a bowling ball fall faster than a soccer ball?

In fact, as Galileo showed (though not with modern bowling and soccer balls) years before Newton was born, this isn't the case. For centuries, most people had assumed that heavier objects fall faster than light objects. Newton was aware of Galileo's results, which was why he was able to figure out how to define force the way he did.

By Newton's explanation, it takes more force to move a heavier object. If you dropped a bowling ball and a soccer ball off a building (which we don't recommend), they would accelerate at the exact same rate (ignoring air resistance) — approximately 9.8 meters per second squared.

The force acting between the bowling ball and Earth would be higher than the force acting on the soccer ball, but because it takes more force to get the bowling ball moving, the actual rate of acceleration is identical for the two balls.

Realistically, if you performed the experiment, there would be a slight difference. Because of air resistance, the lighter soccer ball would probably be slowed down if dropped from a high enough point, while the bowling ball would not. But a properly constructed experiment, in which air resistance is completely neutralized (like in a vacuum), shows that the objects fall at the same rate, regardless of their mass.

In Newton's gravitational theory, the force between two objects is based on the product of their masses, divided by the square of the distance between them. In other words, the heavier the two objects are, the more force there is between them, assuming the distance between them stays the same. (See the sidebar "A matter of mass" for clarification of this relationship.)

TIP

The fact that the force is divided by distance squared means that if the same two objects are closer to each other, the power of gravity increases. If the distance gets wider, the force drops. The inverse square relationship means that if the distance doubles, the force drops to one-fourth of its original intensity. If the distance is halved, the force increases by four times.

If the objects are very far away from each other, the effect of gravity becomes very small. The reason gravity has any impact on the universe is because there's *a lot* of it. Gravity itself is very weak, as forces go.

The opposite is true as well. If two objects get extremely close to each other — and we're talking *extremely* close here — then gravity can become incredibly powerful, even among objects that don't have much mass, like the fundamental particles of physics.

This isn't the only reason gravity is observed so much. Gravity's strength in the universe also comes from the fact that it's always attracting objects together. The electromagnetic force sometimes attracts objects and sometimes repulses them, so on the scale of the universe at large, it tends to counteract itself. Finally, gravity interacts at very large distances, as opposed to some other forces (the nuclear forces) that only work at distances smaller than an atom.

We delve a bit deeper into Newton's work, both in gravity and in related areas, in Chapter 5.

Despite the success of Newton's theory, he had a few nagging problems in the back of his mind. First and foremost among them was the fact that although he had a model for gravity, he didn't know *why* gravity works. The gravity that he described was an almost mystical force (like the Force!), acting across great distances with no real physical connection required. It would take two centuries and Albert Einstein to resolve this problem.

Einstein's law of gravity: Gravity as geometry

Albert Einstein would revolutionize the way physicists see gravity. Instead of thinking of gravity as a force acting between objects, Einstein envisioned a universe in which each object's mass caused a slight bending of space (actually space-time) around it. The movement of an object along the shortest distance in this space-time was the net effect of gravity. Instead of being a force, gravity was actually the result of the geometry of space-time itself.

Einstein proposed that motion in the universe could be explained in terms of a coordinate system with three space dimensions — up/down, left/right, and backward/forward, for example — and one time dimension. This 4-dimensional coordinate system, developed by Hermann Minkowski, Einstein's former professor, was called *space-time* and came out of Einstein's 1905 *theory of special relativity.*

As Einstein generalized this theory, creating the *theory of general relativity* in 1915, he was able to include gravity in his explanation of motion. In fact, the concept of space-time was crucial to it. The space-time coordinate system bent when matter was placed in it. As objects moved within space and time, they naturally tried to take the shortest path through the bent space-time.

We follow our orbit around the sun because it's the shortest path (called a *geodesic* in mathematics) through the curved space-time around the sun.

Einstein's relativity is covered in depth in Chapter 6, and the major implications of relativity to the evolution of the universe are covered in Chapter 9. The space-time dimensions are discussed in Chapter 15.

Describing Matter: Physical and Energy-Filled

Einstein helped to revolutionize our ideas about the composition of matter as much as he changed our understanding of space, time, and gravity. Thanks to Einstein, scientists realize that mass — and therefore matter itself — is a form of energy. This realization is at the heart of modern physics. Because gravity is an interaction between objects made up of matter, understanding matter is crucial to understanding why physicists need a theory of quantum gravity.

Viewing matter classically: Chunks of stuff

The study of matter is one of the oldest physics disciplines, going back to the ancient philosophers who tried to understand what made up the objects around them. Even fairly recently, a physical understanding of matter was elusive as physicists debated the existence of *atoms* — tiny, indivisible chunks of matter that can't be broken up any further.

One key physics principle is that matter can be neither created nor destroyed, but can only change from one form to another. This principle is known as the *conservation of mass.*

Though it can't be created or destroyed, matter can be broken, which led to the question of whether there was a smallest chunk of matter, the atom, as the ancient Greeks had proposed — a question that, throughout the 1800s, seemed to point toward an affirmative answer.

As an understanding of *thermodynamics* — the study of heat and energy, which made things like the steam engine (and the Industrial Revolution) possible — grew, physicists began to realize that heat can be explained as the motion of tiny particles.

The atom had returned, though the findings of 20th-century quantum physics would reveal that it wasn't indivisible, as everyone thought.

Viewing matter at a quantum scale: Chunks of energy

With the rise of modern physics in the 20th century, two key facts about matter became clear:

- As Einstein had proposed with his famous $E = mc^2$ equation, matter and energy are, in a sense, interchangeable.
- Matter is incredibly complex, made up of an array of bizarre and unexpected types of particles that join together to form other types of particles.

The atom, it turns out, is composed of a nucleus surrounded by electrons. The nucleus is made up of protons and neutrons, which are in turn made up of strange new particles called quarks! As soon as physicists thought they had reached a fundamental unit of matter, they seemed to discover that it could be broken open and still smaller units could be pulled out.

Not only that, but even these fundamental particles didn't seem to be enough. It turns out that there are three families of particles, some of which only appear at significantly higher energies than scientists had previously explored.

Today, the Standard Model of particle physics contains 25 distinct fundamental particles, all of which have been experimentally detected (often decades after theoreticians proposed them).

Grasping for the Fundamental Forces of Physics

Even while the types of particles identified by scientists became more bizarre and complex, the ways those objects interacted turned out to be surprisingly straightforward. In the 20th century, scientists discovered that objects in the universe experience only four fundamental types of interactions:

- Electromagnetism
- Strong nuclear force

» Weak nuclear force

» Gravity

Physicists have discovered profound connections between these forces — except for gravity, which seems to stand apart from the others for reasons that physicists still aren't completely certain about. Trying to incorporate gravity with all the other forces — to discover how the fundamental forces are related to each other — is a key insight that many physicists hope a theory of quantum gravity will offer.

Electromagnetism: Super-speedy energy waves

Discovered in the 19th century, the *electromagnetic force* (or *electromagnetism*) is a unification of the electrostatic force and the magnetic force. In the mid-20th century, this force was explained in a framework of quantum mechanics called *quantum electrodynamics*, or *QED*. In this framework, the electromagnetic force is transferred by particles of light, called *photons*.

The relationship between electricity and magnetism is covered in Chapter 5, but the basic relationship comes down to electrical charge and its motion. The electrostatic force causes charges to exert forces on each other in a relationship that's similar to (but more powerful than) gravity — an inverse square law. This time, though, the strength is based not on the mass of the objects but on the electrical charge.

The *electron* is a particle that contains a negative electrical charge, while the *proton* in the atomic nucleus has a positive electrical charge. Traditionally, electricity is seen as the flow of electrons (negative charge) through a wire. This flow of electrons is called an *electrical current*.

A wire with an electrical current flowing through it creates a magnetic field. Alternately, when a magnet is moved near a wire, it causes a current to flow. (This is the basis of most electric power generators.)

This is the way in which electricity and magnetism are related. In the 1800s, physicist James Clerk Maxwell unified the two concepts into one theory, called *electromagnetism*, which depicted the electromagnetic force as waves of energy moving through space.

One key component of Maxwell's unification was a discovery that the electromagnetic force moved at the speed of light. In other words, the electromagnetic waves that Maxwell predicted from his theory were a form of light waves.

Quantum electrodynamics retains this relationship between electromagnetism and light because, in QED, the information about the force is transferred between two charged particles (or magnetic particles) by another particle — a photon. (Physicists say that the electromagnetic force is *mediated* by a photon.)

Nuclear forces: What the strong force joins, the weak force tears apart

In addition to gravity and electromagnetism, 20th-century physicists identified two nuclear forces called the *strong nuclear force* and the *weak nuclear force*. These forces are also mediated by particles. The strong force is mediated by a type of particle called a *gluon* — there are eight gluons, distinguished by their *color* charge. This has nothing to do with actual colors, it's just a name that physicists invented. This is why the theory of strong interactions is called quantum chromodynamics (*chroma* is Greek for color).

The weak force is mediated by three particles: *Z*, W^+, and W^- *bosons*. These are actually closely related to the photon, the particle that mediate electromagnetic interactions. (You can read more about these particles in Chapter 8.)

The strong nuclear force holds quarks together to form protons and neutrons, but it also holds the protons and neutrons together inside the atom's nucleus.

The weak nuclear force, on the other hand, is responsible for radioactive decay, such as when a neutron decays into a proton. The processes governed by the weak nuclear force are responsible for the burning of stars and the formation of heavy elements inside stars.

Infinities: Why Einstein and the Quanta Don't Get Along

Einstein's theory of general relativity, which explains gravity, does an excellent job of explaining the universe on the scale of the cosmos. Quantum physics does an excellent job of explaining the universe on the scale of an atom or smaller. In between those scales, good old-fashioned classical physics usually rules.

Unfortunately, some problems bring general relativity and quantum physics into conflict, resulting in mathematical infinities in the equations. (Infinity is essentially an abstract number that's larger than any other numbers. Though certain cartoon characters like to go "To infinity and beyond," scientists don't like to see infinities come up in mathematical equations.) Infinities arise in quantum physics, but physicists have developed mathematical techniques to tame them in many cases so the results match the experiments. In some cases, however, these techniques don't apply. Because physicists never observe real infinities in nature, these troublesome problems motivate the search for quantum gravity.

Each of the theories works fine on its own, but when you get into areas where both have something specific to say about the same thing — such as what's going on at the border of a black hole — things get very complicated. The quantum fluctuations make the distinction between the inside and outside of the black hole kind of fuzzy, and general relativity needs that distinction to work properly. Neither theory by itself can fully explain what's going on in these specific cases.

This is the heart of why physicists need a theory of quantum gravity. With the current theories, you get situations that don't look like they make sense. Physicists don't see infinities, but both relativity and quantum physics indicate that they should exist. Reconciling this bizarre region in the middle, where neither theory can fully describe what's going on, is the goal of quantum gravity.

Singularities: Bending gravity to the breaking point

Because matter causes a bending of space-time, cramming a lot of matter into a very small space causes a lot of bending of space-time. In fact, some solutions to Einstein's general relativity equations show situations where space-time bends an infinite amount — called a *singularity*. Specifically, a space-time singularity shows up in the mathematical equations of general relativity in the following two situations:

- During the early big bang period of the universe's history
- Inside black holes

These subjects are covered in more detail in Chapter 9, but both situations involve a density of matter (a lot of matter in a small space) that's enough to cause problems with the smooth space-time geometry that relativity depends on.

These singularities represent points where the theory of general relativity breaks down completely. Even talking about what goes on at this point becomes meaningless, so physicists need to refine the theory of gravity to include rules for how to talk about these situations in a meaningful way.

Some believe that this problem can be solved by altering Einstein's theory of gravity (as you see in Chapter 20). String theorists don't usually want to modify gravity (at least at the energy levels scientists normally look at); they just want to create a framework that allows gravity to work without running into these mathematical (and physical) infinities.

Quantum jitters: Space-time under a quantum microscope

A second type of infinity, proposed by John Wheeler in 1955, is the *quantum foam* or, as it's called by string theorist and best-selling author Brian Greene, the *quantum jitters*. Quantum effects mean that space-time at very tiny distance scales (called the *Planck length*) is a chaotic sea of virtual particles being created and destroyed. At these levels, space-time is certainly not smooth, as relativity suggests, but is a tangled web of extreme and random energy fluctuations, as Figure 2-1 shows.

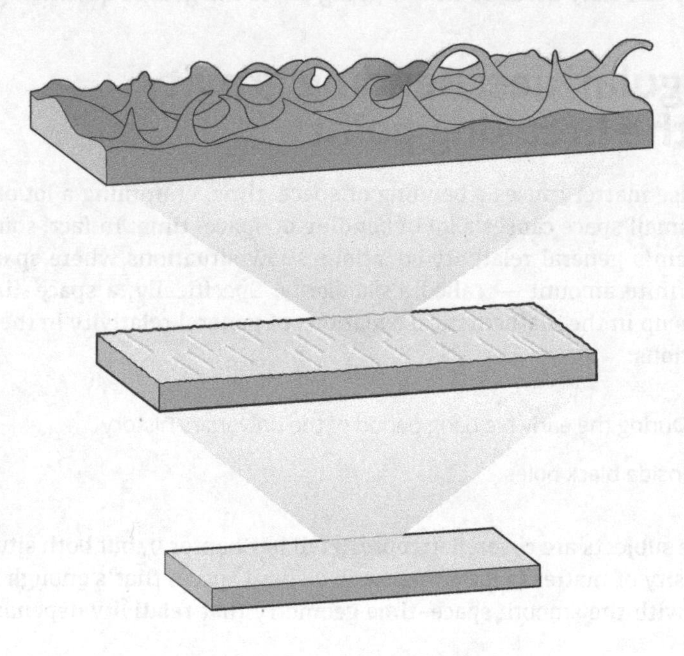

FIGURE 2-1: If you zoom in on space-time enough, you may see a chaotic "quantum foam."

The basis for the quantum jitters is the *uncertainty principle*, one of the key (and most unusual) features of quantum physics. This is explained in more detail in Chapter 7, but the key component of the uncertainty principle is that certain pairs of quantities — for example, position and velocity, or time and energy — are linked together, so that the more precisely one quantity is measured, the more uncertain the other quantity is. This isn't just a statement about measurement, though; it's fundamental uncertainty in nature!

TIP

In other words, nature is a bit "blurry" according to quantum physics. This blurriness only shows up at very small distances, but this problem creates the quantum foam.

One example of the blurriness comes in the form of virtual particles. According to quantum field theory (a *field theory* is one where each point in space has a certain value, similar to a gravitational field or an electromagnetic field), even the empty void of space has a slight energy associated with it. This energy can be used to, very briefly, bring a pair of particles — a particle and its antiparticle, to be precise — into existence. The particles exist for only a moment, and then they destroy each other. It's as if they borrowed enough energy from the universe to exist for just a few fractions of a second.

The problem is that when you look at space-time at very small scales, the effects of these virtual particles become very important. The energy fluctuations predicted by the uncertainty principle take on massive proportions. Without a quantum theory of gravity, there's no way to really figure out what's going on at sizes that small.

Unifying the Forces

The attempt to unite gravity with the other three forces, as well as with quantum physics, was one of the driving forces of physics throughout the 20th century (and it still is). In a way, these sorts of unifications of different ideas are the major discoveries in science throughout the ages.

Quantum electrodynamics successfully created a quantum theory of electromagnetism. Later, the electroweak theory unified this theory with the weak nuclear force. The strong nuclear force is explained by quantum chromodynamics. The current model of physics that explains all three of these forces is called the *Standard Model of particle physics,* which is covered in much more detail in Chapter 8.

Unifying gravity with the other forces would create a new version of the Standard Model and would explain how gravity works on the quantum level. Many physicists hope that string theory will ultimately prove to be this theory.

Einstein's failed quest to explain everything

After Einstein successfully worked the major kinks out of his theory of general relativity, he turned his attention toward trying to unify this theory of gravity with electromagnetism as well as with quantum physics. In fact, he would spend most of the rest of his life trying to develop this unified theory but would die unsuccessful.

Throughout this quest, Einstein looked at almost any theory he could think of. One of these ideas was to add an extra space dimension and roll it up into a very small size. This approach, called *Kaluza-Klein theory* after the men who proposed it, is addressed in Chapter 6. This same approach would eventually be used by string theorists to deal with the pesky extra dimensions that arose in their own theories.

Ultimately, none of Einstein's attempts bore fruit. To the day of his death, he worked feverishly on completing his unified field theory in a manner that many physicists consider a sad end to such a great career.

Today, however, some of the most intense theoretical physics work is in the search for a theory to unify gravity and the rest of physics, mainly in the form of string theory.

A particle of gravity: The graviton

The Standard Model of particle physics explains electromagnetism, the strong nuclear force, and the weak nuclear force as fields that follow the rules of gauge theory. *Gauge theory* is based heavily on mathematical symmetries. Because these forces are quantum theories, the gauge fields come in discrete units (*quantum* means "how much" in Latin) — and these units actually turn out to be particles in their own right, called *gauge bosons.* The forces described by a gauge theory are carried, or *mediated,* by these gauge bosons.

For example, the electromagnetic force is mediated by the photon. When gravity is written in the form of a gauge theory, the gauge boson for gravity is called the *graviton.* (If you're confused about gauge theories, don't worry too much — just remember that the graviton is what makes gravity work and you'll know everything you need to know to understand their application to string theory.)

Physicists have identified some features of the theoretical graviton so that, if it exists, it can be recognized. For one thing, the particle is *massless,* which means it has no rest mass — a graviton is always in motion, and that probably means it travels at the speed of light (although in Chapter 20 you find out about a theory of modified gravity in which gravity and light move through space at different speeds).

Another feature of the graviton is that it has a spin of 2. (*Spin* is a quantum number indicating an inherent property of a particle that acts kind of like angular momentum. Fundamental particles have an inherent spin, meaning they interact with other particles like they're spinning even when they aren't.)

A graviton also has no electrical charge. It's a stable particle, which means it won't decay.

REMEMBER

So physicists are looking for a massless particle moving at an incredibly fast speed, with no electrical charge and a quantum spin of 2. Even though the graviton has never been discovered by experiment, it's the gauge boson that mediates the gravitational force. Given the extremely weak strength of gravity in relation to other forces, trying to identify a single graviton is an incredibly hard task. (It is, however, possible to identify the *gravitational waves* created by many gravitons, which was done very recently, as you find out in Chapter 6.)

The possible existence of the graviton in string theory is one of the major motivations for looking toward the theory as a likely solution to the problem of quantum gravity.

Supersymmetry's role in quantum gravity

Supersymmetry is a principle that says that two types of fundamental particles, bosons and fermions, are connected to each other. The benefit of this type of symmetry is that the mathematical relationships in gauge theory reduce in such a way that unifying all the forces becomes more feasible. (We explain bosons and fermions in greater detail in Chapter 8, while we present a more detailed discussion of supersymmetry in Chapter 10.)

The top graph in Figure 2-2 shows the strengths of the three forces described by the Standard Model modeled at different energy levels. If the three forces met up in the same point, it would indicate that there might be an energy level where these three forces become fully unified into one single force.

However, as the lower graph of Figure 2-2 shows, when supersymmetry is introduced into the equation (literally, not just metaphorically), the three forces meet in a single point. If supersymmetry proves to be true, it's strong evidence that the three forces of the Standard Model unify at high enough energy.

Many physicists believe that all four forces are a manifestation of the same fundamental laws, which in string theory would be the dynamics of quantum strings. This should be apparent at high energy levels, but as the universe reduced into a lower-energy state, the inherent symmetry between the forces began to break down. This broken symmetry caused the creation of four apparently very different forces of nature.

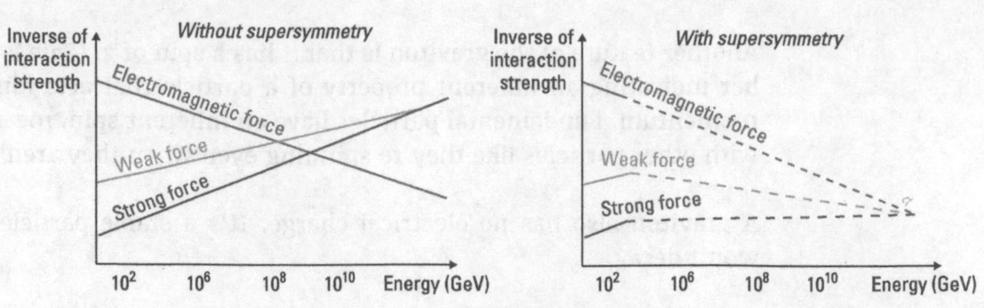

FIGURE 2-2: If supersymmetry is added, the strengths of the forces in the Standard Model scale differently with energy, and they may become equal at high enough energy.

Lisa Randall, 2005. Reproduced by permission of HarperCollins Publishers.

The goal of a theory of quantum gravity is, in a sense, an attempt to look back in time, to when these four forces were unified into a single structure. If successful, it would profoundly affect our understanding of the first few moments of the universe — the last time the forces joined together in this way.

IN THIS CHAPTER

- Embracing string theory's achievements
- Poking holes in string theory
- Wondering what the future of string theory holds

Chapter 3
Accomplishments and Failures of String Theory

String theory is a work in progress, having captured the hearts and minds of much of the theoretical physics community while being apparently disconnected from any realistic chance of definitive experimental proof. Despite this, it has had some successes — unexpected predictions and achievements that may well indicate string theorists are on the right track.

String theory critics would also point out (and many string theorists would probably agree) that the last couple of decades haven't been kind to string theory because the momentum toward a unified theory of everything has slowed, and the latest particle colliders have failed to provide any direct evidence for string theory.

In this chapter, you see some of the major successes and failures of string theory, as well as look at the possibilities for where string theory may go from here. The controversy over string theory rests entirely on how much significance physicists give to these different outcomes.

Celebrating String Theory's Successes

String theory has gone through many transformations since its origins in 1968, when theorists hoped it would be a model of certain types of particle collisions. It initially failed at that goal, but in the 50 years since, string theory has developed into the primary candidate for a theory of quantum gravity. It has driven major developments in mathematics, and theorists have used insights from string theory to tackle other, unexpected problems in physics. In fact, the very presence of gravity within string theory is an unexpected outcome!

Predicting gravity out of strings

The first and foremost success of string theory is the unexpected discovery of objects within the theory that match the properties of the graviton. These objects are a specific type of closed strings that are also massless particles that have a spin of 2, exactly like gravitons. To put it another way, gravitons are a spin-2 massless particle that, under string theory, can be formed by a certain type of vibrating closed string. String theory wasn't created to have gravitons — they're a natural and required consequence of the theory.

One of the greatest problems in modern theoretical physics is that gravity seems to be disconnected from all the other forces of physics that are explained by the Standard Model of particle physics. String theory solves this problem because it not only includes gravity but makes gravity a necessary by-product of the theory.

Explaining what happens to a black hole (sort of)

A major motivating factor for the search for a theory of quantum gravity is to explain the behavior of black holes, and string theory appears to be one of the best methods of achieving that goal. String theorists have created mathematical models of black holes that appear similar to predictions made by Stephen Hawking more than 50 years ago and may be at the heart of resolving a long-standing puzzle within theoretical physics: What happens to matter that falls into a black hole?

Scientists' understanding of black holes has always run into problems, because to study the quantum behavior of a black hole, you need to somehow describe all the *quantum states* (possible configurations, as defined by quantum physics) of the black hole. Unfortunately, black holes are objects in general relativity, so it's not clear how to define these quantum states. (See Chapter 2 for an explanation of the conflicts between general relativity and quantum physics.)

REMEMBER

String theorists have created models that appear to be identical to black holes in certain simplified conditions, and they use that information to calculate the quantum states of black holes. Their results have been shown to match Hawking's predictions, which he made without any precise way to count the quantum states of a black hole.

This is the closest that string theory has come to an experimental prediction. Unfortunately, there's nothing experimental about it because scientists can't directly observe black holes to this level of detail. It's a theoretical prediction that unexpectedly matches another (well-accepted) theoretical prediction about black holes. And, beyond that, the prediction only holds for certain types of black holes and hasn't yet been successfully extended to all black holes.

For a more detailed look at black holes and string theory, check out Chapters 9, 13, and 16.

Explaining quantum field theory using string theory

One of the major successes of string theory is something called the *Maldacena conjecture*, or the *AdS/CFT correspondence*. (We get into what this means in Chapter 13.) Developed in 1997 and later expanded, this correspondence appears to give insights into gauge theories, like those at the heart of quantum field theory, and their relation to gravity. (See Chapter 2 for an explanation of gauge theories.)

The original AdS/CFT correspondence, written by Juan Maldacena, argues that strings (that is, quantum gravity) in certain *D*-dimensional universes are equivalent to certain quantum field theories (without gravity) in a *(D-1)*-dimensional universe. This sounds confusing (it is), but in a nutshell, it means that quantum gravity is a bit like the Standard model (but for a universe in one dimension less), and the Standard model is a bit like quantum gravity (but for a universe in one dimension more). This is a very surprising way to think about quantum gravity (first anticipated by Nobel laureate Gerard 't Hooft), which finds its most precise realization in Maldacena's AdS/CFT correspondence.

More precisely, Maldacena proposed that a certain 3-dimensional (three space dimensions, like our universe) gauge theory, with the most supersymmetry allowed, describes the same physics as a string theory in a 4-dimensional (four space dimensions) world. This means that questions about string theory can be asked in the language of gauge theory, which is a quantum theory that physicists know how to work with!

String theory keeps making a comeback

String theory has suffered more setbacks than probably any other scientific theory in human history, but those hiccups don't seem to last very long. Every time it seems that some flaw in the theory comes along, the mathematical resiliency of string theory seems to not only save it, but bring it back stronger than ever.

When extra dimensions came into the theory in the 1970s, string theory was abandoned by many, but it had a comeback in the first superstring revolution. It then turned out that there were five distinct versions of string theory, but a second superstring revolution was sparked by unifying them. When string theorists realized a vast number of solutions to string theories (each solution to string theory is called a *vacuum*, while many solutions are called *vacua*) were possible, they turned this into a virtue instead of a drawback.

Still, even after so many years, some scientists believe that string theory is failing at its goals. (See "Considering String Theory's Setbacks" later in this chapter.)

Being the most popular theory in town

Many young physicists feel that string theory, as the primary theory of quantum gravity, is the best (or only) avenue for making a significant contribution to our understanding of this topic. Over the last three decades, high-energy theoretical physics (especially in the United States) has become dominated by string theorists. In the high-stakes world of "publish-or-perish" academia, this is a major success.

Why do so many physicists turn toward this field when it offers no experimental evidence? Some of the brightest theoretical physicists of either the 20th or the 21st centuries — Edward Witten, John Henry Schwarz, Leonard Susskind, and others you meet throughout this book — continually return to the same common reasons in support of their interest:

- If string theory were wrong, it wouldn't provide the rich structure that it does, such as with the development of the heterotic string (see Chapter 10), which allows for an approximation of the Standard Model of particle physics within string theory.
- If string theory were wrong, it wouldn't lead to better understanding of quantum field theory, quantum chromodynamics (see Chapter 8), or the quantum states of black holes, as presented by the work of Leonard Susskind, Andrew Strominger, Cumrun Vafa, and Juan Maldacena (see Chapters 11, 13, and 16).

» If string theory were wrong, it would have collapsed in on itself well before now, instead of passing many mathematical consistency checks (such as those discussed in Chapter 10) and providing more and more elaborate ways to be interpreted, such as the dualities and symmetries that allowed for the presentation of M-theory (discussed in Chapter 11).

This is how theoretical physicists think, and it's why so many of them continue to believe that string theory is the place to be. The mathematical beauty of the theory, the fact that it's so adaptable, is seen as one of its virtues. The theory continues to be refined, and it hasn't been shown to be incompatible with our universe. There has been no brick wall where the theory failed to provide something new and — in some eyes, at least — meaningful, so those studying string theory have had no reason to give up and look somewhere else. (The history of string theory in Chapters 10 and 11 offers a better appreciation of these achievements.)

Whether this resilience of string theory will translate someday into proof that the theory is fundamentally correct remains to be seen, but for the majority of those working on the problems, confidence is high.

As you can read in Chapter 18, this popularity is also seen by some critics as a flaw. Physics thrives on the rigorous debate of conflicting ideas, and some physicists are concerned that the driving support of string theory, to the exclusion of all other ideas, isn't healthy for the field. For some of these critics, the mathematics of string theory has indeed already shown that the theory isn't performing as expected (or, in their view, as needed to be a fundamental theory) and string theorists are in denial.

Considering String Theory's Setbacks

Because string theory has made so few specific predictions, it's hard to disprove it, but the theory has fallen short of some of the hype about how it will be a fundamental theory to explain all the physics in our universe, a "theory of everything." This failure to meet that lofty goal seems to be the basis of many (if not most) of the attacks against it.

In Chapter 18, you find more detailed criticisms of string theory. Some of them cut to the very heart of whether string theory is even scientific or whether it's being pursued in the correct way. For now, we leave those more abstract questions and focus on three issues that even most string theorists aren't particularly happy about:

- Because of supersymmetry, string theory requires a large number of particles beyond what scientists have ever observed.
- This new theory of gravity was unable to predict the accelerated expansion of the universe that was detected by astronomers.
- A vastly large number of mathematically feasible string theory *vacua* (solutions) currently exist, so it seems virtually impossible to figure out which could describe our universe.

The following sections cover these dilemmas in more detail.

The universe doesn't have enough particles

For the mathematics of string theory to work, physicists have to assume a symmetry in nature called *supersymmetry*, which creates a correspondence between different types of particles. One problem with this is that instead of the 25 fundamental particles in the Standard Model, supersymmetry requires at least 36 fundamental particles (which means that nature allows 25 more particles that scientists have never seen!). In some ways, string theory does make things simpler — the fundamental objects are *strings* and *branes* or, as predicted by matrix theory, 0-dimensional branes called *partons*. These strings, branes, or possibly partons make up the particles that physicists have observed (or the ones they hope to observe). But that's on a very fundamental level; from a practical standpoint, string theory doubles the number of particles allowed by nature from 25 to 50.

One of the biggest possible successes for string theory would be to experimentally detect these missing supersymmetric partner particles. Many theoretical physicists hoped that when the Large Hadron Collider particle accelerator at the European Organization for Nuclear Research in Switzerland went online, it would detect supersymmetric particles. This hasn't happened — yet.

Even if it's found, proof of supersymmetry doesn't inherently prove string theory, so the debate would continue to rage on, but at least one major objection would be removed. Supersymmetry might well end up being true, whether or not string theory as a whole is shown to accurately describe nature.

Dark energy: The discovery string theory should have predicted

Astronomers found evidence in 1998 that the expansion of the universe was actually accelerating. This accelerated expansion is caused by the *dark energy* that you hear about so often in the news. Not only did string theory not predict the existence of dark energy, but its attempts to use science's best theories to calculate the amount of dark energy come up with a number that's vastly larger than the one observed by astronomers. The theory just absolutely failed to initially make sense of dark energy.

Claiming this as a flaw of string theory is a bit more controversial than the other two problems, but there's some (albeit questionable) logic behind it. The goal of string theory is nothing less than the complete rewriting of gravitational law, so it's not unreasonable to think that string theory should've anticipated dark energy in some way.

When Einstein constructed his theory of general relativity, the mathematics indicated that space could be expanding (later proved to be true). When Paul Dirac formulated a quantum theory of the electron, the mathematics indicated an antiparticle existed (later proved to actually exist). A profound theory like string theory can be expected to illuminate new facts about our universe, not be blindsided by unanticipated discoveries.

Of course, no other theory anticipated an accelerating expansion of the universe either. Prior to the observational evidence, cosmologists (and string theorists) had no reason to assume that the expansion rate of space was increasing. Years after dark energy was discovered, it was shown that string theory could be modified to include it, which string theorists count as a success (although the critics continue to be unsatisfied).

Where did all these "fundamental" theories come from?

Unfortunately, as string theorists performed more research, they had a growing problem (pun intended). Instead of narrowing in on a single *vacuum* (solution) that could be used to explain the universe, it began to look like there were an absurdly large number of vacua. Some physicists' hopes that a unique, fundamental version of string theory would fall out of the mathematics effectively dissolved.

In truth, such hype was rarely justified in the first place. In general relativity, for example, an infinite number of ways to solve the equations exist, and the goal is to find solutions that match our universe. The overly ambitious string theorists (the ones who expected a single vacuum to fall out of the sky) soon realized that they, too, would end up with a rich *string theory landscape*, as Leonard Susskind calls the range of possible vacua (see Chapter 12 for more on Susskind's landscape idea). The goal of string theory has since become to figure out which set of vacua applies to our universe.

STRING THEORY IS . . . INEVITABLE?

A modern take on the study of string theory turns the tables and asks, *what else could be a good theory of quantum gravity?* In this approach, physicists don't try to construct new theories. Instead, they try to list all the properties that *any* reasonable theory of quantum gravity should have.

For instance, a reasonable quantum gravity should contain the graviton, and it should look like ordinary (classical) gravity when the masses and distances involved have the right size. There are additional requirements that come from quantum physics and are a little more technical. They ensure that the various symmetries of the theory are realized and that all probabilities always add up to 100 percent. (This may sound silly, but it's one of the most important concepts in quantum theory, called *unitarity*.)

The hope is that by considering all these constraints, we can pin down what quantum gravity should look like. This approach is sometimes called *bootstrapping* (as in "pulling yourself up by your own bootstraps"). It's not clear if the various requirements that physicists have imposed can nail down a single theory of quantum gravity. However, these constraints actually rule out large families of theories, leaving relatively little space beyond string theory.

If this bootstrap approach can be completed, it might eventually show that any reasonable theory of quantum gravity is *inevitably* string theory. You'll find more about this approach in Chapter 14, in the section "Bootstrapping Our Way into String Theory".

Looking into String Theory's Future

At present, string theory faces two hurdles. The first is the theoretical hurdle, which is whether a model that describes our own universe can be formulated. The second hurdle is the experimental one, because even if string theorists are successful in modeling our universe, they'll then have to figure out how to make a distinct prediction from the theory that's testable in some way.

Right now, string theory falls short on both counts, and it's unclear whether it can ever be formulated in a way that will be uniquely testable. The critics claim that growing disillusionment with string theory is rising among theoretical physicists, while the supporters continue to talk about how string theory is being used to resolve the major questions of the universe.

Only time will tell whether string theory is right or wrong, but regardless of the answer, string theory has driven scientists for years to ask fundamental questions about our universe and explore the answers to those questions in new ways. Even an alternative theory would partly owe its success to the hard work performed by string theorists.

Theoretical complications: Can we figure out string theory?

The current version of string theory is called *M-theory.* Introduced in 1995, M-theory is a comprehensive theory that includes the five supersymmetric string theories and exists in 11 dimensions. There's just one problem: By and large, M-theory remains a mystery.

Scientists are searching for a complete string theory, but they don't have one yet. And until they do, there's no way of knowing that they'll be successful. Until string theorists have a complete theory that describes our universe, the theory could be all smoke and mirrors. Although some aspects of string theory may be shown to be true, it may be that these are only approximations of some more fundamental theory — or it may be that string theory is actually that fundamental theory itself.

String theory, the driving force of 21st-century theoretical physics, *could* prove to be nothing more than a mathematical illusion that provides some approximate insights into science but isn't actually the theory that drives the forces of nature.

It's unclear how long the search for a theory can last without some specific breakthrough. There's a sense (among some) that the most brilliant physicists on the planet have been spinning their wheels for decades, with only a handful of significant insights, and even those discoveries don't seem to lead anywhere specific.

The theoretical implications of string theory are addressed in Chapters 10 and 11, while the criticism of the theory rears its ugly head in Chapter 18.

Experimental complications: Can we prove string theory?

Even if a precise version of string theory (or M-theory) is formulated, the question then moves from the theoretical to the experimental realm. Right now, the energy levels that scientists can reach in experiments are probably way too small to realistically test string theory, although aspects of the theory can be tested today.

Theory moves forward with directions from experiment, but the last input that string theory had from experiment was the realization that it failed as a theory in describing the scattering of particles within particle accelerators. The realm string theory claims to explain involves distances so tiny that it's questionable whether scientists will ever achieve a technology able to probe at that length, so it's possible that string theory is an inherently untestable theory of nature. (Some versions of string theory do make predictions in testable ranges, however, and string theorists hope that these versions of string theory may apply to our universe.)

You find out some ways to possibly test string theory in Chapter 14, although these are only speculative because right now science doesn't even have a theory that makes any unique predictions. The best physicists can hope for are some hints that would give some direction to the theoretical search, such as the discovery of certain types of extra dimensions, new cosmological predictions about the formation of our universe, or evidence of the missing supersymmetric particles.

2 The Physics Upon Which String Theory Is Built

IN THIS PART . . .

Understand the method of science.

Review elements of classical physics.

Apply Einstein's work in relativity.

Grasp quantum theory basics.

Dive into the Standard Model of particle physics.

Explore cosmology and astrophysics.

IN THIS CHAPTER

» Brushing up on scientific theories you know and love

» So you say you want a scientific revolution

» What scientists have unified, let no one put asunder

» Breaking the same old rules to keep things interesting

Chapter 4
Putting String Theory in Context: Understanding the Method of Science

String theory is at the cutting edge of science. It's a mathematical theory of nature that, at present, makes few predictions that are directly testable by empirical experiments. This brings up the question of what it takes for a theory to be scientific.

In this chapter, we look a bit more closely at the methods scientists use to investigate nature's structure. We explore how scientists perform science and some of the ways their work is viewed. We certainly don't solve any big, philosophical issues in this chapter, but our goal is to make it clear that scientists have differing views about how the nature of science is supposed to work. Although we could write volumes on the evolution of scientific thought throughout the ages, we touch on these topics in just enough detail to help you understand some of the arguments in favor of and against string theory.

Exploring the Practice of Science

Before you can figure out whether string theory is scientific, you have to ask, "What is science?"

Science is the methodical practice of trying to understand and predict the consequences of natural phenomena. This is done through two distinct but closely related means: theory and experiment.

Not all science is created equal. Some science is performed with diagrams and mathematical equations. Other science is performed with costly experimental apparatus. Still other forms of science, while also costly, involve observing distant galaxies for clues to the mystery of the universe.

String theory has spent more than 40 years focusing on the theory side of the scientific equation and, sadly, is relatively lacking on the experimental side, as critics never hesitate to point out. Ideally, the theories developed would eventually be validated by experimental evidence. (See the later sections "The need for experimental falsifiability" and "The foundation of theory is mathematics" for more on the necessity of experimentation.)

The myth of the scientific method

In school, many of us were taught that science follows nice, simple rules called the *scientific method.* These rules are a classical model of scientific investigation based on principles of *reductionism* and *inductive logic.* In other words, you take observations, break them down (the reductionism part), and use them to create generalized laws (the inductive logic part). String theory's history certainly doesn't follow this simple model.

The steps of the scientific method students are taught actually change a bit depending on the textbook schools use in a given year, though they generally have mostly common elements. Frequently, they are delineated as a set of bullet points:

- **Observe a phenomenon:** Look at nature.
- **Formulate a hypothesis:** Ask a question (or propose an answer).
- **Test the hypothesis:** Perform an experiment.
- **Analyze the data:** Confirm or reject the hypothesis.

In a way, this scientific method is a myth. You may earn a physics degree from a top college without once being asked a question about the scientific method in a physics course. (Though it may come up if you dabble a bit more in the philosophy of science or similar courses.)

Turns out, there's no single scientific method that all scientists follow. Scientists don't look at a list and think, "Well, I've observed my phenomenon for the day. Time to formulate my hypothesis." Instead, science is a dynamic activity that involves a continuous, active analysis of the world. It's an interplay between the world we observe and the world we conceptualize. Science is a translation between observations, experimental evidence, and the hypotheses and theoretical frameworks that are built to explain and expand on those observations.

Still, the basic ideas of the scientific method do tend to hold. They aren't so much hard-and-fast rules, but they're guiding principles that can be combined in different ways depending on what's being studied.

One of the best situations for a scientist to be in is to observe a pattern or trend in phenomena, and then use that to make a precise prediction about some other phenomenon that hasn't yet been observed. This provides the basis for a new experiment or observation that, if it matches the prediction, provides an excellent foundation for thinking the line of reasoning that led to the prediction was probably on the right track.

BREAKING DOWN NATURE WITH BACON AND GALILEO

The ideas of the scientific method are often traced back to Sir Francis Bacon's 1620 book, *Novum Organum,* and to Galileo Galilei's works in the 1630s. Broadly speaking, the main idea is that reductionism and inductive reasoning could be used to arrive at fundamental truths about the causes of natural events, which can then be compared with experience and experiment. This was quite a revolutionary idea because, at that time, your best bet of convincing anybody of your ideas would have been to argue that they matched Aristotle's theories written 2,000 years earlier!

In the Baconian model, the scientist breaks natural phenomena down into component parts that are then compared to other components based on common themes. These reduced categories are then analyzed using principles of inductive reasoning. *Inductive reasoning* is a logical system of analysis where you start with specific true statements and work to create generalized laws that would apply to all situations by finding commonalities between the observed truths.

The need for experimental falsifiability

Traditionally, the idea has been that an experiment can either confirm or refute a theory. An experimental result yields *positive evidence* if it supports the theory, while a result that contradicts the hypothesis is *negative evidence.*

In the 20th century, a notion arose that the key to a theory — the thing that makes it scientific — is whether it can in some way be shown to be false. This *principle of falsifiability* can be controversial when applied to string theory, which theoretically explores energy levels that can't at present (or possibly ever) be directly explored experimentally. Some claim that because string theory currently fails the test of falsifiability, it's somehow not "real science." (Check out Chapter 18 for more on this idea.)

The focus on this falsifiability is traced back to philosopher Karl Popper's 1934 book, *The Logic of Scientific Discovery.* He was opposed to the reductionist and inductive methods that Francis Bacon had popularized three centuries earlier. In a time that was characterized by the rise of modern physics, it appeared that the old rules no longer applied.

Popper reasoned that the principles of physics arose not merely by viewing little chunks of information, but by creating theories that were tested and repeatedly failed to be proved false. Observation alone could not have led to these insights, he claimed, if they'd never been put in positions to be proven false. In the most extreme form, this emphasis on falsifiability states that scientific theories don't tell you anything definite about the world, but are only the best guesses about the future based on past experience.

For example, if you predict that the sun will rise every morning, you can test your prediction by looking out the window every morning for 50 days. If the sun is there every day, you have not proved that the sun will be there on the 51st day. After you actually observe it on the 51st day, you'll know that your prediction worked out again, but you haven't proved anything about the 52nd day, the 53rd day, and so on.

REMEMBER

No matter how good a scientific prediction is, if you can run a test that shows that it's false, you have to throw out the idea (or at least modify your theory to explain the new data). This led the 19th century biologist Thomas Henry Huxley to define the great tragedy of science as "the slaying of a beautiful hypothesis by an ugly fact."

To Popper, falsifiability was far from tragic, but was instead the brilliance of science. The defining component of a scientific theory, the thing that separates it from mere speculation, is that it makes a falsifiable claim.

Of course, it might not be easy to falsify a scientific claim. The claim may be a prediction at the forefront of current scientific knowledge, perhaps requiring additional technological progress before the experiments that would prove the claim false can be run. This is where things can get tricky for string theory, where much of the benefit is the ability to skirt along the edge of currently testable knowledge.

It's also worth noting that mathematics as a whole doesn't really follow this idea of falsifiability. Instead, it invents some more or less reasonable starting points (axioms) and, through a series of deductions, provides theorems that establish precise relations between mathematical objects. Although math itself is not falsifiable, it cannot be denied that it is incredibly useful to understanding nature, and we'd be hard-pressed to do without it when building a house or flying a plane.

Popper's claim is sometimes controversial, especially when it's being used by one scientist (or philosopher) to discredit an entire field of science. Many still believe that reductionism and inductive reasoning can lead to the creation of meaningful theoretical frameworks that represent reality as it is, even if there's no claim that's expressly falsifiable. The central element of this belief is the idea of *confirmation* — direct positive evidence for a theory, rather than just a lack of negative evidence against it.

String theorist Leonard Susskind and physicist Lee Smolin amicably clashed over exactly this point online in 2004 (you can view the debate at `www.edge.org/3rd_culture/smolin_susskind04/smolin_susskind.html`). To support the idea of confirmation, Susskind lists several theories that have been denounced as unfalsifiable: behaviorism in psychology along with quark models and inflationary theory in physics. Scientists initially believed that certain traits in these areas couldn't be examined, though methods were later developed that allowed them to be tested.

REMEMBER

There's a difference between being unable to falsify a theory in practice and being unable to falsify it in principle.

It may seem as if the debate over confirmation and falsifiability is academic. That's probably true, but some critics of string theory frame the debate around this theory as a battle over the very meaning of physics. Many string theory critics believe that it's inherently unfalsifiable, while some string theorists believe a mechanism to test (and falsify) the prediction of string theory may someday be found.

The foundation of theory is mathematics

Galileo famously wrote that the universe is a book, and the language in which it is written is mathematics. Since his time, we have developed more and more powerful mathematical models that represent the underlying physical laws that nature follows. These mathematical models are the actual theories of physics that physicists can then relate to meaningful events in the real world through experiment and other means.

Science requires both experiment and theory to build explanations of what happens in the world. To paraphrase Einstein, science without theory is lame, while science without experiment is blind.

REMEMBER

If physics is built on a foundation of experimental observation, then theoretical physics is the blueprint that explains how those observations fit together. The insights of theory have to move beyond the details of specific observations and connect them in new ways. Ideally, these connections lead to other predictions that are testable by experiment. String theory has not yet made this significant leap from theory to experiment.

TIP

A large part of the work in theoretical physics is developing mathematical models — frequently including simplifications that aren't necessarily realistic — that can be used to predict the results of future experiments. When physicists "observe" a particle, they're really looking at data that shows a specific trace of that particle's existence. When they look into the heavens, they receive energy readings that fit certain parameters and explanations. To a physicist, these aren't "just" numbers; they're clues to understanding the universe.

High-energy physics (which includes string theory and the physics of fundamental particles) has an intense interplay between theoretical insights and experimental observations. Research papers in this area fall into one of four categories:

- Experiment
- Lattice (computer simulations)
- Phenomenology
- Theory

Phenomenology is the study of phenomena (no one ever said physicists were creative when it comes to naming conventions) and relating them within the framework of an existing theory. In other words, scientists focus on taking the existing theory and applying it to the existing facts or building models describing anticipated facts that may be discovered soon. Then they make predictions about the experimental observations that should be obtained. (Of course, phenomenology has a lot more to it, but this is the gist of what you need to know to understand it in relation to string theory.) It's an intriguing discipline, and one that over the last decades has been focusing on supersymmetry and string theory. When we discuss how to possibly test string theory in Chapter 14, it's largely the work of phenomenologists that tells scientists what they're looking for.

Though scientific research can be conducted with these different methods, there is certainly overlap. Phenomenologists can work on pure theory and can also, of course, prepare a computer simulation. Also, in some ways, a computer simulation can be viewed as a process that's both experimental and theoretical. But what all these approaches have in common is that the scientific results are expressed in the language of science: mathematics.

The rule of simplicity

In science, one goal is to propose the fewest "entities" or rules needed to explain how something works. In many ways, the history of science is seen as a progression of simplifying the complex array of natural laws into fewer and fewer fundamental laws.

REMEMBER

Take Occam's razor, which is a principle developed in the 14th century by Franciscan friar and logician William of Occam. His "law of parsimony" is basically translated (from Latin) as "Entities must not be multiplied beyond necessity." (In other words, keep it simple.) Albert Einstein famously stated a similar rule as "Make everything as simple as possible, but not simpler." Though not a scientific law itself, Occam's razor tends to guide how scientists formulate their theories.

In some ways, string theory seems to violate Occam's razor. For example, in order to be related back to the real world, string theory requires the addition of a lot of odd components that scientists haven't actually observed yet (extra dimensions, new particles, and other features mentioned in Chapters 10 and 11). However, if these components are indeed necessary, then string theory is in accord with Occam's razor.

The role of objectivity in science

Some people believe that science is purely objective. And, of course, science *is* objective in the sense that anyone can apply the principles of science in the same way and get the same empirical results in a specific experimental situation. (At least this is how it usually works in physics. Don't get us started on psychology.) The idea that scientists are themselves inherently objective is a nice thought, but it's about as true as the notion of pure objectivity in journalism. The debate over string theory demonstrates that the discussion isn't always purely objective. At its core, the debate is over different opinions about how to view science.

In truth, scientists continually make choices that are subjective, such as which questions to pursue. For example, when string theorist Leonard Susskind met Nobel Prize winner Murray Gell-Mann, Gell-Mann laughed at the very idea of vibrating strings. Two years later, Gell-Mann wanted to hear more about Susskind's theory.

In other words, physicists are people. They have mastered a difficult discipline, but that doesn't make them infallible or immune to pride, passion, or any other human foible. The motivation for their decisions may be financial, aesthetic, personal, or any other reason that influences human decisions.

The degree to which scientists rely on theory versus experiment in guiding their activities is another subjective choice. Einstein, for example, spoke of the ways in which only the "free inventions of the mind" (pure physical principles, conceived in the mind and aided by the precise application of mathematics) could be used to perceive the deeper truths of nature in ways that pure experiment never could. Of course, had experiments never confirmed his "free inventions," it's unlikely that we (or anyone else) would remember his contributions a century later.

Understanding How Scientific Change Is Viewed

The debates over string theory represent fundamental differences in how to view science. As the first part of this chapter points out, many people have proposed ideas about what the goals of science should be. But over the years, science changes as new ideas are introduced, and it's in trying to understand the nature of these changes where the meaning of science really comes into question.

The methods in which scientists adapt old ideas and adopt new ones can also be viewed in different ways, and string theory is all about adapting old ideas and adopting new ones.

Precision and accuracy: Science as measurement

The transition from early "natural philosophy" to what we now think of as hard science is largely a technological phenomenon that evolved alongside the development of tools that allowed for more precise measurement. The ancient Greeks and Chinese didn't fail to develop Newton's theory of gravity because they weren't inherently intelligent enough. The Greeks are, after all, the same civilization that brought us Socrates, Plato, Aristotle, Euclid, and Archimedes. They were smart enough. What they didn't have, however, were precise clocks.

Even if they had been inclined to try to develop the sort of science that would later have taken shape during the Enlightenment a couple of millennia later, their theories wouldn't have taken shape if they couldn't properly quantify the measurements of what they were talking about. Without precise clocks, it was difficult to make accurate and exact measurements of the time associated with motion.

Within the area where they *could* measure things precisely, these ancient people did some great work. They used shadows and mathematics to calculate with a fair amount of accuracy the radius and circumference of Earth, for example. Their ability to measure length was more refined than their ability to measure time.

In a practical sense, one of the important things that science accomplishes is making precise and accurate measurements of quantities in the physical world. When people talk about a "scientific fact" that's truly objective, they're usually talking about this measurement aspect of science.

Changes in what can be measured, in turn, lead to changes in what can be talked about scientifically. The invention of devices such as the telescope, microscope, and particle accelerator has allowed for measurements and explorations of whole new realms of nature, turning things that had previously been beyond the reach of scientific inquiry into something that can be measured and, in turn, analyzed and better understood.

Old becomes new again: Science as revolution

The interplay between experiment and theory is never so obvious as in those realms where they fail to match up. At that point, unless the experiment contained a flaw, scientists have no choice but to adapt the existing theory to fit the new evidence. The old theory must transform into a new theory. The philosopher of science Thomas Kuhn spoke of such transformations as *scientific revolutions*.

In Kuhn's model (which not all scientists agree with), science progresses along until it accumulates a number of experimental problems that make scientists redefine the theories that science operates under. These overarching theories are *scientific paradigms*, and the transition from one paradigm to a new one is a period of upheaval in science. In this view, string theory would be a new scientific paradigm, and physicists would be in the middle of the scientific revolution where it gains dominance.

A scientific paradigm, as proposed by Kuhn in his 1962 work, *The Structure of Scientific Revolutions*, is a period of business as usual for science. A theory explains how nature works, and scientists work within this framework.

Kuhn viewed the Baconian scientific method — regular puzzle-solving activities — as taking place within an existing scientific paradigm. The scientist gains facts and uses the rules of the scientific paradigm to explain them.

The problem is that there always seems to be a handful of facts that the scientific paradigm can't explain. A few pieces of data don't seem to fit. During the periods of normal science, scientists do their best to explain this data, to incorporate it into the existing framework, but they aren't overly concerned about these occasional anomalies.

That's fine when there are only a few such anomalies, but when enough of them pile up, it can pose serious problems for the prevailing theory.

As these abnormalities begin to accumulate, the activity of normal science becomes disrupted and eventually reaches the point where a full scientific revolution takes place. In a *scientific revolution*, the current scientific paradigm is replaced by a new one that offers a different conceptual model of how nature functions.

At some point, scientists can't just proceed with business as usual anymore, and they're forced to look for new ways to interpret the data. Initially, scientists attempt to do that with minor modifications to the existing theory. They can tack on an exception here or a special case there. But if there are enough anomalies, and if these makeshift fixes don't resolve all the problems, scientists are forced to build a new theoretical framework.

REMEMBER

In other words, during a scientific revolution, scientists are forced not only to amend their theory, but to construct an entirely new paradigm. It isn't just that some factual details were wrong, but that their most basic assumptions were wrong. In a period of scientific revolution, scientists begin to question everything they thought they knew about nature. For example, in Chapter 10 you see that string theorists have been forced to question the number of dimensions in the universe.

Combining forces: Science as unification

Science can be seen as a progressive series of *unifications* between ideas that were, at one point, seen as separate and distinct. For example, biochemistry came about by applying the study of chemistry to systems in biology. Together with zoology, this yields genetics and *neo-Darwinism* — the modern theory of evolution by natural selection, the cornerstone of biology.

In this way, we know that all biological systems are fundamentally chemical systems. And all chemical systems, in turn, come from combining different atoms to form molecules that ultimately follow the assorted laws defined in the Standard Model of particle physics.

Physics, because it studies the most fundamental aspects of nature, is the science most interested in these principles of unification. String theory, if successful, might unify all fundamental physical forces of the universe down to one fundamental object — a string.

Galileo and Newton unified the heavens and Earth in their work in astronomy, defining the motion of heavenly bodies and firmly establishing that Earth follows exactly the same rules as all other bodies in our solar system. Michael Faraday and James Clerk Maxwell unified the concepts of electricity and magnetism into a single concept governed by uniform laws — electromagnetism. (If you want more information on gravity or electromagnetism, you'll be attracted to Chapter 5.)

Albert Einstein, with the help of his old teacher Hermann Minkowski, unified the notions of space and time as dimensions of space-time, through his theory of special relativity. In the same year, as part of the same theory, he unified the concepts of mass and energy as well. Years later, in his general theory of relativity, he unified gravitational force and special relativity into one theory.

Central to quantum physics is the notion that particles and waves aren't the separate phenomena they appear to be. Instead, particles and waves can be seen as the same unified phenomenon, viewed differently in different circumstances.

The unification continued in the Standard Model of particle physics, when electromagnetism was ultimately unified with the "weak" nuclear force (which is responsible for radioactivity) into a single force, which, in line with physicists' lack of imagination in naming things, was dubbed the "electroweak" force.

The process of unification has been astoundingly successful, because nearly everything in nature can be traced back to the Standard Model — except for gravity. String theory, if successful, will be the ultimate unification theory, finally bringing gravity into harmony with the other forces.

What happens when you break it? Science as symmetry

A *symmetry* exists when you can take something and transform it in some way, and nothing seems to change about the situation. The principle of symmetry is crucial to the study of physics and has special implications for string theory in particular. Even when a symmetry used to be there, but is then broken by some other effect, physicists find it extremely useful to use it to describe the world. They call those *spontaneously broken symmetries*.

Symmetries are obvious in geometry. Take a circle and draw a line through its center, as in Figure 4-1. Now picture flipping the circle around that line. The resulting image is identical to the original image when it's flipped about the line. This is *reflection symmetry*. If you were to spin the figure 180 degrees, you'd end up with the same image again. This is *rotational symmetry*. The trapezoid in Figure 4-1, on the other hand, has *asymmetry* (or lacks symmetry) because no rotation or reflection of the shape will yield the original shape.

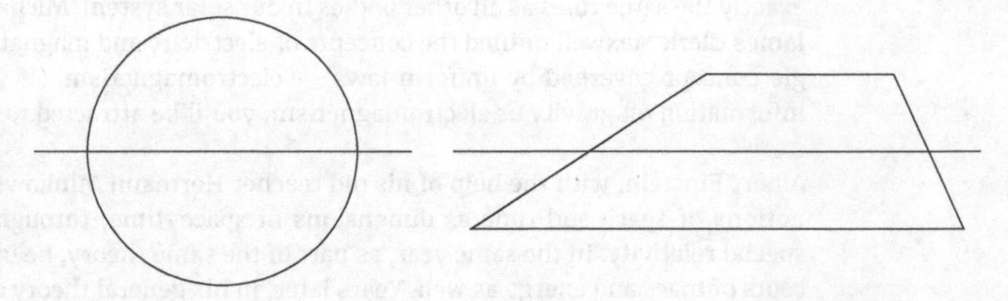

FIGURE 4-1: The circle has symmetry, but the trapezoid doesn't.

The most fundamental form of symmetry in physics is the idea of *translational symmetry*, which is where you take an object and move it from one location in space to another. If we move from one location to another, the laws of physics should be the same in both places. This principle is how scientists use laws discovered on Earth to study the distant universe. Another familiar symmetry is *rotational symmetry*: if we make some experiment on a bench facing south or east, we generally do not expect to see a difference.

In physics, though, symmetry means way more than just taking an object and flipping, spinning, or sliding it through space.

The most detailed studies of energy in the universe indicate that, no matter which direction you look, space is basically the same in all directions. The universe itself seems to have been symmetric from the very beginning, at least to a good approximation.

REMEMBER

The laws of physics don't change over time and across space, as far as we can tell. If we perform an experiment today and perform the same experiment tomorrow, we believe that we will be able to interpret the result according to the same fundamental laws. The same is true if we perform the experiment in New York, in Tokyo, or on Mars.

This does not mean that the outcome of the experiment will be necessarily the same! For instance, while the laws of gravity are, as far as we believe, the same now as they were in the distant past, we know that the universe looked very different back then, compared to now.

Symmetries like rotation, translations, and translations through time are seen as central to the study of science, and in fact, many physicists have stated that symmetry is the single most important concept for physics to grasp.

Indeed, symmetry is so important in physics that we can use it even when it's no longer there. Take a small drop of water, which is a perfect sphere. If we cool it enough, it will become an ice crystal, and as we all know from playing with snowflakes, they are not spherically symmetric. Yet, they do have a beautiful residual symmetry, in many ways richer and more interesting than the original droplet. This is an example of *spontaneous symmetry breaking.*

Another example is the fact that the equations describing the universe are invariant under time-translations (redefining our clocks), while the universe itself is not: We actually believe that it had a beginning with the Big Bang.

In a much more sophisticated way, the study of how the fundamental symmetries of nature are broken is one of the keys to understanding modern physics. An important example that's a bit beyond the scope of our discussion is the famous Higgs boson, which is a cornerstone of particle physics and is intimately related to breaking a symmetry — specifically, the symmetry of the electroweak force mentioned in the previous section.

Another type of symmetry which plays a big role in string theory, and has been getting a lot of attention from theoretical physicists, is called supersymmetry. On top of having a catchy name, it is a required ingredient to make sense of string theory. Supersymmetry makes some very strong predictions on the elementary particles that should exist in the universe, and because of that we know that it is not exactly realized in the universe: It can only be one of the spontaneously broken symmetries. We will tell you all about this in Chapter 10.

IN THIS CHAPTER

» **Matter and energy: Each affects the other**

» **Transferring energy through waves and vibrations**

» **Newton's four revolutionary breakthroughs**

» **Electricity and magnetism: One and the same**

Chapter 5
What You Must Know about Classical Physics

No matter how complex modern physics concepts get, they have their roots in classical ideas. To understand the revolutions leading up to string theory, you need to first understand these basic ideas. You'll then be able to understand how string theory recovers and generalizes them.

In this chapter, we present some physics concepts that you need to be familiar with to understand string theory. First, we discuss three fundamental concepts in physics: matter, energy, and how they interact. Next, we explain waves and vibrations, which are crucial to understanding stringy behavior. Gravity is also key, so Sir Isaac Newton's important discoveries come next. Finally, we give a brief overview of electromagnetic radiation, an important aspect of physics that led directly into the discovery of both relativity and quantum physics — the two theories that together gave birth to modern string theory!

This Crazy Little Thing Called Physics

Physics is the study of matter and its interactions. Physics tries to understand the behavior of physical systems from the most fundamental laws that we can achieve. String theory could provide these most fundamental laws and explain all of the universe in an elegant way.

One other key principle of physics is the idea that many of the laws that work in one location also work in another location — a principle known as *symmetry* (we cover this in more detail later in this section and also in Chapter 4). The connection between physics in different locations is just one sort of symmetry, allowing physics concepts to be related to each other. Science has progressed by taking diverse concepts and unifying them into cohesive physical laws.

That's a very broad definition of physics, but then physics is the broadest science. Because everything you see, hear, smell, touch, taste, or in any way interact with is made of matter and interacts according to some sort of rules, that means physics is literally the study of anything that happens. In a way, chemistry and all the other sciences are approximations of the fundamental laws of physics.

Even if string theory (or some other "theory of everything") were to be verified experimentally, there would still be a need for other sciences. Trying to figure out every single physical system from string theory would be as absurd as trying to study the weather by analyzing every single atom in the atmosphere.

No laughing matter: What we're made of

One of the traits of matter (the "stuff" that everything is made of) is that it requires force to do something. (There are some exceptions to this, but as a rule, a *force* is any influence that produces a change, or prevents a change, in a physical quantity.) *Mass* is the property that allows matter to resist a change in motion (in other words, the ability to resist force). Another key trait of matter is that it's *conserved*, meaning it can't be created or destroyed but can only change forms. (Einstein's theory of relativity showed this wasn't entirely true, as you see in Chapter 6.)

Without matter, the universe would be a pretty boring place. Matter is all around you. The book you're reading, as you lean back comfortably in your matter-laden chair, is made of matter. You yourself are made of matter. But what, exactly, is this stuff called matter?

A matter of inertia, and a matter of the utmost gravity

When we think of any type of matter, a couple of unrelated images come up. First, we cannot go "through" matter: We cannot move through walls, and even wading in water, we encounter resistance. Second, all matter falls down, thanks to gravity.

Newton's key insight was to understand that these are truly fundamental properties of all matter. The first property can be summarized by saying that all matter resists any attempt to put it into motion (we explain this in more detail later in this chapter in "Force, mass, and acceleration: Putting objects into motion"). This is called *inertia*, and the degree to which an object resists this change is its *inertial mass*.

The second property is that all matter is subject to the force of gravity. In fact, in Newton's law of universal gravitation, the mass of an object plays the role of the charge of the gravitational attraction, just like the electric charge does for the electric force. Not that Newton could have known this: The electric force was discovered by Charles-Augustin de Coulomb a century after Newton. To emphasize that the mass is the source of the gravitational attraction, we talk of *gravitational mass*.

If you think that using two different words — inertial and gravitational — to refer to mass is a little odd, you are spot on: It turns out that the two concepts of mass are one and the same! This is absolutely not obvious at first look. It wasn't until the early 1900s that a Hungarian gentleman named Loránd Eötvös came up with a clever experiment that convinced people that this was the right way to think about Newton's discovery. This fact is a crucial ingredient in Einstein's general relativity theory and in string theory.

Scientists discover that mass can't be destroyed

Antoine-Laurent Lavoisier's work in the 18th century provided physics with another great insight into matter. Lavoisier and his wife, Marie Anne, performed extensive experiments that indicated matter can't be destroyed; it merely changes from one form to another. This principle is called the *conservation of mass*.

This isn't an obvious property. If you burn a log, when you look at the pile of ash, it certainly looks like you have a lot less matter than you started with. But, indeed, Lavoisier found that if you're extremely careful that you don't misplace any of the pieces — including the pieces that normally float away during the act of burning — you end up with as much mass at the end of the burning as you started with.

Over and over again, Lavoisier showed this unexpected trait of matter to be the case, so much so that we now take it for granted as a familiar part of our universe. Water may boil from liquid into gas, but the particles of water continue to exist and can, if care is taken, be reconstituted back into liquid. Matter can change form, but can't be destroyed This picture becomes more complicated when one starts looking at chemical and nuclear reactions, when one can trade a tiny amount of mass for quite a lot of energy. But this was not yet understood until well after Lavoiser's time.

As the study of matter progresses through time, things grow stranger instead of more familiar. In Chapter 8, we discuss the modern understanding of matter, which is that we are composed mostly of tiny particles that are linked together with invisible forces across vast (from their scale) empty distances. In fact, as string theory suggests, it's possible that even those tiny particles aren't really there — at least not in the way we normally picture them.

Add a little energy: Why stuff happens

The matter in our universe would never do anything interesting if it weren't for the addition of energy. There would be no change from hot to cold or from fast to slow. Energy is also conserved, as scientists discovered throughout the 1800s as they explored the laws of thermodynamics, but the story of energy's conservation is more elusive than that of matter. You can see matter, but tracking energy proves to be trickier.

Kinetic energy is the energy involved when an object is in motion. *Potential energy* is the energy contained within an object, waiting to be turned into kinetic energy. It turns out that the *total* energy — kinetic energy plus potential energy — is conserved any time a physical system undergoes a change.

String theory makes new predictions about physical systems that contain a *large* amount of energy packed into a very small space. The energies needed for string theory predictions are so large that it may never be possible to construct a device able to generate that much energy and directly test the theory's new predictions. String theory also works when the energies are small, in which case it reproduces the "usual" laws of gravity, as we expect.

The energy of motion: Kinetic energy

Kinetic energy is most obvious in large objects, but it's present in objects of all sizes, down to molecules and atoms. Heat (or *thermal energy*) is really just a bunch of atoms moving rapidly, representing a form of kinetic energy. When water is heated, the particles accelerate until they break free of the bonds with other water molecules and become a gas. The motion of particles can cause energy to be emitted in different forms, such as when a burning piece of coal glows white hot.

Sound is another form of kinetic energy. If two billiard balls collide, the particles in the air will be forced to move, resulting in a noise carried by a wave. Normally this is a tiny amount of energy, but it can actually reach destructive levels when we consider the blast of an explosion or the waves in a fluid other than air — for instance, a tsunami in the ocean. It turns out that light and radiation are also described by waves, and they also carry energy, but they do so in a different way than sound waves and ocean waves do. We will come back to this in Chapter 7.

Stored energy: Potential energy

Potential energy is stored energy. Potential energy takes many more forms than kinetic energy and can be a bit trickier to understand.

A spring, for example, has potential energy when it's stretched out or compressed. When the spring is released, the potential energy transforms into kinetic energy as the spring moves into its least energetic length.

Moving an object in a gravitational field changes the amount of potential energy stored in it. A penny held out from the top of the Empire State Building has a great deal of potential energy due to gravity, which turns into a great deal of kinetic energy when it's dropped (although not, as demonstrated on an episode of *MythBusters*, enough to kill an unsuspecting pedestrian on impact).

It may sound a bit odd, talking about something having more or less energy just because of where it is, but the environment is part of the physical system described by the physics equations. These equations tell us exactly how much potential energy is stored in different physical systems, and they can be used to determine outcomes when the potential energy gets released.

Symmetry: Why some laws were made to be broken

A change in location or position that retains the properties of the system is called a *geometric symmetry* (or sometimes *translational symmetry*). Another form of symmetry is an *internal symmetry*, which is when something within the system can be swapped for something else and the system (as a whole) doesn't change. When a symmetrical situation at high energy collapses into a lower energy *ground state* that is asymmetrical, it's called *spontaneous symmetry breaking*. An example would be when a roulette wheel spins and slows into a "ground state." The ball ultimately settles into one slot in the wheel — and the gambler either wins or loses.

String theory goes beyond the symmetries we observe to predict even more symmetries that aren't observed in nature. It predicts a necessary symmetry that's not observed in nature called *supersymmetry*. At the energies we observe, supersymmetry is an example of a broken symmetry, though physicists believe that in high-energy situations, supersymmetry would no longer be broken (which is what makes it so interesting to study). We cover supersymmetry in Chapters 2 and 10.

Translational symmetry: Same system, different spot

If an object has *translational symmetry*, you can move it and it continues to look the same (for a detailed explanation of this, flip to Chapter 4). Moving objects in space doesn't change the physical properties of the system.

Now, didn't we just say in the last section that the potential energy due to gravity changes depending on where an object is? Yes, we did. Moving an object's location in space can have an impact on the physical system, but the laws of physics themselves don't change (so far as we can tell). If the Empire State Building, Earth, and the penny held over the edge (the entire "system" in this example) were all shifted by the same amount in the same direction, there would be no noticeable change to the system.

Internal symmetry: The system changes, but the outcome stays the same

In an *internal symmetry*, some property of the system can undergo a change without changing anything that we may measure in an experiment.

For example, changing every particle with its antiparticle — changing positive charges to negative and negative charges to positive — leaves the electromagnetic forces involved completely identical. This is a form of internal symmetry called *charge conjugation symmetry*.

Spontaneous symmetry breaking: A gradual breakdown

Physicists believe that the laws of the universe used to be even more symmetric but have gone through a process called *spontaneous symmetry breaking*, where the symmetry falls apart in the universe we observe.

If everything were perfectly symmetric, the universe would be a very boring place. The slight differences in the universe — the broken symmetries — are what make the natural world so interesting, but when physicists look at the physical laws, they tend to find that the differences are fairly small in comparison to the similarities.

TIP

To understand spontaneous symmetry breaking, consider a pencil perfectly balanced on its tip. The pencil is in a state of perfect balance, of equilibrium, but it's unstable. Any tiny disturbance will cause it to fall over. However, no law of physics says *which way* the pencil will fall. The situation is perfectly symmetrical because all directions are equal.

As soon as the pencil starts to fall, however, definite laws of physics dictate the direction it will continue to fall. The symmetrical situation spontaneously (and, for all intents and purposes, randomly) begins to collapse into one definite, asymmetrical form. As the system collapses, the other options are no longer available to the system.

The Standard Model of particle physics, as well as string theory (which includes the Standard Model in an appropriate limit), predicts that some properties of the universe were once highly symmetrical but have undergone spontaneous symmetry breaking into the universe we observe now.

All Shook Up: Waves and Vibrations

In string theory, the most fundamental objects are tiny strings of energy that vibrate or oscillate in simple, regular patterns. In physics, such systems are called harmonic oscillators. Harmonic oscillators are the simplest (and in many ways most universal) physical system that you will ever encounter.

TIP

Though the strings of string theory are different, understanding the vibrations of classical objects — like air, water, jump ropes, springs — can help you understand the behavior of these exotic little creatures when you encounter them. These classical objects can carry what are called *mechanical waves.*

Catching the wave

Waves (as we usually think of them) move through some sort of medium. Like in the examples we discussed when talking about kinetic energy, tidal waves can move through the water, and sound waves through the air, with those materials acting as the medium for the motion. Similarly, if you flick the end of a jump rope or string, a wave moves along the rope or string. In classical physics, waves transport energy, but not matter, from one region to another. One set of water molecules transfers its energy to the nearby water molecules, which means that the wave moves through the water, even though the actual water molecules don't really travel all the way from the start of the wave to the end of the wave.

This is even more obvious if we take the end of a jump rope and shake it, causing a wave to travel along its length. Clearly, the molecules at our end of the jump rope aren't traveling along it. Each group of jump-rope molecules is nudging the next group of jump-rope molecules, and the end result is the wave motion along its length.

There are two types of mechanical waves, as shown in Figure 5-1:

- **Transverse wave:** A wave in which the displacement of the medium is perpendicular to the direction of travel of the wave along the medium, like the flicking of a jump rope
- **Longitudinal wave:** A wave that moves in the same direction in which the wave travels, like a piston pushing on a cylinder of water

The highest point on a transverse wave (or the densest point in a longitudinal wave) is called a *crest.* The lowest point on a transverse wave (or the least dense point in a longitudinal wave) is called a *trough.*

The displacement from the resting point to the crest — in other words, how high the wave gets — is called the *amplitude.* The distance from one crest to another (or one trough to another) is called the *wavelength.* These values are shown on the transverse wave in Figure 5-1. The wavelength is shown on the longitudinal wave as well, although the amplitude is hard to show on that type of wave, so it isn't included.

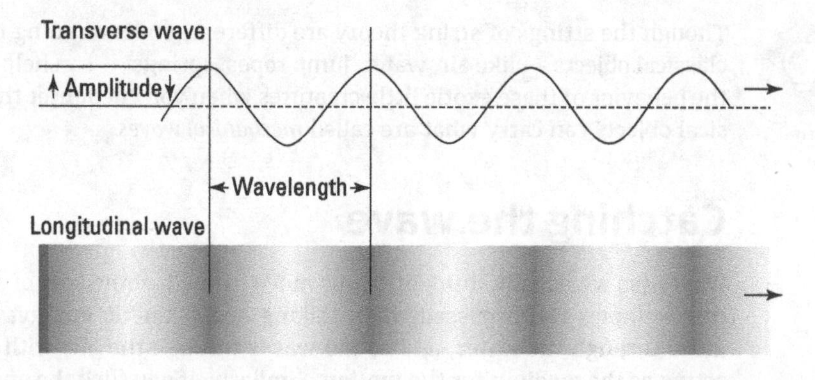

FIGURE 5-1: Waves come in two types: transverse, shown on top, and longitudinal, shown on the bottom.

Another useful thing to consider is the *velocity* (speed and direction) of the wave. This can be determined by its wavelength and *frequency,* which is a measure of how many times the wave passes a given point per second. If you know the frequency and the wavelength, you can calculate the velocity. This, in turn, allows you to calculate the energy contained within the wave.

Another trait of many waves is the *principle of superposition,* which states that when two waves overlap, the total displacement is the sum of the individual displacements, as shown in Figure 5-2. This property is also referred to as *wave interference.*

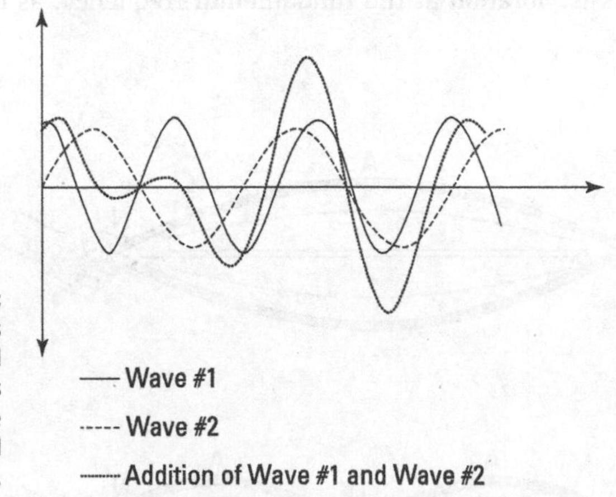

FIGURE 5-2: When two waves overlap, the total displacement is the sum of the two individual displacements.

Consider waves when two ships cross each other's path. The waves made by the ships cause the water to become choppier, and as the waves add height to each other, they cause massive swells.

Similarly, sometimes waves can cancel each other out. If the crest of wave 1 overlaps with the trough of wave 2, they cancel each other out at that point. This sort of interference plays a key role in one of the quantum physics problems we discuss in Chapter 7 — the double slit experiment.

Getting some good vibrations

String theory depicts strings of energy that vibrate, but the strings are so tiny that you never perceive the vibrations directly, only their consequences. To understand these vibrations, you have to understand a classical type of wave called a *standing wave* — a wave that doesn't appear to be moving.

In a standing wave, certain points, called *nodes*, don't appear to move at all. Other points, called *antinodes*, have the maximum displacement. The arrangement of nodes and antinodes determines the properties of various types of standing waves.

The simplest example of a standing wave is one with a node on each end, such as a string that's fixed in place on the ends and plucked. When there is a node on each end and only one antinode in between them, the wave is said to vibrate at the *fundamental frequency*.

Consider a jump rope that is held at each end by a child. The ends of the rope represent nodes because they don't move much. The center of the rope is the antinode, where the displacement is the greatest and where another child will attempt to jump in. This is vibration at the fundamental frequency, as demonstrated in Figure 5-3a.

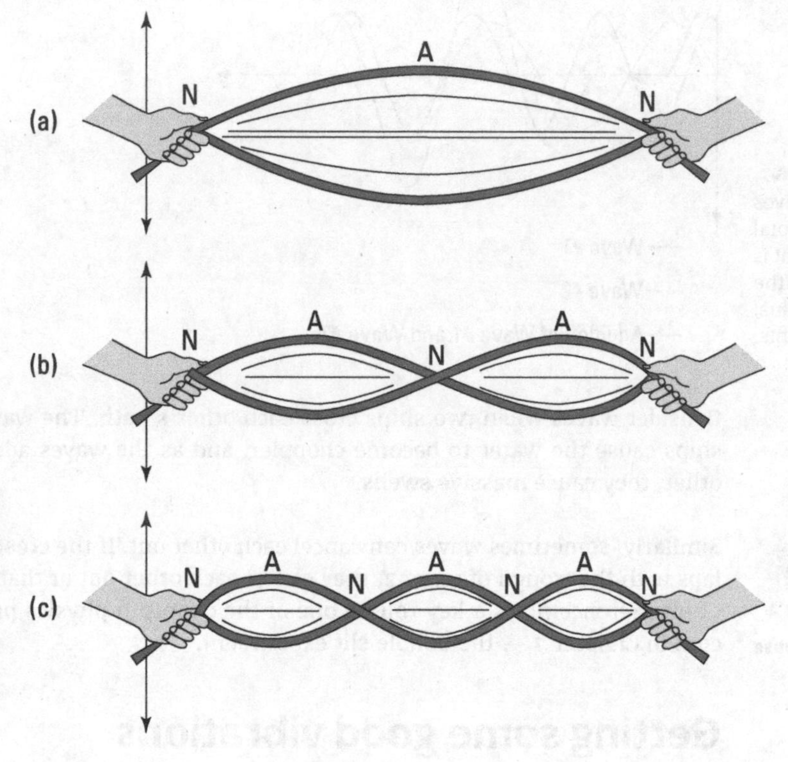

FIGURE 5-3: Examples of standing waves, demonstrating the first three normal modes of a string fixed at both ends. The top wave represents the fundamental frequency.

If the children get ambitious, however, and begin putting more energy into the wave motion of their jump rope, a curious thing happens. Eventually, the children will pump enough energy into the rope that instead of one large antinode, two smaller antinodes are created, and the center of the rope seems to be at rest, as shown in Figure 5-3b. It's almost as if someone grabbed the middle of the rope and is gingerly, but firmly, holding it in place! (If you are a musician, you may recognize this as the *second harmonic* or *first overtone* of the rope.)

Another type of standing wave can be considered if instead of a child holding each end of the rope, one end is mounted on a ring around a pole. The child holding one end begins the wave motion, but the end on the pole is now unconstrained and moves up and down. Instead of having a node on each end, one end is a node (held by the child), and the other is an antinode (moving up and down on the pole).

A similar situation happens when a musician uses a pipe that's closed at one end and open at the other, like pipes in an organ. A node forms at the closed end of the pipe, but the open end of the pipe is always an antinode.

A third type of standing wave has an antinode at each end. This would be represented by either a pipe that's open on both ends or a rope that's free to move on both ends.

The more energy that's pumped into the standing wave, the more nodes form (see Figure 5-3c). The series of frequencies that cause new nodes to form are called *harmonics* or, in music, *overtones.* The waves that correspond to harmonics are called *normal modes* or *vibrational modes.*

Music works because of the manipulation and superposition of harmonic overtones created by these normal modes of vibration. The first three normal modes are shown in Figure 5-3, where a string is fixed on both ends.

REMEMBER

In string theory, the vibrational modes of strings (and other objects) are similar to the vibrating waves that we are talking about in this chapter. In fact, matter itself is seen as the manifestation of standing waves on strings. Different vibrational modes give rise to different particles! We perceive the particles from the lowest vibrational modes, but with higher energies, we may be able to detect other, higher-energy particles.

Newton's Revolution: How Physics Was Born

Many see Sir Isaac Newton's discoveries as the start of modern physics (along with a bit of help from his predecessor Galileo Galilei). Newton's discoveries dominated two centuries of physics, until Albert Einstein took his place at the apex of scientific greatness.

Newton's accomplishments are diverse, but he's known largely for four crucial discoveries that define the realm of physics even today:

- Three laws of motion
- Law of universal gravitation
- Optics
- Calculus

Each of these discoveries has elements that will prove important as you attempt to understand the later discoveries of string theory.

Force, mass, and acceleration: Putting objects into motion

Newton formulated three laws of motion, which showed his understanding of the real meaning of motion and how it relates to force. Under his laws of motion, a force creates a proportional acceleration on an object.

This understanding was a necessary foundation upon which his law of gravity was built (see the next section). In fact, both were introduced in his 1686 book, *Philosophiae Naturalis Principia Mathematica*, a title that translates to *Mathematical Principles of Natural Philosophy*. This book has become known in physics circles by the shorter title *Principia*.

The second law of motion says that the force required to accelerate an object is the product of the mass and acceleration, expressed by the equation $F = ma$, where F is the total force, m is the object's mass, and a is the acceleration. To figure out the total acceleration on an object, you figure out the total forces acting on it and then divide by the mass.

TECHNICAL STUFF

Strictly speaking, Newton said that force is equal to the change in momentum of an object. In calculus, this is the derivative of momentum with respect to time. Momentum is equal to mass times velocity. Because mass is assumed to be constant and the derivative of velocity with respect to time yields the acceleration, the popular $F = ma$ equation is a simplified way of looking at this situation.

REMEMBER

This equation can also be used to define *inertial* mass. If we take a force and divide it by the acceleration it causes on an object, we can determine the mass of the object. One question string theorists hope to answer is *why* some objects have mass and others (such as the photon) do not.

NEWTON MAKES SOME LAWS ABOUT MOTION

Newton's second law of motion, and the way it relates force, acceleration, and mass, is the only law of motion relevant to a string theory discussion. However, for true Newton-o-philes, here are the other two laws of motion, paraphrased for ease of understanding:

- **Newton's first law of motion:** An object at rest remains at rest, or an object in motion remains in motion, unless acted upon by an external force. In other words, it takes a force to cause motion to change.
- **Newton's third law of motion:** When two objects interact through a force, each object exerts a force on the other object that is equal and opposite. In other words, if we exert a force on a wall with a hand, the wall exerts an equal force back on the hand.

Gravity: A great discovery

With the laws of motion in hand, Newton was able to perform the action that would make him the greatest physicist of his age: explaining the motion of the heavens and Earth. His proposal was the *law of universal gravitation*, which defines a force acting between two objects based on their masses and the distance separating them.

The more massive the objects are, the higher the gravitational force is. The relationship with distance is an *inverse* relationship, meaning that as the distance increases, the force drops off. (It actually drops off with the square of the distance — so it drops off very quickly as objects are separated.) The closer two objects are, the higher the gravitational force is.

The strength of the gravitational force determines a value in Newton's equation called the *universal constant of gravitation* or *Newton's constant*. A striking property of the universal constant of gravitation is that it is . . . well, universal. This means that its value is the same in New York, Tokyo, Mars, Alpha Centauri, or the Andromeda galaxy. This value is obtained by performing laboratory experiments and astronomical observations, and calculating what the constant should be. One question still open to physics and string theory is why gravity is so weak compared to other forces.

REMEMBER

Gravity seems fairly straightforward, but it actually causes quite a few problems for physicists because it won't behave itself and get along with the other forces of the universe. Newton himself wasn't comfortable with the idea of a force acting at a distance, without understanding the mechanism involved. But the equations, even without a thorough explanation for what causes gravity, worked. In fact, the equations worked well enough that for more than two centuries, until Einstein, no one could figure out what was missing from the theory. More on this in Chapter 6.

Optics: Shedding light on light's properties

Newton also performed extensive work in understanding the properties of light, a field known as *optics*. Newton supported a view that light moves as tiny particles, as opposed to a theory that light travels as a wave. He performed all his work in optics assuming that light moves as tiny balls of energy flying through the air.

For nearly a century, Newton's view of light as particles dominated, until Thomas Young's experiments in the early 1800s demonstrated that light exhibits the properties of waves — namely, the principle of superposition (see the earlier "Catching the wave" section for more on superposition and the later "Light as a wave: The ether theory" section for more on light waves).

The understanding of light, which began with Newton, would lead to the revolutions in physics by Albert Einstein and, ultimately, to the ideas at the heart of string theory. In string theory, both gravity and light are caused by the behavior of strings.

Calculus and mathematics: Enhancing scientific understanding

To study the physical world, Newton had to develop new mathematical tools. One of the tools he developed was a type of math that we call *calculus*. At the same time Newton invented it, philosopher and mathematician Gottfried Leibniz also created calculus completely independently! Newton needed calculus to perform his analysis of the natural world. Leibniz, on the other hand, developed it mainly to explain certain geometric problems.

REMEMBER

Think for a moment how amazing this really is. A purely mathematical construct, calculus, provided key insights into the physical systems that Newton explored. Alternately, the physical analysis that Newton performed led him to create calculus. In other words, this is a case where mathematics and physics seemed to help build upon each other! One of the major successes of string theory is that it has provided motivation for important mathematical developments that have gone on to be useful in other realms.

The Forces of Light: Electricity and Magnetism

In the 19th century, the physical understanding of the nature of light changed completely. Experiments began to show strong cases where light acts like waves instead of particles, which contradicted Newton (see the "Optics: Shedding light on light's properties" section for more on Newton's findings). During the same time, experiments into electricity and magnetism began to reveal that these forces behave like light, except we can't see them.

By the end of the 19th century, it became clear that electricity and magnetism are different manifestations of the same force: *electromagnetism*. One of the goals of string theory is to develop a single theory that incorporates both electromagnetism and gravity.

Light as a wave: The ether theory

Newton had treated light as particles, but experiments in the 19th century began to show that light acts like a wave. The major problem with this was that waves require a medium. Something has to do the waving. Light seemed to travel through empty space, which contained no substance at all. What was the medium that light used to move through space? What was waving?

To explain the problem, physicists proposed that space is filled with a substance. When looking for a name for this hypothetical substance, they turned back to Aristotle and named it *luminous aether*, or *ether*. (No relation to the gas used in old-timey surgeries as anesthetic.)

Even with this hypothetical ether, though, there were problems. Newton's optics still worked, and his theory described light in terms of tiny balls moving in straight lines, not as waves. It seemed that sometimes light acts like a wave and sometimes it acts as a particle.

Most physicists of the 19th century believed in the wave theory, largely because the study of electricity and magnetism helped support the idea that light is a wave, but they were unable to find solid evidence of the ether.

Invisible lines of force: Electric and magnetic fields

Electricity is the study of how charged particles affect each other. *Magnetism*, on the other hand, is the study of how magnetized objects affect each other. In the 19th century, research began to show that these two seemingly separate phenomena are, in fact, different aspects of the same thing. The physicist Michael Faraday proposed that invisible fields transmit the force.

Electricity and magnetism are linked together

An electrical force acts between two objects that contain a property called *electrical charge*, which can be either positive or negative. Positive charges repel other positive charges, and negative charges repel other negative charges, but positive and negative charges *attract* each other, as in Figure 5-4.

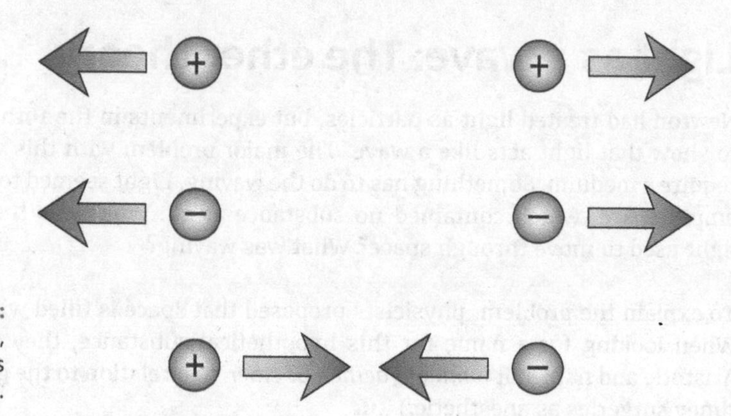

FIGURE 5-4: Like repels like, but opposites attract.

Coulomb's Law, which describes the simplest behavior of the electric force between charged particles (a field called *electrostatics*), is an inverse square law, similar to Newton's law of gravity. This provided some of the first inklings that gravity and electrostatic forces (and, ultimately, electromagnetism) may have something in common.

When electrical charges move, they create an electrical current. These currents can influence each other through a magnetic force. This was discovered by Hans Christian Oersted, who found that a wire with an electrical current running through it can deflect the needle of a compass.

Later experimentation by Michael Faraday and others showed that this works the other way as well — a magnetic force can influence an electrical current. As demonstrated in Figure 5-5, moving a magnet toward a conducting loop of wire causes a current to run through the wire.

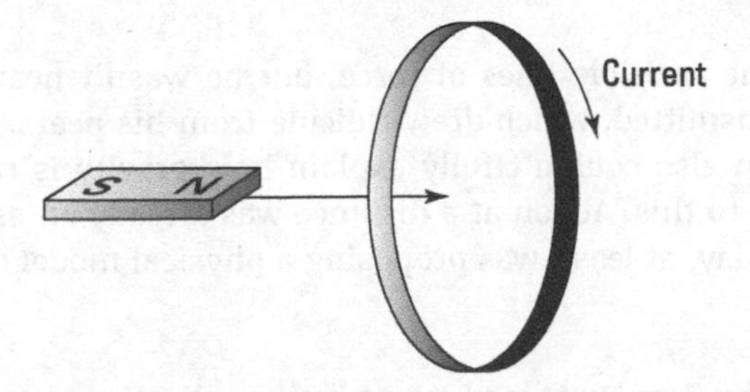

FIGURE 5-5: A magnet moving toward a metal ring creates a current in the ring.

Faraday proposes force fields to explain these forces

In the 1840s, Michael Faraday proposed the idea that invisible lines of force are at work in electrical currents and magnetism. These hypothetical lines make up a *force field* that has a certain value and direction at any given point and can be used to calculate the total force acting on a particle at that point. This concept was quickly adapted to apply to gravity in the form of a *gravitational field.*

REMEMBER

According to Faraday, these invisible lines of force are responsible for electrical force (as shown in Figure 5-6) and magnetic force (as shown in Figure 5-7). They result in an electric field and a magnetic field that can be measured.

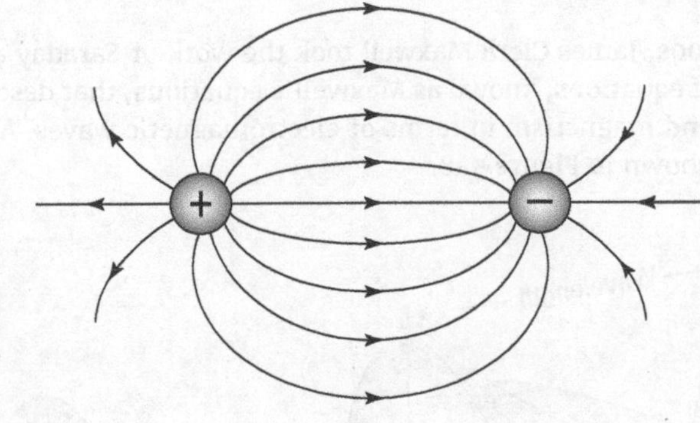

FIGURE 5-6: Positive and negative charges are connected by invisible lines of force.

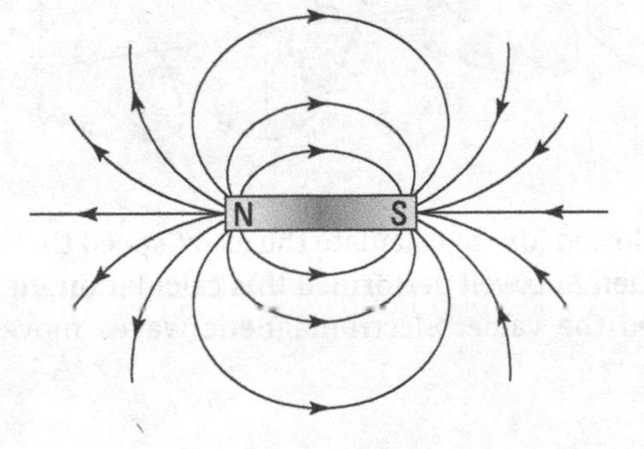

FIGURE 5-7: The north and south poles of a bar magnet are connected by invisible lines of force.

Faraday proposed the invisible lines of force, but he wasn't nearly as clear on how the force is transmitted, which drew ridicule from his peers. Keep in mind, though, that Newton also couldn't fully explain how gravity is transmitted, so there was precedent to this. Action at a distance was already an established part of physics, and Faraday, at least, was proposing a physical model of how it could take place.

TECHNICAL STUFF

The fields proposed by Faraday turned out to have applications beyond electricity and magnetism. Gravity, too, can be written in a field form. The benefit of a force field is that every point in space has a value and direction associated with it. If you can calculate the value of the field at a point, you know exactly how the force will act on an object placed at that point. Today, every law of physics can be written in the form of fields.

Maxwell's equations bring it all together: Electromagnetic waves

Physicists now know that electricity and magnetism are both aspects of the same *electromagnetic force*. This force travels in the form of *electromagnetic waves*. We see a certain range of this electromagnetic energy in the form of visible light, but there are other forms, such as X-rays and microwaves, that we don't see.

In the mid-1800s, James Clerk Maxwell took the work of Faraday and others and created a set of equations, known as Maxwell's equations, that describe the forces of electricity and magnetism in terms of electromagnetic waves. An electromagnetic wave is shown in Figure 5-8.

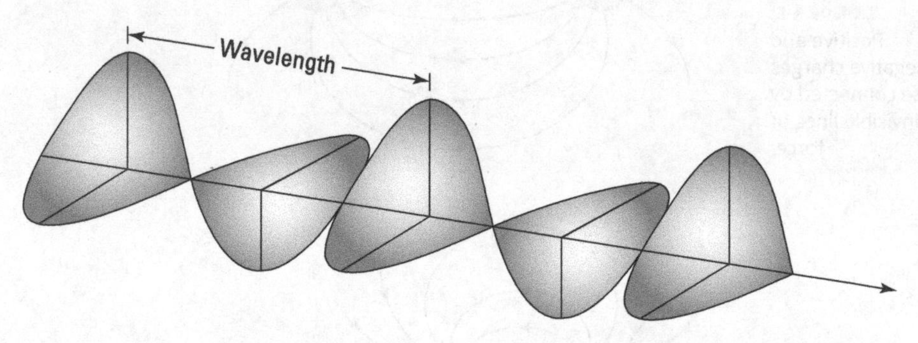

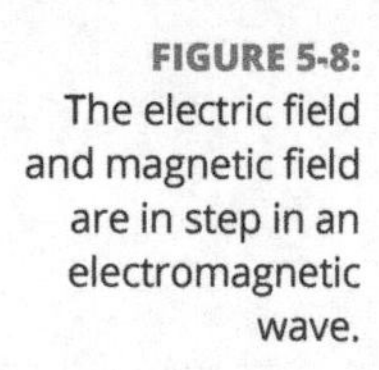

FIGURE 5-8: The electric field and magnetic field are in step in an electromagnetic wave.

Maxwell's equations allowed him to calculate the exact speed that an electromagnetic wave traveled. When Maxwell performed this calculation, he was amazed to find that he recognized the value. Electromagnetic waves move at exactly the speed of light!

Maxwell's equations showed that visible light and electromagnetic waves are different manifestations of the same underlying phenomena. In other words, we see only a small range of the entire spectrum of electromagnetic waves that exist in our universe. Extending this unification to include all the forces of nature, including gravity, would ultimately lead to theories of quantum gravity such as string theory.

Two dark clouds and the birth of modern physics

Two significant unanswered questions about the electromagnetic theory remained. The first problem was that the ether hadn't been detected, while the second involved an obscure problem about energy radiation, called the blackbody problem (described in Chapter 7). What's amazing, in retrospect, is that physicists didn't see these problems (or *dark clouds*, as British scientist Lord Kelvin called them in a 1900 speech) as especially significant, but instead believed they were minor issues that would soon be resolved. As you see in Chapters 6 and 7, resolving these two problems would introduce the great revolutions of modern physics: relativity and quantum physics.

IN THIS CHAPTER

- **Looking for something that allows light waves to travel**
- **Nailing down the relationships between space and time, and mass and energy**
- **Grasping the forces behind gravity**
- **Meeting one of string theory's forerunners**

Chapter 6 Revolutionizing Space and Time: Einstein's Relativity

Albert Einstein introduced his theory of relativity to explain the issues arising from the electromagnetic concepts introduced in Chapter 5. The theory has had far-reaching implications, altering our understanding of time and space. It provides a theoretical framework that tells us how gravity works, but it has left open certain questions that string theory hopes to answer.

In this book, we give you only a glimpse of relativity — the glimpse needed to understand string theory. For a more in-depth look at the fascinating concepts of Einstein's theory of relativity, we suggest *Einstein For Dummies* by Carlos I. Calle, PhD (Wiley).

In this chapter, we explain how the ether model failed to match experimental results and how Einstein introduced special relativity to resolve the problem. We discuss Einstein's theory of general relativity, including a brief look at a rival theory of gravity and how Einstein's theory was confirmed. We then point out some issues arising from general relativity. Finally, we introduce a theory that tried to unify relativity and electromagnetics and is seen by many as a predecessor of string theory.

What Waves Light Waves? Searching for the Ether

In the latter part of the 19th century, physicists were searching for the mysterious *ether* — the medium they believed existed for light waves to wave through. Their inability to discover this ether, despite good experiments, was frustrating, to say the least. Their failure paved the way for Einstein's explanation, in the form of the theory of relativity.

As we explain in Chapter 5, mechanical waves have to pass through a medium, a substance that actually does the waving. Following this logic, light waves passing through the "empty space" of a *vacuum* (a space without any air or other regular matter) would also need a medium. Hence, physicists had predicted a luminous ether that must exist everywhere and be some sort of substance that scientists had never before encountered. In other words, the "empty space" was not (in the view of the time) really empty because it contained ether.

Some things could be predicted about the ether, though. For example, if there was a medium for light, the light was moving through it, like a swimmer moving through the water. And, like a swimmer, the light should travel slightly faster when going in the same direction as the current than when it was trying to go against the current.

REMEMBER

This doesn't mean that the ether itself was moving. Even if the ether was completely still, Earth was moving within the ether, which is effectively the same thing. If you walk through a still body of water, it feels basically the same as if you are walking in place and the water is flowing around you. (In fact, small pools that use this exact principle now exist. You can swim for hours in a pool that's only a few feet long. Because a powerful current is pumping through it, you swim against the current and never go anywhere.)

Physicists wanted to construct an experiment based on this concept that would test whether light traveled different speeds in different directions. This sort of variation would support the idea that light was traveling through an ether medium.

In 1881, physicist Albert Michelson created a device called an *interferometer* designed to do just that. With the help of his colleague Edward Morley, he improved the design and precision of the device in 1887. The Michelson-Morley interferometer is shown in Figure 6-1.

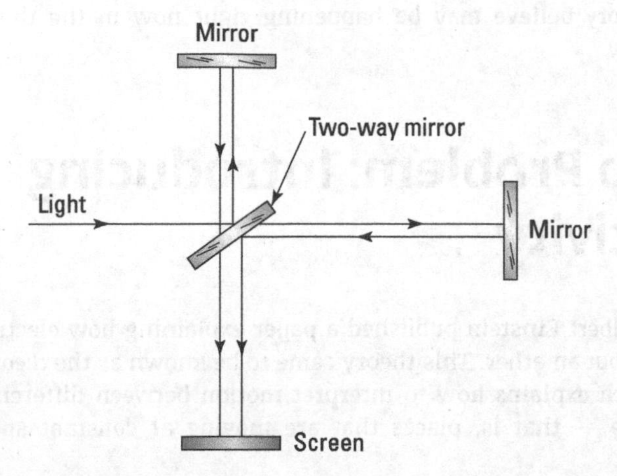

FIGURE 6-1: The Michelson-Morley interferometer sends light beams along two different paths to meet up on a screen.

The interferometer used mirrors that were only partially reflective, so they let half the light pass through and reflected half the light. The mirrors were set at an angle, splitting a single beam of light so it ended up traveling two different paths. The paths traveled perpendicular to each other, but ended up hitting the same screen.

In 1887, Michelson and Morley ran a series of tests with the improved interferometer to discover the ether. They thought that the light traveling along one of the paths should be slightly faster than the light traveling along the other path because one of them would be going either with or against the ether and the other path would be perpendicular to the ether. When the light hit the screen, each beam would have traveled the exact same distance. If one had traveled a slightly different speed, the two beams would be slightly out of phase with each other, which would show distinctive wave interference patterns — light and dark bands would appear — on the screen.

No matter how many times Michelson and Morley conducted the experiment, they never found a difference in speed for the two light beams. They always found the same speed of approximately 670 million miles per hour, regardless of the direction the light traveled.

Physicists didn't immediately dismiss the ether model; instead, they (including Michelson and Morley) considered it a failed experiment, even though it *should* have worked had there been an ether. In 1900, when Lord Kelvin gave his "two dark clouds" speech, 13 years had passed without physicists being able to detect the ether's motion, but it was still assumed that the ether existed.

Sometimes scientists are reluctant to give up on a theory that they've devoted years to, even if the evidence turns against them — something that the critics of string theory believe may be happening right now in the theoretical physics community.

No Ether? No Problem: Introducing Special Relativity

In 1905, Albert Einstein published a paper explaining how electromagnetics can work without an ether. This theory came to be known as the *theory of special relativity*, which explains how to interpret motion between different *inertial frames of reference* — that is, places that are moving at constant speeds relative to each other.

The key to special relativity was that Einstein explained the laws of physics when two objects are moving at a constant speed as the *relative motion* between the two objects, instead of appealing to the ether as an absolute frame of reference that defined what was going on. If you and another astronaut, Amber, are moving in different spaceships and want to compare your observations, all that matters is how fast you and Amber are moving with respect to each other.

Einstein's 1905 paper that introduced special relativity, "On the Electrodynamics of Moving Bodies," was based on two key principles:

- **The principle of relativity:** The laws of physics take the same form for all objects moving in inertial (constant speed) frames of reference.
- **The principle of the speed of light:** The speed of light is the same for all observers, regardless of their motion relative to the light source. (Physicists write this speed using the symbol *c*.)

The genius of Einstein's discoveries is that he looked at the experiments and assumed the findings were true. That was the exact opposite of what other physicists seemed to be doing. Instead of assuming that the theory was correct and the experiments had failed, he assumed that the experiments were correct and the theory had failed.

The ether had made a mess of things, in Einstein's view, by introducing a medium that caused certain laws of physics to work differently depending on how the observer moved relative to the ether. Einstein just removed the ether entirely and assumed that the laws of physics, including the speed of light equal to c, worked the same way regardless of how you were moving — exactly as experiments and mathematics showed them to be!

This came at a conceptual price that physicists had been unwilling to pay before Einstein: The notion of time is no longer absolute in Einstein's theory. For instance, if you and a friend are making a measurement in two corners of the lab, *it does not make sense* to say that you make the two measurements simultaneously — not even in principle. Counterintuitive as this may sound, we now know that it is a fundamental feature of nature.

GIVING CREDIT WHERE CREDIT IS DUE

No physicist works in a vacuum, and that was certainly true of Albert Einstein. Though he revolutionized the world of physics, he did so by resolving the biggest issues of his day, which means he was tackling problems that many other physicists were also working on. He had a lot of useful research to borrow from. Some have accused Einstein of plagiarism or implied that his work wasn't truly revolutionary because he borrowed so heavily from the work of others.

For example, his work in special relativity was largely based on the work of Hendrik Lorentz, George FitzGerald, and Jules Henri Poincaré, who had developed mathematical transformations that Einstein would later use in his theory of relativity. Essentially, they did the heavy lifting of creating special relativity, but they fell short in one important way: They thought it was a mathematical trick, not a true representation of physical reality.

The same is true of the discovery of the photon. Max Planck introduced the idea of energy in discrete packets but thought it was only a mathematical trick to resolve a specific odd situation. Einstein took the mathematical results literally and created the theory of the photon.

The accusations of plagiarism are largely dismissed by the scientific community because Einstein never denied that the work was done by others and, in fact, gave them credit when he was aware of their work. Physicists tend to recognize the revolutionary nature of Einstein's work and know that others contributed greatly to it.

Unifying space and time

Einstein's theory of special relativity created a fundamental link between space and time. The universe can be viewed as having three space dimensions — up/down, left/right, forward/backward — and one time dimension. This 4-dimensional space is referred to as the *space-time continuum.*

The necessity of treating space and time on the same footing is a consequence of the lack of an absolute reference clock, as we are about to see. If you move fast enough through space, the observations that you make about space and time differ somewhat from the observations that other people who are moving at different speeds make. The formulas Einstein used to describe these changes were developed by Hendrik Lorentz (see the nearby sidebar, "Giving credit where credit is due").

REMEMBER

String theory introduces many more space dimensions, so grasping how the dimensions in relativity work is a crucial starting point to understanding some of the confusing aspects of string theory. The extra dimensions are so important to string theory that they get their own chapter, Chapter 15.

Following the bouncing beam of light

The reason for this space-time link comes from applying the principles of relativity and the speed of light very carefully. The speed of light is the distance light travels divided by the time it takes to travel this path, and (according to Einstein's second principle) all observers must agree on this speed. Sometimes, though, different observers disagree on the distance a light beam has traveled, depending on how they're moving through space.

This means that to get the same speed, those observers must *disagree* about the time the light beam travels the given distance.

You can picture this for yourself by understanding the thought experiment depicted in Figure 6-2. Imagine that you're on a spaceship and holding a laser so it shoots a beam of light directly up, striking a mirror you've placed on the ceiling. The light beam then comes back down and strikes a detector.

The spaceship is traveling at a constant speed of half the speed of light (0.5c, as physicists would write it). According to Einstein, this makes no difference to you — you can't even tell that you're moving. However, if astronaut Amber were spying on you, as in the bottom of Figure 6-2, it would be a different story.

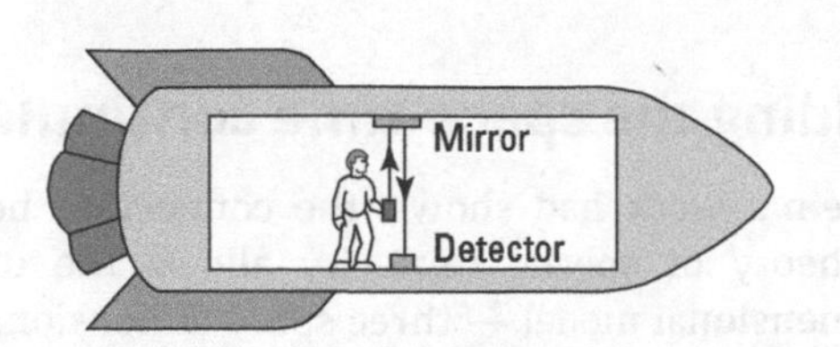

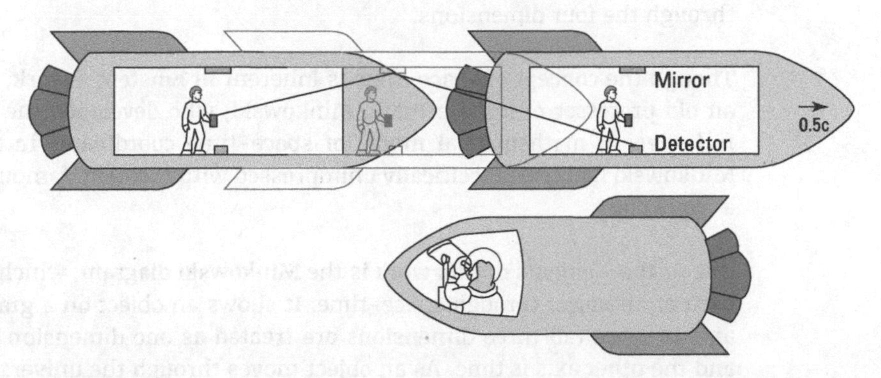

FIGURE 6-2: (Top) You see a beam of light go up, bounce off the mirror, and come straight down. (Bottom) Amber sees the beam travel along a diagonal path.

Amber would see your beam of light travel upward along a diagonal path, strike the mirror, and then travel downward along a diagonal path before striking the detector. In other words, you and Amber would see *different* paths for the light, and more important, those paths aren't even the same length. This means that the time the beam takes to go from the laser to the mirror to the detector must also be different for you and Amber so that you both agree on the speed of light.

This phenomenon is known as *time dilation*, where the time on a ship moving very quickly appears to pass slower than on Earth. In Chapter 17, we explain some ways that this aspect of relativity can be used to allow time travel. In fact, it allows the only form of "time travel" that scientists know for sure is physically possible — though more than travelling through time, we are just experiencing it at different rates.

As strange as it seems, this example (and many others) demonstrates that in Einstein's theory of relativity, space and time are intimately linked together. If you apply Lorentz's transformation equations, they work out so that the speed of light is perfectly consistent for both observers.

REMEMBER

This strange behavior of space and time is evident only when you're traveling close to the speed of light, so no one had ever observed it before. Experiments carried out since Einstein's discovery have confirmed that it's true — time and space are perceived differently, in precisely the way Einstein described, for objects moving near the speed of light.

Building the space-time continuum

Einstein's work had shown the connection between space and time. In fact, his theory of special relativity allows the universe to be represented as a 4-dimensional model — three space dimensions and one time dimension. In this model, any object's path through the universe can be described by its *worldline* through the four dimensions.

Though the concept of space-time is inherent in Einstein's work, it was actually an old professor of his, Hermann Minkowski, who developed the concept into a full, elegant mathematical model of space-time coordinates in 1907. Actually, Minkowski had been specifically unimpressed with Einstein, famously calling him a "lazy dog."

One of the elements of this work is the Minkowski diagram, which illustrates the path of an object through space-time. It shows an object on a graph, where one axis is space (all three dimensions are treated as one dimension for simplicity) and the other axis is time. As an object moves through the universe, its sequence of positions represents a line or curve on the graph, depending on how it travels. This path is called the object's *worldline,* as shown in Figure 6-3. In string theory, the idea of a worldline becomes expanded to include the motion of strings, into objects called *worldsheets.* (See Chapter 17 for more information. A worldsheet can be seen in Figure 17-1.)

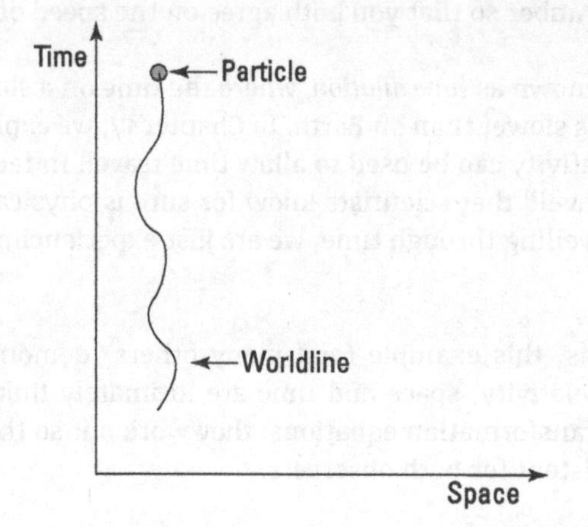

FIGURE 6-3: The path a particle takes through space and time creates its worldline.

Unifying mass and energy

The most famous work of Einstein's life also dates from 1905 (a very busy year for him), when he applied the ideas of his relativity paper to come up with the equation $E=mc^2$, which represents the relationship between mass *(m)* and energy *(E).*

The reason for this connection is a bit involved, but essentially it relates to the concept of kinetic energy discussed in Chapter 5. Einstein found that as an object approaches the speed of light (c), the mass of the object increases. The object goes faster, but it also gets heavier. In fact, if it were actually able to move at c, the object's mass and energy would both be infinite. A heavier object is harder to speed up, so it's impossible to ever actually get the particle up to a speed of c.

In this 1905 paper — "Does the Inertia of a Body Depend on its Energy Content?" — Einstein explained this work and extended it to stationary matter, showing that mass at rest contains an amount of energy equal to mass times c^2.

REMEMBER

Until Einstein, the concepts of mass and energy were viewed as completely separate. He proved that the principles of conservation of mass and conservation of energy are part of the same larger, unified principle, *conservation of mass-energy.* Matter can be turned into energy and energy can be turned into matter because a fundamental connection exists between the two substances.

TIP

If you're interested in greater detail on the relationship of mass and energy, check out *Einstein For Dummies* (Wiley) or the book *E=mc²: A Biography of the World's Most Famous Equation* by David Bodanis (Walker & Company).

Changing Course: Introducing General Relativity

General relativity was Einstein's theory of gravity, published in 1915. With respect to special relativity, it makes one extra assumption: that the laws of physics take the same in all frames of reference, whether inertial or not. (Non-inertial frames of reference are accelerating with respect to each other.) It turns out that this is enough to give a geometric meaning to the force of gravity, and make some predictions that go beyond Netwon's theory of gravitation. General relativity takes the form of field equations, describing the curvature of space-time and the distribution of matter throughout space-time. The effects of matter and space-time on each other are what we perceive as gravity.

Gravity as acceleration

Several years after proposing special relativity, Einstein eventually grasped hold of the principle that would prove crucial to developing his general theory of relativity. He called it the *principle of equivalence*, and it states that an accelerated system is completely physically equivalent to a system inside a gravitational field.

As Einstein later explained the discovery, he was sitting in a chair thinking about the problem when he realized that if someone fell from the roof of a house, they wouldn't feel their own weight. This realization gave him a sudden understanding of the equivalence principle.

As with most of Einstein's major insights, he introduced the idea as a thought experiment. If a group of scientists in an accelerating spaceship performs a series of experiments, they would get exactly the same results as if they were sitting still on a planet whose gravity provides the same acceleration, as shown in Figure 6-4.

FIGURE 6-4: (Left) Scientists performing experiments in an accelerating spaceship. (Right) The scientists get the same results after landing on a planet.

Importantly, this is true locally, meaning when you only have access to a small area of space. On the whole earth for instance this would be wrong because the acceleration or opposite sides of the earth are in opposite directions, whereas this would not be the case for an accelerating spaceship.

Einstein's brilliance was that after he realized an idea applied to reality, he applied it uniformly to every physics situation he could think of.

For example, if a beam of light enters an accelerating spaceship, then the beam will appear to curve slightly, as in the left picture of Figure 6-5. The beam is trying to go straight, but the ship is accelerating, so the path, as viewed inside the ship, will be a curve.

FIGURE 6-5: Both acceleration and gravity bend a beam of light.

By the principle of equivalence, this meant that gravity should also bend light, as shown in the right picture of Figure 6-5. When Einstein first realized this in 1907, he had no way to calculate the effect, other than to predict that it would probably be very small. Ultimately, though, this exact effect would be the one used to give general relativity its strongest support.

Gravity as geometry

The theory of the space-time continuum already existed, but under general relativity Einstein was able to describe gravity as the bending of space-time geometry. He defined a set of *field equations*, which represent the way gravity behaves in response to matter in space-time. These field equations can be used to represent the geometry of space-time at the heart of the theory of general relativity.

As Einstein developed his general theory of relativity, he had to refine Minkowski's notion of the space-time continuum into a more precise mathematical framework (see the earlier "Building the space-time continuum" section for more on this concept). He also introduced another principle, *the principle of covariance*, which states that the laws of physics must take the same form in all coordinate systems. (These coordinate systems are a generalization of the Cartesian coordinates that you are probably familiar with.)

TIP

In other words, all space-time coordinates are treated the same by the laws of physics — in the form of Einstein's field equations. This is similar to the relativity principle, which states that the laws of physics are the same for all observers moving at constant speeds. In fact, after general relativity was developed, it was clear that the principles of special relativity were a special case.

REMEMBER

Einstein's basic principle was that no matter where you are — Toledo, Mount Everest, Jupiter, or the Andromeda galaxy — the same laws apply. This time, though, the laws were the field equations, and your motion could very definitely impact the solutions that came out of the field equations.

Applying the principle of covariance meant that the space-time coordinates in a gravitational field had to work exactly the same way as the space-time coordinates on a spaceship that was accelerating. If you're accelerating through empty space (where the space-time field is flat, as in the left picture of Figure 6-6), the geometry of space-time would appear to curve. This means that if there's an object with mass generating a gravitational field, it has to curve the space-time field as well (as shown in the right picture of Figure 6-6).

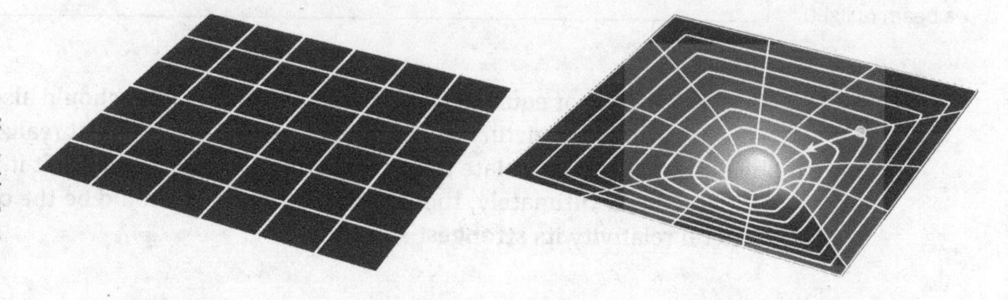

FIGURE 6-6: Without matter, space-time is flat (left), but it curves when matter is present (right).

REMEMBER

In other words, Einstein had succeeded in explaining the Newtonian mystery of where gravity came from! Gravity results from massive objects bending space-time geometry itself.

Because space-time curves, the objects moving through space will follow the "straightest" path along the curve, which explains the motion of the planets. They follow a curved path around the sun because the sun bends space-time around it.

TIP

Again, you can think of this by analogy. If you're flying by plane on Earth, you follow a path that curves around the planet. In fact, if you take a flat map and draw a straight line between the start and end points of a trip, that would *not* be the shortest path to follow. The shortest path is actually the one formed by a "great circle" that you'd get if you cut Earth directly in half, with both points along the outside of the cut. Traveling from New York City to northern Australia involves flying up along southern Canada and Alaska — nowhere close to a straight line on the flat maps we're used to.

Similarly, the planets in the solar system follow the shortest paths — those that require the least amount of energy — and that results in the motion we observe.

Testing general relativity

For most purposes, the theory of general relativity matched the predictions of Newton's gravity, and it also incorporated special relativity — it was a relativistic theory of gravity. But no matter how impressive a theory is, it still has to be confirmed by experiment before the physics community fully embraces it. Today, scientists have seen extensive evidence of general relativity.

One stunning modern example of applying relativity is the global positioning system (GPS). The GPS satellite system sends carefully synchronized beams around the planet. This is what allows military and commercial devices to know their location to within a few meters or better. But the entire system is based upon the synchronization of satellites that had to be programmed with corrections to take into account the curvature of space-time near Earth. (We said earlier that relativity messes with time and clocks, and this is also true for general relativity!) Without these corrections, minor timing errors would accumulate day after day, causing the system to completely break down.

Of course, such equipment wasn't available to Einstein when he published his theory of general relativity in 1915, so the theory had to gain support in other ways.

One solution that Einstein immediately arrived at was to explain an anomaly in the orbit of Mercury. For years, it had been known that Newtonian gravity wasn't quite matching up with astronomers' observations of Mercury's path around the sun. By taking into account the effects of relativity's curved space-time, Einstein's solution precisely matched the path observed by astronomers.

Still, that wasn't quite enough to win over all the critics because another theory of gravity had its own appeal.

Pulled in another direction: Einstein's competition for a theory of gravity

A couple years before Einstein completed his theory of general relativity, Finnish physicist Gunnar Nordström introduced his metric theory of gravity, which also combined gravity with special relativity. He went further, taking James Clerk Maxwell's electromagnetic theory and applying an extra space dimension, which meant that the electromagnetic force was also included in his theory. It was simpler and more comprehensive than Einstein's general relativity, but ultimately wrong (in a way that most physicists then and today see as fairly obvious). But it was the first attempt to use an extra dimension in a unification theory, so it's worth investigating a bit.

Einstein himself was supportive of Nordström's work to incorporate special relativity with gravity. In a 1913 speech on the state of unifying the two, he said that only his work and that of Nordström met the necessary criteria. In 1914, though, Nordström introduced a mathematical trick that increased the stakes of unification. He took Maxwell's electromagnetic equations and formulated them in four space dimensions, instead of the usual three that Einstein had used. The resulting equations included an equation that seemed like it may describe the force of gravity.

REMEMBER

Including the dimension of time, this made Nordström's theory a 5-dimensional space-time theory of gravity. He treated our universe as a 4-dimensional portion of a 5-dimensional space-time. (This is kind of similar to how your shadow on a wall is a 2-dimensional projection of your 3-dimensional body.) By adding an extra dimension to an established physical theory, Nordström unified electromagnetics and gravity! This is an early example of a principle from string theory — that the addition of extra dimensions can provide a mathematical means for unifying and simplifying physical laws.

When Einstein published his complete theory of general relativity in 1915, Nordström jumped ship on his own theory because Einstein could explain Mercury's orbit, while his theory could not.

Nordström's theory had a lot going for it, though, because it was much simpler than Einstein's theory of gravity. In 1917, a year after Nordström himself had given up on it, some physicists considered his metric theory a valid alternative to general relativity. Nothing noteworthy came out of these scientists' efforts, though, so clearly they had backed the wrong theory.

The eclipse that confirmed Einstein's life work

One major difference between Einstein's and Nordström's theories was that they made different predictions about light's behavior. Under Nordström's theory, light always traveled in a straight line. According to general relativity, a beam of light would curve within a gravitational field.

In fact, as early as the late 1700s, physicists had predicted that light would curve under Newtonian gravity. Einstein's equations showed that these earlier predictions were off by a factor of 2.

The deflection of light predicted by Einstein is due to the curvature of space-time around the sun. Because the sun is so massive that it causes space-time to curve, a beam of light that travels near the sun will travel along a curved path — the "shortest" path along the curved space-time, as shown in Figure 6-7.

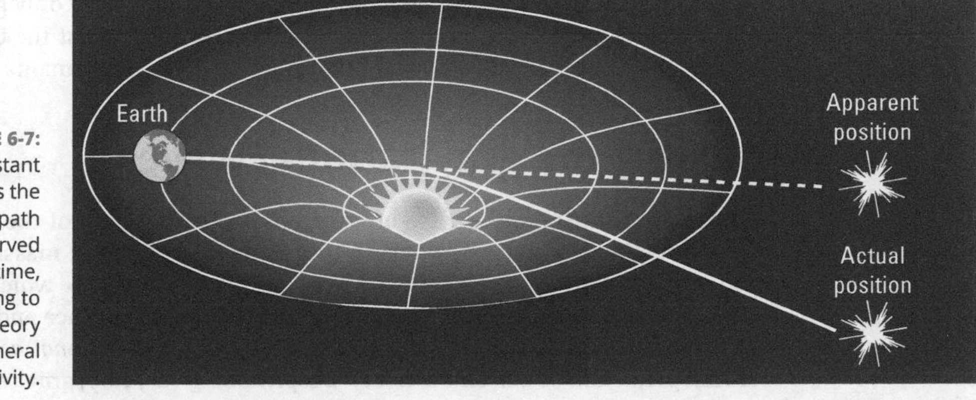

FIGURE 6-7: Light from distant stars follows the shortest path along curved space-time, according to Einstein's theory of general relativity.

In 1911, Einstein had done enough work on general relativity to predict how much the light should curve in this situation, which should be visible to astronomers during an eclipse.

Astronomers on an expedition to Russia in 1914 attempted to observe the deflection of light by the sun, but the team ran into one little snag: World War I. Arrested as prisoners of war and released a few weeks later, the astronomers missed the eclipse that would have tested Einstein's theory of gravity.

That turned out to be great news for Einstein because his 1911 calculations contained an error! Had the astronomers been able to view the eclipse in 1914, the negative results might have caused Einstein to give up his work on general relativity.

When he published his complete theory of general relativity in 1915, he'd corrected the problem, making a slightly modified prediction for how the light would be deflected. In 1919, another expedition set out, this time to the West African Island of Principe. The expedition leader was British astronomer Arthur Eddington, a strong supporter of Einstein.

Despite hardships on the expedition, Eddington returned to England with the pictures he needed, and his calculations showed that the deflection of light precisely matched Einstein's predictions. General relativity had made a prediction that matched observation.

Albert Einstein had successfully created a theory that explained the gravitational forces of the universe and had done so by applying a handful of basic principles. To the degree possible, the work had been confirmed, and most of the physics world agreed with it.

Already well-known from his 1905 work, Einstein's popularity only grew as word spread about general relativity. In 1921, he traveled throughout the United States to a media circus that probably wasn't matched until the Beatlemania of the 1960s.

Surfing the gravitational waves

According to Einstein, the force of gravity is a manifestation of the curvature of the space-time continuum, which in turn is generated by the masses of planets and stars. But all those planets and stars are moving around, which means the curvature is also changing. As a result, you should see that space and time themselves wobble as heavy objects move around you. This *gravitational wave* is another crucial prediction of Einstein's theory. Despite being very tiny, gravitational waves can be observed, providing another spectacular test of general relativity.

To create a wave, it's enough to make a periodic motion in some medium, like air (for instance, sound waves) or water. Light itself is a wave, though, as we note above in "No Ether? No Problem: Introducing Special Relativity," it doesn't need any ether to travel — it's just an oscillation of the electromagnetic field, created by the periodic motion of an electrical charge (that's what happens in an antenna). To generate a gravitational wave, it's enough to take any mass and make it oscillate — for instance, on a pendulum.

However, the strength of the gravitational interaction is really tiny. Although it's easy to make an antenna for electromagnetic radiation, we would need to move an enormous amount of mass to create a gravitational wave big enough to be measured. In practice, you will need to ask your astrophysicist friend for some help on this one: Observable gravitational waves come only from the rapid movement of very heavy objects, like black holes and perhaps very dense stars.

A binary star system is, quite simply, a system of two stars that rotate around each other. It's not much different from a pendulum or an antenna, except for the fact that the rotation moves along an orbit in a plane rather than on a line. If the two stars are emitting gravitational waves, they must be spending some of their energy to do so (exactly like a radio transmitter). As a result, the orbit of the two stars decays, and they spiral closer and closer over time. The measurement of this spiraling motion by Russell Hulse and Joseph Taylor was the first indirect observation of gravitational waves, which won them the 1993 Nobel Prize.

It's harder to observe a gravitational wave directly. We need to take an object of a fixed length, like a rod, and see that it shrinks and stretches as the gravitational wave passes through it, shrinking and stretching the space-time continuum along its way. But this effect is very tiny. Once again, physicists resorted to building an interferometer, like the one created by Michelson and Morley, only much bigger. In fact, they built several: the Laser Interferometer Gravitational-Wave Observatory, or LIGO (in the United States), Virgo (in Italy), and Kamioka Gravitational Wave Detector, or KAGRA (in Japan), with 3-kilometer-long arms.

Using these interferometers, physicists performed the first direct observations of gravitational waves, with the very first wave (from the merger of two black holes) seen by LIGO in 2015, resulting in another Nobel Prize, to Rainer Weiss, Barry Barish, and Kip Thorne in 2017. This was a further confirmation not only of the existence of gravitational waves but also of black holes, some of the most mysterious objects in general relativity — and something about which string theory has much to say.

Applying Einstein's Work to the Mysteries of the Universe

Einstein's work in developing the theory of relativity had shown amazing results, unifying key concepts and clarifying important symmetries in the universe. Still, there are some cases where relativity predicts strange behavior, such as *singularities*, where the curvature of space-time becomes infinite and the laws of relativity seem to break down. String theory continues this work today by trying to extend the concepts of relativity into these areas, hoping to find predictions that work in these regions.

With relativity in place, physicists could look to the heavens and begin a study of how the universe evolved over time, a field called *cosmology*. However, Einstein's field equations also allow for some strange behavior — such as black holes and time travel — that caused great distress to Einstein and others over the years.

If you haven't read about relativity before, this chapter may seem like a whirlwind of strange, exotic concepts — and these new theories certainly felt like that to the physicists of the time. Fundamental concepts — motion, mass, energy, space, time, and gravity — were transformed in a period of only 15 years!

Motion, instead of being just some incidental behavior of objects, was now crucial to understanding how the laws of physics manifested themselves. The laws don't change — this was key to all of Einstein's work — but they can manifest in different ways, depending on where you are and how you're moving — or how space-time is moving around you.

In Chapter 9, we cover the ideas of modern cosmology and astrophysics arising from Einstein's work, such as the black holes that can form when massive quantities of mass cause space-time to curve around itself, and similar problems that come up when trying to apply relativity to the early universe. And, as you see in Chapter 17, we explore how some solutions to Einstein's equations allow time travel.

Einstein himself was extremely uncomfortable with these unusual solutions to his equations. To the best of his ability, he tried to disprove them. When he failed, he would sometimes violate his own basic belief in the mathematics and claim that these solutions represented physically impossible situations.

Despite the strange implications, Einstein's theory of general relativity has been around for nearly a century and has met every challenge — at least when applied to objects larger than a molecule. As we point out in Chapter 2, at very small scales quantum effects become important, and the description using general relativity begins to break down. The usual way to incorporate quantum effects by using the formalism of quantum field theory and renormalization (more on this in Chapter 8) does not work for gravity, and the resulting theory is useless at very short distances. String theory (hopefully) represents one way of reconciling gravity at this realm, as we explain in Chapters 10 and 11.

Kaluza-Klein Theory — String Theory's Predecessor

One of the earliest attempts to unify gravity and electromagnetic forces came in the form of *Kaluza-Klein theory*, a short-lived theory that again unified the forces by introducing an extra space dimension. In this theory, the extra space dimension was curled up to a microscopic size. Though it failed, many of the same concepts were eventually applied in the study of string theory.

Einstein's theory had proved so elegant in explaining gravity that physicists wanted to apply it to the other force known at the time — the electromagnetic force. Was it possible that this other force was also a manifestation of the geometry of space-time?

In 1915, even before Einstein completed his general relativity field equations, the British mathematician David Hilbert said that research by Nordström and others indicated "that gravitation and electrodynamics are not really different." Einstein responded, "I have often tortured my mind in order to bridge the gap between gravitation and electromagnetism."

One theory in this regard was developed and presented to Einstein in 1919 by German mathematician Theodor Kaluza. In 1914, Nordström had written Maxwell's equations in five dimensions and had obtained the gravity equations (see the section "Pulled in another direction: Einstein's competition for a theory of gravity"). Kaluza took the gravitational field equations of general relativity and wrote them in five dimensions, obtaining results that included Maxwell's equations of electromagnetism!

When Kaluza wrote to Einstein to present the idea, the founder of relativity replied by saying that increasing the dimensions "never dawned on me" (which means he must have been unaware of Nordström's attempt to unify electromagnetism and gravity, even though he was clearly aware of Nordström's theory of gravity).

In Kaluza's view, the universe could be viewed as a 5-dimensional cylinder, and our 4-dimensional world was a projection on its surface. Einstein wasn't quite ready to take that leap without any evidence for the extra dimension. Still, he incorporated some of Kaluza's concepts into his own unified field theory, which he published and almost immediately recanted in 1925.

A year later, in 1926, Swedish physicist Oskar Klein dusted off Kaluza's theory and reworked it into the form that has come to be known as *Kaluza-Klein theory.* Klein introduced the idea that the fourth space dimension was rolled up into a tiny circle, so small that there was essentially no way for us to detect it directly.

In Kaluza-Klein theory, the geometry of this extra, hidden space dimension dictated the properties of the electromagnetic force — the size of the circle, and a particle's motion in that extra dimension, related to the electrical charge of a particle. The physics fell apart on this level because the predictions of an electron's charge and mass never worked out to match the true value. Also, many physicists initially intrigued by Kaluza-Klein theory became far more intrigued with the growing field of quantum mechanics, which had actual experimental evidence (as you see in Chapter 7).

Another problem with the theory is that it predicted a particle with zero mass, zero spin, and zero charge. This particle corresponded to the radius of the extra dimension, which was free to oscillate. However, the particle was never observed, despite the fact that it should have been even with the technology of the time. It didn't make sense to have a theory with extra dimensions and then find that the extra dimensions effectively didn't exist.

Whatever the ultimate reason for its failure, Kaluza-Klein theory lasted for only a short time, although there are indications that Einstein continued to tinker with it off and on until the early 1940s, incorporating elements into his various failed unified field theory attempts.

In the 1970s, as physicists began to realize that string theory contained extra dimensions, the original Kaluza-Klein theory served as an example from the past. Physicists once again curled up the extra dimensions, as Klein had done, so they were essentially undetectable (we explain this in more detail in Chapter 10). Such theories are called Kaluza-Klein theories.

IN THIS CHAPTER

- **Exploring the early days of quantum physics**
- **Riding a wave of new ideas about light and particles**
- **Realizing that some quantities can't be measured precisely**
- **Considering different models of quantum physics**
- **Scaling Planck units down (or up) to size**

Chapter 7 Brushing Up on Quantum Theory Basics

As strange as relativity may have seemed to you (see Chapter 6), it's a cakewalk compared to understanding quantum physics. In this strange realm of physics — the realm of the extremely small — particles don't have definite positions or energies. They can exist not only as particles but also as waves, but only when you don't look at them. One hope scientists have is that string theory will account for the many unusual results in quantum physics and reconcile them with general relativity. Particle physics, on the other hand, is at the heart of string theory's origins and is a direct consequence of this early work in quantum physics (see Chapter 8). Without quantum physics, string theory could not exist.

As in the other chapters in this part, the goal of this chapter is not to provide a complete overview of all of quantum physics — there are other books that do a fine job of that, including *Quantum Physics For Dummies* by Steven Holzner (Wiley). Our goal here is to give you the background you need to know about quantum physics so you can understand certain aspects of string theory. It may not seem that these ideas relate directly to string theory, but being familiar with these concepts will be handy down the road when we finally explain string theory itself.

In this chapter, we give you a brief introduction to the history and principles of quantum physics, just enough so you can understand the later concepts related to string theory. We explain how quantum theory allows objects to act as both particles and waves. We explore the implications of the uncertainty principle and probability in quantum physics (dead cat not required). Finally, we discuss the idea that special natural units can be used to describe reality.

Unlocking the First Quanta: The Birth of Quantum Physics

Quantum physics traces its roots back to 1900, when German physicist Max Planck proposed a solution to a *thermodynamics* problem — a problem having to do with heat. He resolved the problem by introducing a mathematical trick: If he assumed that energy was bundled in discrete packets, or *quanta*, the problem went away. (It proved to be brilliant because it worked. There was no theoretical reason for doing this until Einstein came up with one five years later, as discussed in the next section.) In the process of resolving the heat issue, Planck used a quantity known as *Planck's constant*, which has proved essential to quantum physics — and string theory.

The blackbody radiation problem, which Planck was trying to solve, is a basic thermodynamics problem where you have an object that is so hot, it glows. This light can be measured and studied as we let the temperature of the body vary. The problem was that in the 1800s, experiments and theories in this area didn't match up.

A hot object radiates heat in the form of light, like you see with hot charcoals or the metal rings on an electric stove. The best way to describe a blackbody is to consider an old-time coal stove. The radiation (heat) of the coals keeps bouncing around inside the stove until it thoroughly and uniformly heats all of it. At this point, the only interesting property of this object is its temperature. This sort of object is called a *blackbody* — because the object itself doesn't reflect light, but only radiates heat — and throughout the 1800s, much theoretical work in thermodynamics had examined the way heat behaves inside a blackbody.

Now assume that there's a small opening — like a window — in the stove through which light can escape. Studying this light reveals information about what's going on within the blackbody.

Essentially, the heat inside a blackbody takes the form of electromagnetic waves, and because the oven in our example is metal, they're standing waves, with nodes where they meet the side of the oven (see Chapter 5 for details about waves). This fact — along with an understanding of electromagnetics and thermodynamics — can be used to calculate the relationship between light's intensity (or brightness) and wavelength.

REMEMBER

The result of this (classical) description is that as the wavelength of light gets very small (the ultraviolet range of electromagnetic energy), the intensity is supposed to increase dramatically, approaching infinity.

In nature, scientists never actually observe infinities, and this was no exception (see Chapter 2 for more about infinities). The research showed there were maximum intensities in the ultraviolet range that completely contradicted the theoretical expectations, as shown in Figure 7-1. This discrepancy came to be known as the *ultraviolet catastrophe.*

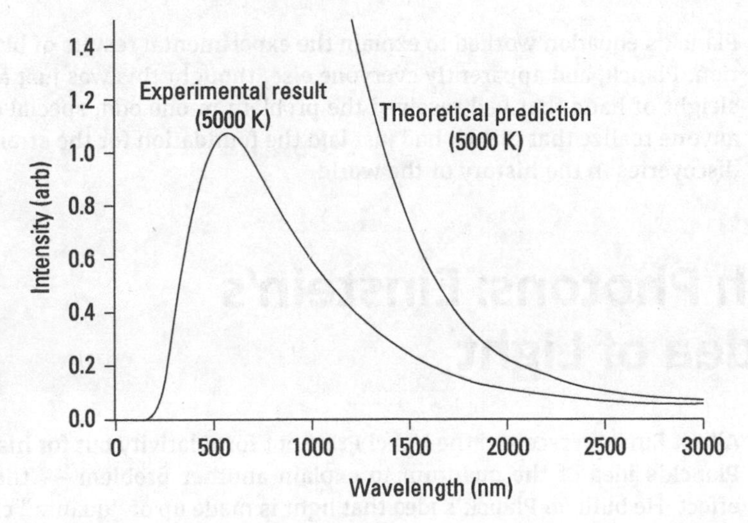

FIGURE 7-1: The ultraviolet catastrophe occurred when theory and experiment didn't match in studying blackbody radiation.

The ultraviolet catastrophe threatened to undermine the theories of electromagnetics and/or thermodynamics. Clearly, if they didn't match the experiment, then one or both of the theories contained errors.

When Planck resolved the ultraviolet catastrophe in 1900, he did so by introducing the idea that the atom could only absorb or emit light in *quanta* (or discrete bundles of energy). One implication of this radical assumption was that there would be less radiation emitted at higher energies. By introducing the idea of discrete energy packets — by quantizing energy — Planck produced a solution that resolved the situation without having to dramatically revise the existing theories (at least at that time).

Planck's insight came when he looked at the data and tried to figure out what was going on. Clearly, the long wavelength predictions were close to matching with the experimental results, but the short wavelength light was not. The theory was over-predicting the amount of light that would be produced at short wavelengths, so he needed a way to limit this short wavelength.

Planck knew that wavelength and frequency were inversely related. So if you're talking about waves with short wavelength, you're also talking about waves with high frequency. All he had to do was find a way to lower the amount of radiation at high frequencies.

Planck reworked the equations, assuming that the atoms could only emit or absorb energy in finite quantities. The energy and frequency were related by a proportion called *Planck's constant*. Physicists use the variable h to represent Planck's constant in his resulting physics equations.

Planck's equation worked to explain the experimental results of blackbody radiation. Planck, and apparently everyone else, thought this was just a mathematical sleight of hand that had resolved the problem in one odd, special case. Little did anyone realize that Planck had just laid the foundation for the strangest scientific discoveries in the history of the world.

Fun with Photons: Einstein's Nobel Idea of Light

Albert Einstein received the Nobel Prize not for relativity but for his work in using Planck's idea of the quantum to explain another problem — the photoelectric effect. He built on Planck's idea that light is made up of "quanta" carrying energy proportional to their frequency. Light, Einstein said, moves not in waves, but in packets of energy. These packets of energy became known as *photons*. Photons are one of the fundamental particles of physics that physicists hope to explain using string theory.

POWERED BY THE PHOTOELECTRIC EFFECT

Modern solar cells work off the same principle as the photoelectric effect. Composed of photoelectric materials, they take electromagnetic radiation in the form of sunlight and convert it into free electrons. Those free electrons then run through wires to create an electric current that can power devices such as ornamental lights in your flowerbed or NASA's Martian rovers. Improvements in photoelectric materials in recent years have resulted in the rapid growth of solar energy as a sector of the economy, so that solar energy panels are now seen as a viable alternative, at some scales, to more traditional electrical power generated from other methods.

The *photoelectric effect* occurs when light shines on certain materials that then emit electrons. It's almost as if the light knocks the electrons loose, causing them to fly off the material. The photoelectric effect was first observed in 1887 by Heinrich Hertz, but it continued to puzzle physicists until Einstein's 1905 explanation.

At first, the photoelectric effect didn't seem that hard to explain. Electrons absorbed light's energy, which caused the electrons to fly off certain materials, such as a metal plate. Physicists still knew very little about electrons — and virtually nothing about the atom — but this explanation made sense.

As expected, if you increased the light's *intensity* (the total energy per second carried by the beam), more electrons definitely were emitted (see the top of Figure 7-2). There were two unexpected problems, though:

- Above a certain wavelength, no electrons are emitted — no matter how intense the light is (as shown in the bottom of Figure 7-2).
- When you increase the light's intensity, the speed of the electrons doesn't change.

Einstein saw a connection between the first problem and the ultraviolet catastrophe faced by Max Planck (see the preceding section for more about Planck's work), but in the opposite direction. The longer wavelength light (or light with lower frequency) failed to do things that were being achieved by the shorter wavelength light (light with higher frequency).

Planck had created a proportional relationship between energy and frequency. Einstein once again did what he was best at — he took the mathematics at face value and applied it consistently. The result was his discovery that high frequency light has higher energy photons, so it's able to transfer enough energy into an electron to knock it loose. Lower frequency photons don't have enough energy to help any electrons escape. The photons must have energy above a certain threshold to knock the electrons loose.

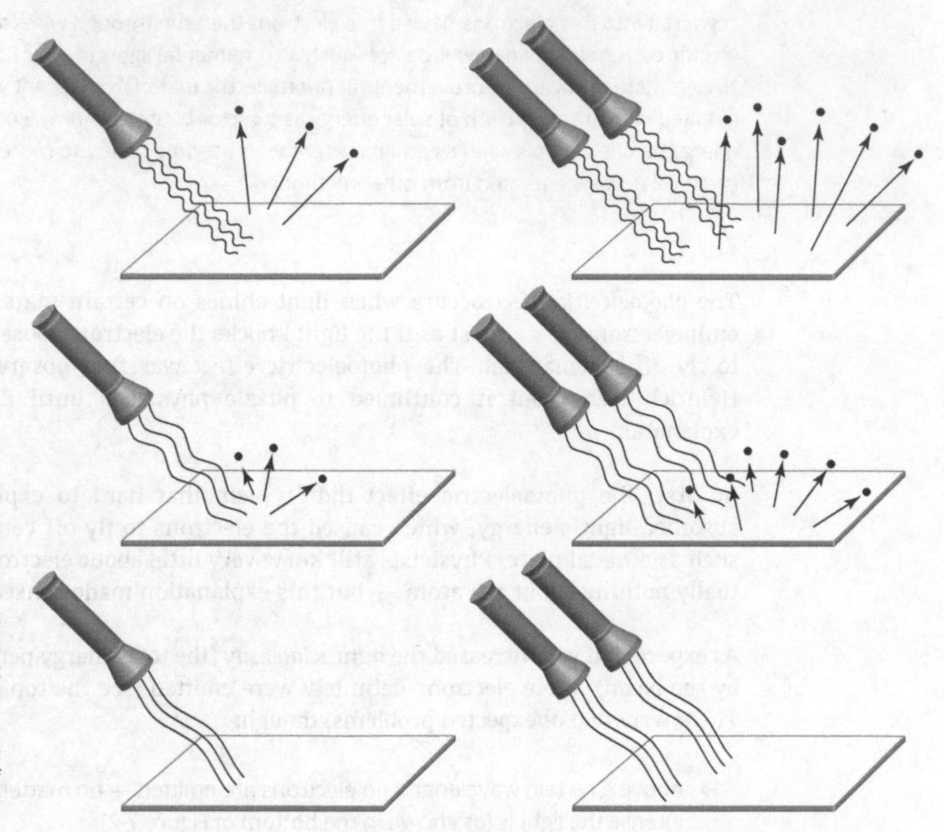

FIGURE 7-2: The photoelectric effect occurs when light collides with a metal plate, causing the release of electrons.

Similarly, the second problem of the noneffect of light's intensity on an electron's speed is also solved by Einstein's quantum view of light. Each photon's energy is based on its frequency (or wavelength), so increasing the intensity doesn't change the energy of each photon; it only increases the total number of photons. This is why increasing the intensity causes more electrons to be emitted, but each electron maintains the same speed. The individual photon knocks out an electron with the same energy as before, but more photons are doing the same job. No single electron gets the benefit of the increase in intensity.

REMEMBER

Based on the principle that the speed of light is constant (the basis of his special theory of relativity), Einstein knew that photons would always move at the same velocity, *c*. Their energy would be proportional to the frequency of the light, based on Planck's equation.

Waves and Particles Living Together

Within quantum physics, two alternate explanations of light work, depending on the circumstances. Sometimes light acts like a wave, and sometimes it acts like a particle, the photon. As quantum physics continued to grow, this *wave particle duality* would come up again and again, as even particles seemed to begin acting like waves. The explanation for this strange behavior lies in the *quantum wave-function*, which describes the behavior of individual particles in a wave-like way. The strange quantum behavior of particles and waves is crucial to understanding quantum theories such as string theory.

Einstein's theory of special relativity had seemingly destroyed the theory of an ether medium, and with his theory of the photon, he proved how light could work without it. The problem was that for more than a century, there had been proof that light does indeed act like a wave.

Light as a wave: The double slit experiment

The experiment that proved light acts like a wave was the *double slit experiment*, which showed a beam of light passing through two slits in a barrier, resulting in light and dark interference bands on a screen. This sort of interference is a hallmark of wave behavior, meaning that light had to be in the form of waves.

These interference patterns in light had been observed in Isaac Newton's time, in the work of Francesco Maria Grimaldi. British physicist and physician Thomas Young vastly improved upon these experiments in 1802.

For the experiment to work, the light passing through the two slits needed to have the same wavelength. Today, you can accomplish this with lasers, but they weren't available in Young's day, so he came up with an ingenious way to get a single wavelength. He made a single slit in a barrier and let light pass through it, and then the light passed through two slits in another barrier. Because the light passing through the two slits came from the same source, the crests of one wave perfectly matched the crests of the other (or, as physicists say, the waves were *in phase*) and the experiment worked. This experimental setup is shown in Figure 7-3.

As you can see in the figure, the end result is a series of bright and dark bands on the final screen. This comes from the interference of the light waves, shown back in Chapter 5 (Figure 5-2). Recall that *interference* means you add the amplitude of the waves. Where high and low amplitudes overlap, they cancel each other out, resulting in dark bands. If high amplitudes overlap, the amplitude of the total wave is the sum of the overlapping amplitudes, and the same happens with low amplitudes, resulting in the bright bands.

FIGURE 7-3: In the double slit experiment, light creates bright and dark bands on a screen.

This dual behavior was the problem facing Einstein's photon theory of light because although a photon has a wavelength, according to Einstein, it's still a particle! How could a particle possibly have a wavelength? Conceptually, it made no real sense until a young Frenchman offered a resolution to the situation.

Particles as a wave: The de Broglie hypothesis

In 1923, Frenchman Louis de Broglie proposed a bold new theory: Particles of matter also have wavelengths and can behave as waves, just as photons do.

Here was de Broglie's line of reasoning: Under special relativity, matter and energy are different manifestations of the same thing. The photon, a particle of energy, has a wavelength associated with it. Therefore, particles of matter, such as electrons, should also have wavelengths. His PhD dissertation set out to calculate what that wavelength (and other wave properties) should be.

Two years later, two American physicists demonstrated de Broglie's hypothesis by performing experiments that showed interference patterns with electrons, as illustrated in Figure 7-4. (The 1925 experiment wasn't actually a double slit experiment, but it showed the interference clearly. The double slit experiment with electrons was conducted in 1961.)

REMEMBER

This behavior showed that whatever quantum law governs photons also governs particles. The wavelength of particles such as the electron is very small compared to the one of visible light. For larger objects, the wavelength is even smaller still, quickly becoming so small as to become unnoticeable. That's why this sort of behavior doesn't show up for larger objects. If you were to fling baseballs through the two slits, you'd never notice an interference pattern.

Still, this left open the question of what was causing the wave behavior in these particles of energy or matter. The answer would be at the core of the new field of quantum mechanics. (String theory will later re-interpret both types of particles — matter and energy — are manifestations of vibrating strings, but that's about 50 years down the road from de Broglie's time.)

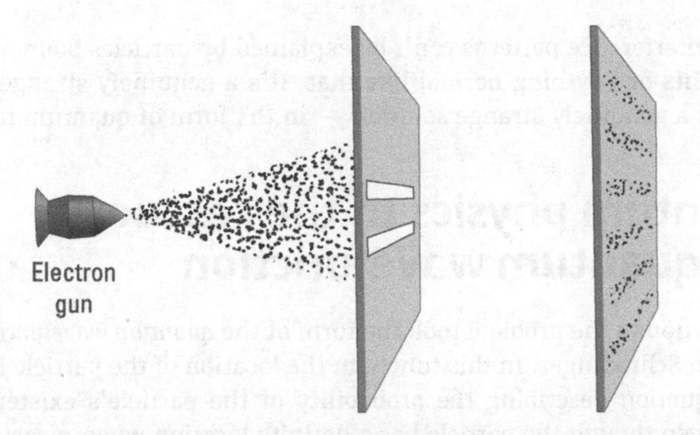

FIGURE 7-4: Electrons demonstrate interference in the double slit experiment.

You can picture the problem if you look at the way the experiment is set up in Figure 7-5. The light wave passes through *both* slits, and that's why the waves interfere with each other. But electrons — or photons, for that matter — *cannot* pass through both slits at the same time if you think of them the way we're used to thinking of them; they have to pick a slit. In this classical case (where the photon is a solid object that has a certain position), there shouldn't be any interference. The beam of electrons should hit the screen in one general spot, just as if you were throwing baseballs through a hole against a wall. (This is why quantum physics challenges our classical thinking about objects and was deemed so controversial in its early years.)

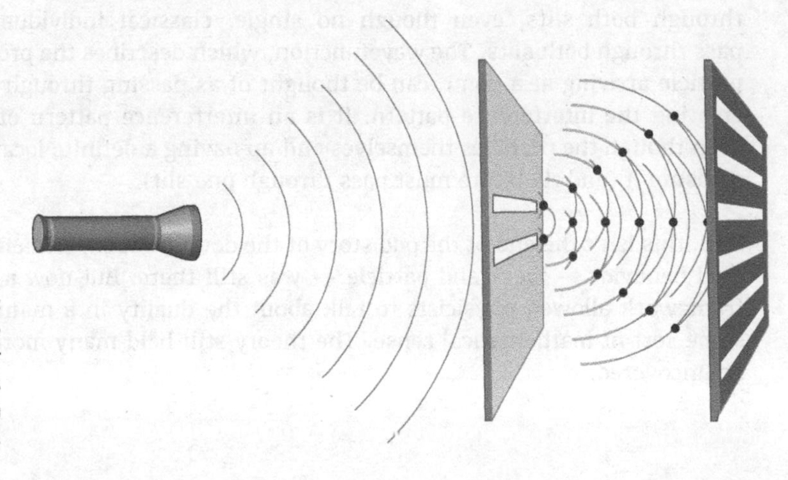

FIGURE 7-5: Interference patterns occur when waves pass through both slits.

In fact, if you close one of the slits, that is exactly what happens. When a slit is closed, the interference pattern goes away — the photons or electrons collect in a single band that spreads out from the brightest spot at the center.

So the interference patterns can't be explained by particles bouncing off the side of the slits or anything normal like that. It's a genuinely strange behavior that requires a genuinely strange solution — in the form of quantum mechanics.

Quantum physics to the rescue: The quantum wavefunction

The solution to the problem took the form of the *quantum wavefunction*, developed by Erwin Schrödinger. In this function, the location of the particle is dictated by a wave equation describing the probability of the particle's existence at a given point, even though the particle has a definite location when measured.

Schrödinger's wavefunction was based in part on his reading of de Broglie's hypothesis about matter having a wavelength. He used this behavior to analyze atomic models created by Niels Bohr (which we cover in Chapter 8). The resulting wavefunction explained the behavior of the atoms in terms of waves. (Bohr's student, Werner Heisenberg, had come up with a different mathematical representation to solve the atomic problem. Heisenberg's matrix method was later shown to be mathematically equivalent to Schrödinger's wavefunction. This sort of parallel work comes up often in physics, as you'll see in Chapters 10 and 11, which cover the development of string theory.)

TIP

The wavefunction created the wave behavior. In this viewpoint, the wave passes through both slits, even though no single, classical individual particle can pass through both slits. The wavefunction, which describes the probability of the particle arriving at a point, can be thought of as passing through both slits and creating the interference pattern. It is an interference pattern of probabilities, even though the particles themselves end up having a definite location when you measure it (and therefore must pass through one slit).

Still, this isn't the end of the odd story of the double slit experiment. The strange dual behavior — wave and particle — was still there. But now a mathematical framework allowed physicists to talk about the duality in a manner that made some sort of mathematical sense. The theory still held many more mysteries to be uncovered.

Why We Can't Measure It All: The Uncertainty Principle

Werner Heisenberg is best known in quantum physics for his discovery of the *uncertainty principle*, which states that the more precisely you measure one quantity, the less precisely you can know another associated quantity. The quantities sometimes come in set pairs that can't both be completely measured. Heisenberg argued that the observation of a system in quantum mechanics disturbs the system enough that you can't know everything about the system. The more precisely you measure the position of a particle, for example, the less it's possible to precisely measure the particle's momentum. The degree of this uncertainty is related directly to Planck's constant — the same value Max Planck had calculated in 1900 in his original quantum calculations of thermal energy. (You'll shortly see that Planck's constant has a lot of unusual implications.)

Heisenberg found that certain complementary quantities in quantum physics are linked by this sort of uncertainty:

- Position and momentum (momentum is mass times velocity)
- Energy and time

This uncertainty is a very odd and unexpected result from quantum physics. Until Heisenberg, no one had ever made any sort of prediction that knowledge was somehow inaccessible on a fundamental level. Sure, there were technological limitations to how well a measurement was made, but Heisenberg's uncertainty principle went further, saying that nature itself doesn't allow you to make measurements of both quantities beyond a certain level of precision.

One way to think about this is to imagine that you're trying to observe a particle's position very precisely. To do so, you have to look at the particle. But you want to be very precise, which means you need to use a photon with a very short wavelength, and a short wavelength relates to a high energy. If the photon with high energy hits the particle — which is exactly what you need to happen if you want to observe the particle's position precisely — then it's going to give some of its energy to the particle. This means any measurement you also try to make of the particle's momentum will be off. The more precisely you try to measure the particle's position, the more you throw off your momentum measurement!

Similar explanations work if you observe the particle's momentum precisely, so you throw off the position measurement. The relationship of energy and time has a similar uncertainty. These are mathematical results that come directly out of analyzing the wavefunction and the equations de Broglie used to describe his waves of matter.

How does this uncertainty manifest in the real world? For that, we'll return to your favorite quantum experiment: the double slit. The double slit experiment has continued to grow odder over the years, yielding increasingly stranger results. For example:

- If you send the photons (or electrons) through the slits one at a time, the interference pattern shows up over time (recorded on film), even though each photon (or electron) has seemingly nothing to interfere with.
- If you set up a detector near either (or both) slits to detect which slit the photon (or electron) went through, the interference pattern goes away.
- If you set up the detector but leave it turned off, the interference pattern comes back.
- If you set up a means of determining later which slit the photon (or electron) went through but do nothing to impact it right now, the interference pattern goes away.

REMEMBER

What does all that have to do with the uncertainty principle? The common denominator among the cases where the interference pattern goes away is that the experiment measured which slit the photons (or electrons) passed through.

When no slit measurement is made, the uncertainty in position remains high, and the wave behavior appears dominant. As soon as a slit measurement is made, the uncertainty in position drops significantly, and the wave behavior vanishes. (There's also a case where you observe *some* of the photons or electrons. Predictably, in this case, you get both behaviors, in exact ratio to how many particles you're measuring.)

Dead Cats, Live Cats, and Probability in Quantum Physics

In the traditional interpretation of quantum physics, the wavefunction is seen as a representation of the probability that a particle will be in a given location. After a measurement is made, the wavefunction collapses, giving the particle a definite value for the measured quantity.

In the double slit experiments, the wavefunction splits between the two slits, and this split wavefunction results in an interference of probabilities on the screen. When the measurements are made on the screen, the probabilities are distributed so that it's more likely to find particles in some places and less likely to find them in other places, resulting in the light and dark interference bands. The particle never splits, but the probability of where the particle will be does split. Until the measurement is made, the distribution of probabilities is all that exists.

This interpretation was developed by the physicist Max Born and grew to be the core of the Copenhagen interpretation of quantum mechanics (which we explain toward the end of this chapter). For this explanation, Born received (three decades later) the 1954 Nobel Prize in physics.

Almost as soon as the explanation of probabilities was proposed, Erwin Schrödinger came up with a morbid thought experiment intended to show how absurd it was. It's become one of the most important, and most misunderstood, concepts in all of physics: Schrödinger's cat experiment.

In this experiment, Schrödinger hypothesized that a radioactive particle has a 50 percent chance of decaying within an hour. He proposed that the radioactive material be placed within a closed box next to a Geiger counter that would detect the radiation. When the Geiger counter detects the radiation from the decay, it will break a glass of poison gas. Also inside the box is a cat. If the glass breaks, the cat dies. (We told you it was morbid.)

Now, according to Born's interpretation of the wavefunction, after an hour the atom is in a quantum state where it is both decayed and not decayed — there's a 50 percent chance of each result. This means the Geiger counter is in a state where it's both triggered and not triggered. The glass containing the poison gas is both broken and not broken. The cat is both dead and alive!

That may sound absurd, but it's the logical extension of the particle being both decayed and not decayed. Schrödinger believed that quantum physics couldn't describe such an insane world, but that the cat had to be either completely alive or completely dead even before the box is opened and the outcome is observed.

After you open the box, according to this interpretation, the cat's state becomes well-defined one way or the other, but in the absence of a measurement, it's in both states. Though Schrödinger's cat experiment was created to oppose Born's interpretation of quantum mechanics, it has become the most dramatic example used to illustrate the strange quantum nature of reality.

Does Anyone Know What Quantum Theory Means?

Quantum physics is based on experimental evidence, much of which was obtained in the first half of the 20th century. The odd behavior of particles has been seen in laboratories around the world, continually backing up the theory, despite all common sense. Precisely because the rules of quantum physics allow us to make specific predictions and match experiments, most physicists are relatively unfazed by how strange these rules are, or by *why* we have such rules to begin with.

Besides, the really strange behavior occurs only on small scales; when you get to the size of cats, the quantum phenomena seem to always take on a definite value, so we don't really expect quantum surprises in our everyday life. Still, even today, the exact meaning of this strange quantum behavior is up in the air — and some physicists are actively trying to understand it better. For instance, all physicists agree that certain objects are quantum (like an electron) and others are not (like a cat), but it's a mystery how exactly we should transition from the quantum world to the classical one.

Some physicists hope that a "theory of everything," perhaps even string theory, may provide a fundamental explanation for the underlying physical meaning of quantum physics. Among them, Lee Smolin has cited string theory's failure to explain quantum physics as a reason to look elsewhere for a fundamental theory of the universe — a view that is certainly not maintained by the majority of string theorists. Most string theorists, and indeed most physicists, believe what matters is that quantum physics works (that is, it makes predictions that match experiment) and the philosophical concerns of why it works are less important. Indeed, explaining quantum mechanics was never really a motivation behind the development of string theory. Einstein spent the last 30 years of his life railing against the scientific and philosophical implications of quantum physics. It was a lively time of debate in physics, as he and Niels Bohr sparred back and forth. "God does not play dice with the universe," Einstein was quoted as saying. Bohr replied, "Einstein, stop telling God what to do!"

REMEMBER

A similar era may be upon us now, as theoretical physicists attempt to uncover the fundamental principles that guide string theory. Unlike quantum theory, there are few (if any) experimental results to base new work on, but there are many Einsteinian critics — again, on both scientific and philosophical grounds. (We get to them in Part 5.)

Even with a firm theory that clearly works, physicists continue to question what quantum physics really means. What is the physical reality behind the mathematical equations? What actually happens to Schrödinger's cat? Where does the quantum world become classical again? Some physicists hope string theory may provide an answer to those questions, though this is far from the dominant view. Still, any successful attempt to extend quantum physics into a new realm could lead to unexpected insights that may resolve the questions.

Quantum Units of Nature: Planck Units

Physicists occasionally use a system of natural units, called *Planck units*, which are calculated based on fundamental constants of nature like Planck's constant, the gravitational constant, and the speed of light.

The idea is simple: Because all of these are universal constants, we can use them to express other quantities in a universal way. For instance, if you say that your spaceship travels at 1 percent of the speed of light, everyone in the universe would know exactly what you mean. If you express the same velocity in meters per second, it would be trickier to communicate (not to mention if you used miles per hour!). Clearly, you can play this game with other universal constants too.

Among these fundamental constants, Planck's constant comes up often in discussing quantum physics. In fact, if you were to perform the mathematics of quantum physics, you'd find that little h variable all over the place. Physicists have found that you can define many useful quantities in terms of Planck's constant and other fundamental constants, such as the speed of light, the gravitational constant, and the charge of an electron.

These Planck units come in a variety of forms. There is Planck length, time, mass, and temperature. You can use various Planck units to derive other units such as the Planck area and volume (the square and cube of the Planck length), the Planck frequency (the inverse of the Planck time), the Planck force . . . well, you get the idea.

For the purposes of the discussion of string theory, only a few Planck units are relevant. They are created by combining the gravitational constant, the speed of light, and Planck's constant, which makes them the natural units to use when talking about phenomena that involve those three constants, such as quantum gravity. The exact values aren't important, but here are the general scales of the relevant Planck units:

- **Planck length:** 10^{-35} meters (if a hydrogen atom were as big as our galaxy, the Planck length would be the size of a human hair)
- **Planck time:** 10^{-43} seconds (the time light takes to travel the Planck length — a very, *very* short period of time)

» **Planck mass:** 10^{-8} kilograms (about the same as a large bacteria or very small insect)

» **Planck energy:** 10^{9} joule or 10^{28} electronvolts (roughly equivalent to a ton of TNT explosive)

TIP

Keep in mind that the exponents represent the number of zeroes, so the Planck energy is a 1 followed by 28 zeroes, in electronvolts. The most powerful particle accelerator on Earth, the Large Hadron Collider (LHC), can produce energy only in the realm of TeV — that is, a 1 followed by 12 zeroes, in electronvolts.

The negative exponents, in turn, represent the number of decimal places in very small numbers, so the Planck time has 42 zeroes between the decimal point and the first nonzero digit. It's a very small amount of time!

Some of these units were first proposed in 1899 by Max Planck himself, before either relativity or quantum physics. Such proposals for *natural units* — units based on fundamental constants of nature — had been made at least as far back as 1881. Planck's constant makes its first appearance in the physicist's 1899 paper. The constant would later show up in his paper on the quantum solution to the ultraviolet catastrophe.

Planck units can be calculated in relation to each other. For example, it takes exactly the Planck time for light to travel the Planck length. The Planck energy is calculated by taking the Planck mass and applying Einstein's $E = mc^2$ (meaning that the Planck mass and the Planck energy are basically two ways of writing the same value).

In quantum physics and cosmology, these Planck units sneak up on us all the time. The Planck mass represents the amount of mass needed to be crammed into the Planck length to create a black hole. A field in quantum gravity theory would be expected to have a vacuum energy with a density roughly equal to 1 Planck energy per cubic Planck length — in other words, it's 1 Planck unit of energy density.

Why are these quantities so important to string theory?

REMEMBER

The Planck length represents the distance where the smoothness of relativity's space-time continuum and the quantum nature of reality begin to rub up against each other. This is the quantum foam we mentioned in Chapter 2. It's the distance where we cannot understand either theory without understanding how they fit together. At those short length-scales, the *quantum behavior* of gravity, whatever it may be, will also dramatically affect any experiment we may attempt. This is the realm where a theory of quantum gravity, such as string theory, is absolutely needed to explain what's going on.

PLANCK UNITS AND ZENO'S PARADOX

If the Planck length represents the shortest distance allowed in nature, it could be used to solve the ancient Greek puzzle called *Zeno's paradox*. Here it is:

> You want to cross a river, so you get in your boat. To reach the other side, you must cross half the river. Then you must cross half of what's left. Now cross half of what's left. No matter how close you get to the other side of the river, you will always have to cover half that distance, so it will take you forever to get across the river because you have to cross an infinite number of halves.

The traditional way to solve this problem is with calculus, where you can show that even though there are an infinite number of halves, it's possible to cross them all in a finite amount of time. (Unfortunately for generations of stymied philosophers, calculus was invented by Newton and Leibnitz 2,000 years after Zeno posed his problem.)

If we take the view that the Plank length is the smallest possible length, however, Zeno's paradox is a nonstarter: When your distance from the opposite shore reaches the Planck length, you *can't* go half anymore. Although this is a rather sophisticated way to answer a 2,500-year-old puzzle, taking the existence of a minimal length too seriously may result in other, far more confusing, paradoxes. Hence, it's best to proceed with care!

In some sense, Planck units are sometimes considered to be *fundamental quanta* of time and space. The idea goes like this: Suppose that you want to make an experiment to determine what goes on at much shorter distances than the Planck length. You have the best possible machines at your disposal, so you don't need to worry about any practical issue (unlike the good people at the LHC in Geneva!). Because of the Heisenberg principle, you know that you need a vast amount of energy to probe such short lengths, larger than the Planck energy and packed in a tiny space. But as soon as you pack a Planckian amount of energy in a Planck volume . . . you get a black hole! (We talk more about black holes in Chapters 9 and 16.) Because of that, there is a *fundamental* limit on the shortest distance you can resolve, and this limit is a direct consequence of quantum physics and general relativity.

The fact that you (probably) cannot make an experiment that allows you to look at a distance shorter than the Planck length doesn't necessarily mean that the spacetime continuum breaks down. Although some strange quantum effects may show up, it may still be possible to treat space as a continuum (this is what string theory does, at least as a first approximation). Still, it's tempting to take the Planck length as a minimal length. You can read an amusing consequence of this idea in the nearby sidebar "Planck units and Zeno's paradox."

REMEMBER

In most string theories, the length of the strings is assumed to be very small. Typically, it is much smaller than the wavelength of the elementary particles that we have so far discovered and, depending on the precise string model, it can get small as the Planck length. The problem with this is that a small length means a very large energy, because energy and length are inversely related. To explore anything in the vicinity of the Planck length you would need an enormous amount of energy.

The Planck energy (the energy related to the Planck length) is 16 orders of magnitude (add 16 zeroes!) larger than the newest, most powerful particle accelerator on Earth can reach. Directly exploring such small distances requires a vast amount of energy, far more energy than we can produce with current technology. It is possible that the string length ends up being quite a bit larger than the Planck length, which would mean we need less energy to observe strings. But most scenarios require energies well beyond our current capabilities.

IN THIS CHAPTER

» Accepting the atom and examining its parts

» Applying quantum physics to the atom's bits and pieces

» Categorizing particles into bosons and fermions

» Using the Standard Model to reveal four forces of physics

» Pondering the range of energies and masses observed

Chapter 8

The Standard Model of Particle Physics

During the mid-1900s, physicists further explored the foundations of quantum physics and the components of matter. They focused on the study of particles in a field that became known as *particle physics.* More itty-bitty particles seemed to spring up every time physicists looked for them! By 1974, physicists had determined a set of rules and principles called the *Standard Model of particle physics* — a model that includes all interactions except gravity. While some of the finer features of the Standard Model may need to be corrected, by and large it captures most of the features of elementary particles.

Here we explore the Standard Model of particle physics and how it relates to string theory. Any complete string theory will have to include all the nice features of the Standard Model and extend beyond it to include gravity as well. In this chapter, we describe the structure of the atom, including the smaller particles contained within it, and the physical models used to explain the interactions holding matter together. We identify the two categories of particles that exist in our universe, fermions and bosons, and the different rules they follow. Finally, we point out the problems that remain with the Standard Model, which string theory hopes to resolve.

The topics related to the development of the Standard Model of particle physics are detailed and fascinating in their own right, but this book is about string theory. So our review of the material in this chapter is necessarily brief and is in no way intended to be a complete look at the subject. Many of the initial topics regarding the discovery of the structure of the atom are recounted in *Einstein For Dummies* (Wiley), and many other popular books explore some of the more involved concepts of particle physics that come along later.

Atoms, Atoms, Everywhere Atoms: Introducing Atomic Theory

Physicist Richard P. Feynman once said that if he could boil down the most important principles of physics to a single sentence, it would be, "All things are made of atoms." (He actually expanded on that, meaning he boiled physics down to a single compound sentence. For our immediate purposes, this first bit is enough.) The structure of atoms determines fundamental properties of matter in our universe, such as how atoms interact with one another in chemical combinations. The study of physics at the scale of an atom is called *atomic theory*, or atomic physics. Though this is several scales above the scale that string theory operates on, understanding the smaller structure of matter requires some understanding of its atomic-level structure.

Ancient Greeks considered the question of whether you can divide an object forever. Some — such as the fifth century BCE philosopher Democritus — believed that you would eventually reach a small chunk of matter that can't be divided any more, and they called these smallest chunks *atoms*.

Our modern understanding of atoms began in 1738 when Swiss mathematician David Bernoulli explained how pressurized gas behaved by assuming that gas was made up of tiny particles. Bernoulli hypothesized that the heat of a gas was related to the speed of the particles. (This built on the work of Robert Boyle, nearly a century earlier.)

In 1808, British chemist John Dalton tried to explain the behavior of *elements* — substances that can't be chemically broken down into simpler substances — by assuming that they were made up of atoms.

REMEMBER

According to Dalton, each atom of an element was identical to other atoms of the same element, and they combined in specific ways to form the more complex substances we see in our universe.

Over the next century, evidence for atomic theory mounted (see the sidebar "Einstein's contribution to atomic theory"). The Russian chemist Dmitry Ivanovich Mendeleyev (often written as Dmitri Mendeleev) formalized the properties of the atoms for the various elements into the periodic table of the elements. The complex structures formed by different atomic elements were called *molecules*, though the exact mechanism for how atoms formed molecules was still unclear.

Mendeleyev's periodic table provided a structure for looking at the atomic elements, grouped together by common behaviors and physical properties. Each element was designated by an atomic number, for example, even though scientists didn't yet have a clear sense of the physical meaning of the atomic numbers. It wouldn't be until years later, with the discovery that the atomic nucleus contained protons, that the physical basis for these atomic numbers would be fully understood.

REMEMBER

It took more than 150 years from the time of Bernoulli for physicists to fully adopt the atomic model. Then, as you find out in the next section, after it was finally adopted, it was found to be incomplete! The complications arising in the study of string theory may well prove to take just as long to be resolved. Perhaps ultimately the theory will be just as incomplete! But that doesn't mean string theory is necessarily "wrong," any more than atomic theory is "wrong."

EINSTEIN'S CONTRIBUTION TO ATOMIC THEORY

As if he weren't credited with enough, Albert Einstein is also frequently cited as the person who provided some of the last definitive support for the atomic theory of matter in two of his 1905 papers.

One of the papers was his PhD thesis, in which he calculated the approximate mass of an atom and the size of sugar molecules. This work earned him his doctorate from the University of Zurich.

The other paper involved analyzing random motion in smoke and liquids. This type of motion, called *Brownian motion,* had puzzled physicists for some time. Einstein pictured the motion as the result of atoms of smoke or liquid being jostled around by atoms of the surrounding gas or liquid, which explained the phenomenon perfectly. His predictions were supported by experimental findings.

Popping Open the Atomic Hood and Seeing What's Inside

Today, scientists know that atoms are not, as the Greeks imagined, the smallest chunks of matter. Scientists eventually realized that atoms have multiple parts inside them:

- Negatively charged electrons circling the nucleus
- A positively charged nucleus

The particles that compose the nucleus (it's made up of smaller pieces too) and electrons are among the particles that the Standard Model of particle physics explains — and that ultimately string theory should also explain.

Discovering the electron

An *electron* is a negatively charged particle contained within an atom. It was discovered in 1897 by British physicist J.J. Thomson, though scientists had hypothesized about charged particles (including the name "electron") earlier.

Some physicists had already hypothesized that units of charge might be flowing around in electrical apparatus. (Benjamin Franklin proposed such an idea as early as the 1700s.) Technology only caught up to this idea in the late 1800s, with the creation of the cathode ray tube, shown in Figure 8-1.

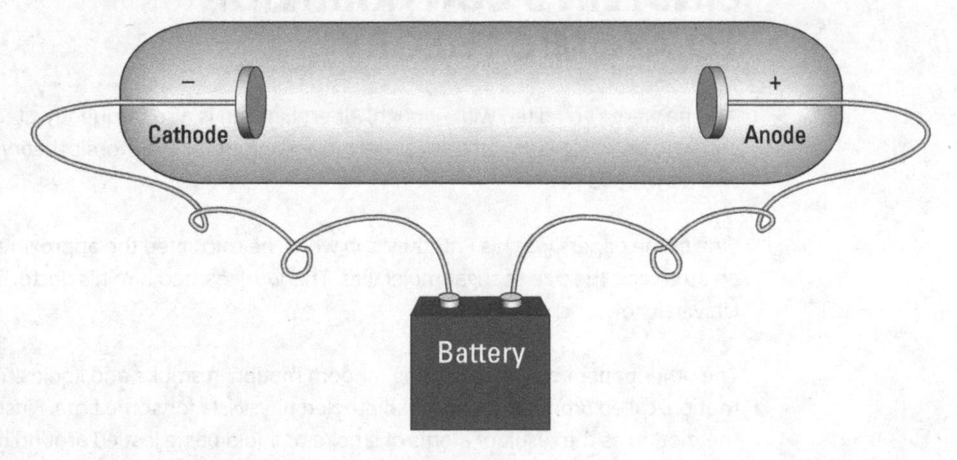

FIGURE 8-1: Cathode ray tubes allow charged particles to be studied in a vacuum.

In a cathode ray tube, a pair of metal disks is connected to a battery. The metal disks are placed inside a sealed glass tube that contains no air — a vacuum tube. The electrical voltage causes one of the metal plates to become positively charged (an *anode*) and one to become negatively charged (the *cathode*, from which the device gets its name). Cathode ray tubes are the basis of traditional television and computer monitor tubes.

When the electrical current is switched on, the tube begins to glow green. In 1897, Thomson was head of the Cavendish laboratory in Cambridge, England, and set about to test the properties of this cathode ray tube glow. He discovered that the glow was due to a beam of negatively charged particles flying between the plates. These negatively charged particles later came to be called electrons. Thomson also figured out that electrons are incredibly light — 2,000 times lighter than a hydrogen atom.

Thomson not only discovered the electron but also theorized that the electron was part of the atom (atoms weren't a completely accepted idea at the time) that somehow got knocked free from the cathode and flowed through the vacuum to the anode. With this discovery, scientists began finding ways to explore the inside of atoms.

The nucleus is the thing in the middle

In the center of an atom is a *nucleus*, a dense ball of matter with a positive electrical charge. Shortly after electrons were discovered, it became clear that if you extracted an electron from an atom, the atom was left with a slightly positive electrical charge. For a while, the assumption was that the atom was a positively charged mass that contained negative electrons, like pieces of negatively charged fruit in a positively charged fruitcake. The entire fruitcake would be neutral unless you extracted some fruit from it. (Scientists of the day, being of a different dietary constitution than most of us today, explained it as plum pudding instead of fruitcake. Plum pudding or fruitcake — it amounts to roughly the same unappetizing picture.)

In 1909, however, an experiment by Hans Geiger (of Geiger counter fame) and Ernest Marsden, working under Ernest Rutherford, challenged this picture. These scientists fired positively charged particles at a thin sheet of gold foil. Most of the particles passed straight through the foil, but every once in a while, one of them bounced back sharply. Rutherford concluded that the positive charge of the gold atom wasn't spread throughout the atom, as hypothesized in the fruitcake model, but was concentrated in a small positively charged nucleus, and that the rest of the atom was empty space. The particles that bounced were the ones that hit this nucleus.

Watching the dance inside an atom

As they tried to figure out the atom's structure, a natural model for scientists to look to was the planetary model, shown in Figure 8-2. In this model, the electrons move around the nucleus in orbits. Physicist Niels Bohr determined that these orbits were governed by the same quantum rules that Max Planck had originally applied in 1900 — that energy had to be transferred in discrete packets.

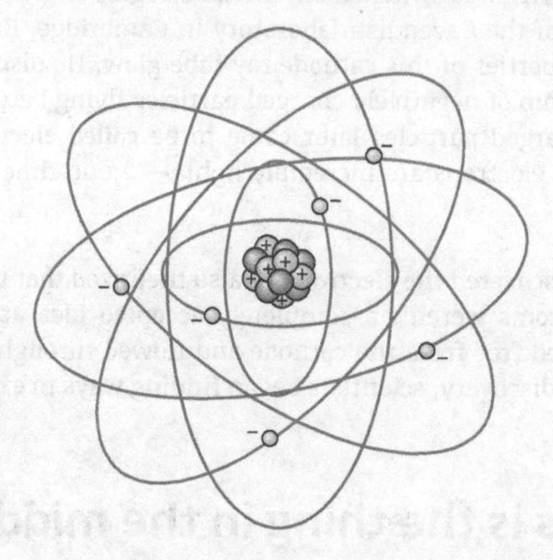

FIGURE 8-2: The Rutherford-Bohr model of the atom has electrons moving in orbits around a positively charged nucleus.

In astronomy, Earth and the sun are attracted to each other by gravity, but because Earth is in motion around the sun, they never come into contact. A similar model could explain why the negative and positive portions of the atom never came into contact.

The first planetary model was proposed in 1904 by Nobel Prize winner Hantaro Nagaoka. It was based on the rings of Saturn and called the Saturnian model. Certain details of the model were disproved by experiment, and Nagaoka abandoned the model in 1908, but Ernest Rutherford revised the concept to create his own planetary model in 1911, which was more consistent with experimental evidence.

When atoms emit electrons, the electrons' energy follows certain precise patterns. Bohr realized in 1913 that this meant Rutherford's model required some revision. To fit the patterns, he applied the idea that energy was *quantized*, or bundled together in certain quantities, which allowed for stable orbits (instead of the collapsing orbits predicted by electromagnetism). Each electron could exist only in a certain, precisely defined energy state within its orbit. Going from one orbit to a different orbit required the electron to have enough energy to jump from one energy state to another.

TIP

Because of the quantum nature of the system, adding half the amount of energy to go from one orbit to another didn't move the electron halfway between those orbits. The electron remained in the first orbit until it received enough energy to kick it all the way into the higher-energy state. This is yet more of the strange behavior you've (hopefully) come to expect from quantum physics.

The Rutherford-Bohr model works pretty well in describing the hydrogen atom, but as atoms get more complex, the model begins to break down. Still, the basic principles hold for all atoms:

- A nucleus is at the center of an atom.
- Electrons move in orbits around the nucleus.
- The electron orbits are quantized (they have discrete energy levels) and are governed by the rules of quantum physics (though it would take several years for those rules to become developed, as described in Chapter 7).

The Quantum Picture of the Photon: Quantum Electrodynamics

The development of the theory of *quantum electrodynamics (QED)* was one of the great intellectual achievements of the 20th century. Physicists were able to redefine electromagnetism by using the new rules of quantum mechanics, unifying quantum theory and electromagnetic theory. Quantum electrodynamics was one of the first quantum approaches to a quantum field theory (described in the next section), so it introduced many features found in string theory (which is also built on the framework of quantum field theory).

Quantum electrodynamics began with the attempt to describe particles in terms of quantum fields, starting in the late 1920s. In the 1940s, QED was completed three distinct times: by the Japanese physicist Sin-Itiro Tomonaga during World War II and later by American physicists Richard Feynman and Julian Schwinger. These three physicists shared the 1965 Nobel Prize in physics for their work.

Dr. Feynman's doodles explain how particles exchange information

Though the principles of quantum electrodynamics were worked out by three individual physicists, the most famous founder of QED was undeniably Richard

Phillips Feynman. Feynman was equally good at the mathematics and the explanation of a theory, which resulted in his creation of *Feynman diagrams* — a visual representation of the mathematics in QED.

Feynman is one of the most interesting characters in 20th century physics, easily ranking with Einstein in personality, if not in pure fame. Early on in his career, Feynman made the conscious decision to work only on problems he found interesting, something that certainly served him well. Fortunately for the world of physics, one of these problems was quantum electrodynamics.

Because electromagnetism is a field theory, the result of QED was a *quantum field theory* — a quantum theory for the electron and photon fields. Electrons and photons are ripples of these quantum fields across time and space. You can imagine that the mathematics of such a theory was intimidating, to say the least, even to those trained in physics and mathematics.

Feynman was brilliant not only with physical theory and mathematics, but also with explanation. One way he simplified things was through the application of his Feynman diagrams. Though the math was still complex, the diagrams meant you could begin talking about the physics without needing all the complexity of the equations. And when you did need the actual numbers, the diagrams helped you organize your computations.

In Figure 8-3, you can see a Feynman diagram of two electrons approaching each other. The Feynman diagram is set on a Minkowski space, introduced in Chapter 6, which depicts events in space-time. The electrons are the solid lines (called *propagators*), and as they get near each other, a photon (the squiggly propagator; see Chapter 7 for the basics of photons) is exchanged between the two electrons.

In other words, in QED two particles communicate their electromagnetic information by emitting and absorbing a photon. A photon that acts in this manner is called a *virtual photon* or a *messenger photon* because it's created solely for the purpose of exchanging this information. This was the key insight of QED, because without this exchange of a photon, there was no way to explain how the information was communicated between the two electrons.

But things can get quite a bit more complicated because that virtual photon may also emit virtual particles (in this case, an electron and a positron) and you have to consider all these contributions to find the correct interaction. At first glance, this isn't a big issue: Contributions with more and more virtual particles should become smaller and smaller, so they should change the final result only by a little. (This idea is called *perturbation theory,* and it's important in all of physics, including string theory. You'll find more on perturbation theory in Chapter 11.) However, there's a snag: We have to consider all possible configurations of virtual particles, including those where particles get infinitely close to each other. This makes our computation blow up in our face.

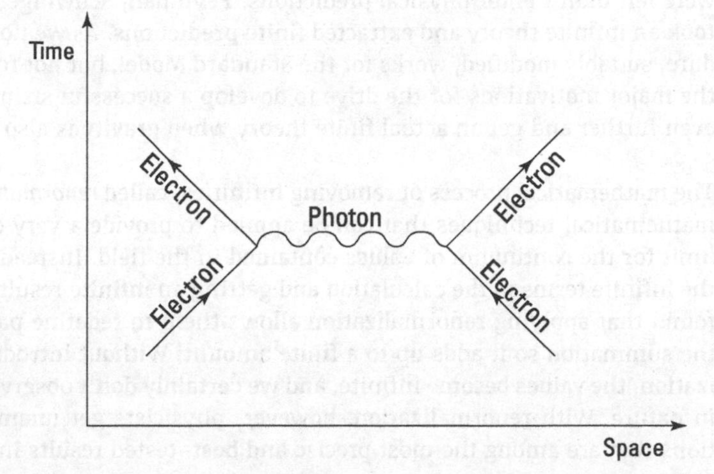

FIGURE 8-3: A Feynman diagram demonstrates how particles interact with each other.

In quantum field theory, particles are just points, with no physical extension. When thought of this way, two particles can get as close to each other as they want. But, as we expect from classical physics, the strength of the particles' interaction becomes very large — infinite! — when their distance gets to zero. One way out of this pickle is to say that particles have a finite radius, and we shouldn't consider any contributions coming from the region where particles are too close to each other. This is called a *cutoff* because we are cutting off some of the possible interactions from what we consider.

The advantage of the cutoff is that the infinities immediately disappear. The disadvantage is that now all our computations depend on the radius we used for the cutoff. It isn't a physical quantity, but just an invention we use to give some sense to our computations. In fact, not only is there no way to determine the cutoff experimentally, but giving real credence to the idea that particles are solid spheres would seriously mess with Einstein's special relativity.

The genius of Feynman, Schwinger, and Tomonaga was to show that there is a way to introduce the cutoff so that the result of the computations eventually doesn't depend on it. They called this process *renormalization*. The whole procedure seems like a technical complication (and to some extent, it is), but it's a crucial part of our understanding of all quantum physics, and it works only for a very special class of nice theories. Crucially, renormalization can be used to study all of the Standard Model except gravity.

REMEMBER

By quantizing electromagnetics, as QED does, Feynman, Schwinger, and Tomonaga were able to use the theory despite the infinities. The infinities were still present, but because the virtual photon meant that the electrons didn't need to get so close to each other, there weren't as many infinities, and the ones that were left didn't enter physical predictions. Feynman, Schwinger, and Tomonaga took an infinite theory and extracted finite predictions. As we noted, their procedure, suitably modified, works for the Standard Model, but not for gravity. One of the major motivations for the drive to develop a successful string theory is to go even further and get an actual finite theory when gravity is also in the mix.

TECHNICAL STUFF

The mathematical process of removing infinities, called *renormalization*, is a set of mathematical techniques that can be applied to provide a very carefully defined limit for the continuum of values contained in the field. Instead of adding up all the infinite terms in the calculation and getting an infinite result, physicists have found that applying renormalization allows them to redefine parameters within the summation so it adds up to a finite amount! Without introducing renormalization, the values become infinite, and we certainly don't observe these infinities in nature. With renormalization, however, physicists get unambiguous predictions that are among the most precise and best-tested results in all of science.

Discovering that other kind of matter: Antimatter

Along with the understanding of quantum electrodynamics came a growing understanding of the existence of *antimatter*, a different form of matter that was identical to known matter, but with opposite charge. Quantum field theory indicated that for each particle, there existed an antiparticle. The antiparticle of the electron is called the *positron*. Sometimes, a particle can be its own antiparticle – for instance, the photon is its own anti-particle.

In 1928, physicist Paul Dirac was creating the quantum theory of the electron (a necessary precursor to a complete QED theory) when he realized that the equation worked only if you allowed extra particles — identical to electrons but with opposite charge — to exist. Just four years later, the first positrons were discovered and named by Carl D. Anderson while he was analyzing cosmic rays.

REMEMBER

The mathematics of the theory implied a symmetry between the known particles and identical particles with opposite charge, a prediction that eventually proved to be correct. The theory demanded that antimatter exists. String theory implies another type of symmetry, called supersymmetry (see Chapter 10), which has yet to experimentally observed, but which many physicists believe will eventually be discovered in nature.

When antimatter comes in contact with ordinary matter, the two types of matter annihilate each other in a burst of energy in the form of a photon. This can also be depicted in QED with a Feynman diagram, as shown on the left side of Figure 8-4. In this view, the positron is like an electron that moves backward through time (as indicated by the direction of the arrow on the propagator).

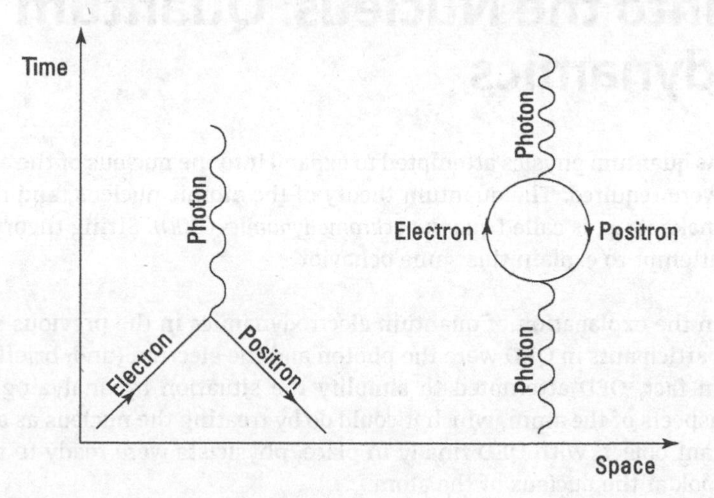

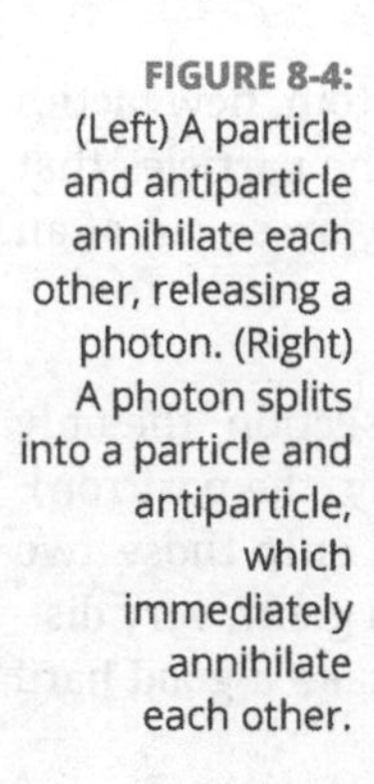

FIGURE 8-4: (Left) A particle and antiparticle annihilate each other, releasing a photon. (Right) A photon splits into a particle and antiparticle, which immediately annihilate each other.

Sometimes a particle is only virtual

In quantum electrodynamics, *virtual particles* can exist briefly, arising from the energy fluctuations of the quantum fields that exist at every point in space. Some virtual particles — such as the photon in Figure 8-3 — exist just long enough to communicate information about a force. Other virtual particles spring into existence seemingly for no purpose other than to make the lives of physicists more interesting.

REMEMBER

The existence of virtual particles is one of the strangest aspects of physics, but it's a direct consequence of quantum physics. Virtual particles can exist because the uncertainty principle, in essence, allows them to carry a large fluctuation of energy as long as they exist for only a brief period of time.

The right side of Figure 8-4 shows a pair of virtual particles — this time an electron and a positron. In some cases, a photon can actually split into an electron and a positron and then recombine back into a photon.

The problem is that even though these particles are virtual, their effects have to be taken into account when you perform calculations about what is occurring in a given area. So no matter what you're doing, an infinite number of strange virtual particles are springing into and out of existence all around you, wreaking havoc with the smooth, orderly calculations you'd like to perform! (If this sounds familiar, it's because this is the quantum foam discussed in Chapter 2.)

Digging into the Nucleus: Quantum Chromodynamics

As quantum physics attempted to expand into the nucleus of the atom, new tactics were required. The quantum theory of the atomic nucleus, and the particles that make it up, is called *quantum chromodynamics (QCD)*. String theory arose out of an attempt to explain this same behavior.

In the explanation of quantum electrodynamics in the previous section, the only participants in QED were the photon and the electron (and, briefly, the positron). In fact, QED attempted to simplify the situation by analyzing only those two aspects of the atom, which it could do by treating the nucleus as a giant, very distant object. With QED finally in place, physicists were ready to take a good hard look at the nucleus of the atom.

The pieces that make up the nucleus: Nucleons

The nucleus of an atom is composed of particles called *nucleons*, which come in two types: positively charged *protons* and noncharged *neutrons*. Protons were discovered in 1919, while neutrons were discovered in 1932.

The proton is about 1,836 times as massive as the electron. The neutron is about the same size as the proton, so the pair of them is substantially larger than the electron. Despite this difference in size, the proton and the electron have identical electrical charges, but of opposite signs: The proton is positive, while the electron is negative.

The growth of technology allowed for the design and construction of larger and more powerful *particle accelerators*, which physicists use to smash particles into each other and see what comes out. With great delight, physicists began smashing protons into each other, in hopes of finding out what was inside them.

In fact, this work on trying to uncover the secrets of nucleons would lead directly to the first insights into string theory. A young physicist at the European Organization for Nuclear Research in Geneva, known as CERN, applied an obscure mathematical formula to describe the behavior of particles in a particle accelerator. This is seen by many as the starting point of string theory. (These events are covered in more detail in Chapter 10.)

The pieces that make up the nucleon's parts: Quarks

Today, nucleons are known to be types of *hadrons,* which are particles made up of even smaller particles called *quarks.* The concept of quarks was independently proposed by Murray Gell-Mann and George Zweig in 1964 (though the name, taken from James Joyce's *Finnegan's Wake,* is pure Gell-Mann), which in part earned Gell-Mann the 1969 Nobel Prize in physics. Quarks are held together by still other particles, called *gluons.*

In this model, both the proton and the neutron are composed of three quarks. These quarks have quantum properties, such as mass, electrical charge, and spin (see the next section for an explanation of spin). There are a total of six *flavors* (or types) of quarks, all of which have been experimentally observed:

- Up quark
- Down quark
- Charm quark
- Strange quark
- Top quark
- Bottom quark

The properties of the proton and neutron are determined by the specific combination of quarks that compose them. For example, a proton's charge is reached by adding up the electrical charge of the three quarks inside it — two up quarks and one down quark. In fact, every proton is made of two up quarks and one down quark, so they're all exactly alike. Every neutron is identical to every other neutron (composed of one up quark and two down quarks).

In addition to standard quantum mechanical properties (charge, mass, and spin), quarks have another property, which came out of QCD, called *color charge.* This is somewhat similar to electrical charge in principle, but it's an entirely distinct property of quarks. It comes in three varieties: *red, green,* and *blue.* (Quarks don't actually have these colors because they're much, much smaller than the wavelength of visible light. These are just names to keep track of the types of charge.)

While QED describes the quantum theory of the electrical charge, QCD describes the quantum theory of the color charge. The color charge is the source of the name quantum chromodynamics because "chroma" is Greek for "color." In addition to quarks, there exist particles called *gluons.* The gluons bind the quarks together, kind of like rubber bands (in a very metaphoric sense). Gluons are the gauge bosons for the strong nuclear force, just as photons are the gauge bosons for electromagnetism (see the later section on gauge bosons for more on these particles).

Looking into the Types of Particles

Physicists have discovered a large number of particles, and one thing that proves useful is that they can be broken down into categories based on their properties. Physicists have found a lot of ways to do that, but in the following sections we briefly discuss some of the categories most relevant to string theory.

According to quantum mechanics, particles have a property known as *spin.* This isn't an actual motion of the particle, but in a quantum mechanical sense, it means that the particle always interacts with other particles as if it's rotating in a certain way. In quantum physics, spin has a numerical value that can be either an integer (0, 1, 2, and so on) or a half-integer (½, 3⁄2, and so on). Particles that have integer spin are called *bosons,* while particles that have half-integer spin are called *fermions.*

Particles of force: Bosons

Bosons, named after Satyendra Nath Bose, are particles that have an integer value of quantum spin. The bosons that are known act as carriers of forces in quantum field theory, as the photon does in Figure 8-3. The Standard Model of particle physics predicts several fundamental bosons.

Some of these bosons are the mediators of fundamental forces:

- Photon, for the electromagnetic force
- Z boson and W boson (actually two particles — the W^+ and W^- bosons), for the "weak force"
- Gluons (of which there are eight) for the "strong" force

In addition, many physicists believe that there probably exists a boson called the *graviton,* which is related to gravity. The relationship of these bosons to the forces of physics is covered in the "Gauge Bosons: Particles Holding Other Particles Together" section later in this chapter.

Additionally, there is another Boson, which does not mediate a force, but is a crucial ingredient of the standard model:

» Higgs boson

Composite bosons, formed by combining together an even number of different fermions, can also exist. For example, a carbon-12 atom contains six protons and six neutrons, all of which are fermions. The nucleus of a carbon-12 atom is therefore a composite boson. *Mesons*, on the other hand, are particles made up of exactly two quarks, so they are also composite bosons.

Particles of matter: Fermions

Fermions, named after Enrico Fermi, are particles that have a half-integer value of quantum spin. Unlike bosons, they obey the *Pauli exclusion principle*, which means that multiple fermions can't exist in the same quantum state.

While bosons are seen as mediating the forces of nature, fermions are particles that are a bit more "solid" and are what we tend to think of as matter particles. Quarks are fermions.

In addition to quarks, there is a second family of fermions called *leptons*. Leptons are elementary particles that can't (so far as scientists know) be broken down into smaller particles. The electron is a lepton, but the Standard Model of particle physics tells us that there are actually three generations of particles, each heavier than the last. (The three generations of particles were predicted by theoretical considerations before they were discovered by experiment, an excellent example of how theory can precede experiment in quantum field theory.)

Also within each generation of particles are two flavors of quarks. Table 8-1 shows the 12 types of fundamental fermions, all of which have been observed. The numbers shown in Table 8-1 are the masses, in terms of energy, for each of the known particles. (Neutrinos have virtually, but not exactly, zero mass.)

REMEMBER

The mass of a particle could be given in kilograms (or pounds if the metric system scares you), but for such minuscule particles this is not very handy. Instead, physicists use a measure of *energy* (remember, mass and energy are related by $E = mc^2$), which is the electron-volt (eV). This is the energy that you get by accelerating something with the electric charge of an electron by one volt. One electron-volt divided by c^2 is about $10^{(-36)\text{kg}}$ – very tiny! Sometimes, physicists use multiples of the eV, like the kilo, mega, giga and tera eV (KeV, MeV, GeV, TeV), which correspond to 10^3, 10^6, 10^9 and 10^{12} eV. Even the largest unit, the TeV, is absolutely tiny compared to the scales to which we are accustomed.

TABLE 8-1: Elementary Particle Families for Fermions

	Quarks		Leptons	
First Generation	Up Quark 2-3 MeV	Down Quark 4-5 MeV	Electron Neutrino	Electron 0.5 MeV
Second Generation	Charm Quark 1.2 GeV	Strange Quark 90-100 MeV	Muon Neutrino	Muon 106 MeV
Third Generation	Top Quark 172 GeV	Bottom Quark 4.2 GeV	Tau Neutrino	Tau 1.8 GeV

There are also, of course, composite fermions, made when an odd number of fermions combine to create a new particle, similar to how protons and neutrons are formed by combining quarks.

In the original formulation of the standard model the masses of the neutrinos were assumed to be exactly zero. Today, there is rather strong evidence that they are not zero, but very small, of the order of a single eV (almost a million times smaller than the electron's!).

Gauge Bosons: Particles Holding Other Particles Together

In the Standard Model of particle physics, the forces can be explained in terms of gauge theories, which possess certain mathematical properties. These forces transmit their influence through particles called *gauge bosons*. The structure of gravity is not that different from that of a gauge theory, and this becomes more apparent in string theory. In fact, within string theory it is possible to show that a duality exists between gravity and gauge theories. We will talk about this in detail in Chapter 13.

Throughout the development of the Standard Model, it became clear that all the forces (or, as many physicists prefer, *interactions*) in physics can be broken down into four basic types:

- Electromagnetism
- Gravity
- Weak nuclear force
- Strong nuclear force

The electromagnetic force and weak nuclear force were consolidated in the 1960s by Sheldon Lee Glashow, Abdus Salam, and Steven Weinberg into a single force called the *electroweak force*. This force, in combination with quantum chromodynamics (which defined the strong nuclear force), is what physicists mean when they talk about the Standard Model of particle physics.

One key element of the Standard Model of particle physics is that it's a *gauge theory*, which means that the theory is described in terms of a gauge redundancy, but that the physical results of any experiment should not be independent from the redundancy. A force that operates through a gauge field is transmitted with a *gauge boson*. The following gauge bosons have been observed by scientists for three of the forces of nature:

- Electromagnetism — photon
- Strong nuclear force — gluons (actually, 8 of them)
- Weak nuclear force — Z, W^+, and W^- bosons

REMEMBER

In addition, gravity can be written as a gauge theory, which means that there should exist a gauge boson that mediates gravity. The name for this theoretical gauge boson is the *graviton*. (In Chapter 10, you see how the discovery of the graviton in the equations of string theory led to its development as a theory of quantum gravity.)

Exploring the Theory of Where Mass Comes From

In the Standard Model of particle physics, particles get their mass through something called the *Higgs mechanism*, which is an example of scientists cleverly exploiting symmetries to make predictions (in this case, about the mass of fundamental particles). The Higgs mechanism is based on the existence of a *Higgs field*, which permeates all of space. The Higgs field creates a type of particle called a *Higgs boson*.

It takes a lot of energy for the Higgs field to create a Higgs boson, and physicists have only recently managed to observe it at the Large Hadron Collider (LHC) at the CERN in Geneva. The search for the Higgs boson, together with attempts to find new particles and physical phenomena, such as those motivated by string theory, are among the major reasons scientists have been building more and more powerful particle colliders.

What is the Higgs field?

The weak nuclear force falls off very rapidly above short distances. According to quantum field theory, this means that the particles mediating the force — the W and Z bosons — must have a mass (as opposed to the gluons and photons, which are massless).

The problem is that the gauge theories described in the preceding section are mediated only by massless particles. If the gauge bosons have mass, then a gauge theory can't be sensibly defined. The Higgs mechanism avoids this problem by introducing a new field called the Higgs field.

At high energies, where the gauge theory is defined, the gauge bosons are massless, and the theory works as anticipated. At low energies, the field triggers a spontaneous breaking of the symmetry that allow the particles to have mass. A colloquial way to say this is, at low energy, the W and Z bosons feel the average value of the Higgs field across space, and they are affected by it, acquiring a mass of their own in the process. The particle associated to the Higgs field is the Higgs boson. The mass of the Higgs boson isn't something the theory tells us outright, but theoretical physicists have long expected it to fall in a narrow window — if not, the Standard Model would be in big trouble. For a few decades, physicists expected the Higgs boson to show up in the range between 114 GeV (below that mass, nothing had been found in earlier experiments) and about 200 GeV (higher than that, they would have struggled to explain other predictions of the Standard Model).

To be sure, it was necessary to build a new, more powerful particle collider. This was the LHC, where the Higgs boson was finally discovered.

TECHNICAL STUFF

By the way, the Higgs mechanism, Higgs field, and Higgs boson are named after Scottish physicist Peter Higgs. Though he wasn't the first to propose these concepts (Robert Brout and François Engelert, among others, certainly also deserve credit), he's the one whose name is attached to them, which is just one of those things that sometimes happens in physics.

Discovering the Higgs boson at the LHC

The LHC is an enormous particle collider, consisting of a circle with a circumference of 27 km in which particles are accelerated by powerful magnets. The hunt for the Higgs boson was one of the main motivations for building the collider, which took ten years, from 1998 to 2008. Fortunately, physicists could exploit the infrastructure of a previous CERN collider, the Large Electron-Positron collider (LEP). The two colliders had the same size, but the LHC is much more powerful than the LEP was because electrons and positrons are much lighter than hadrons, which is what the LHC is using. (In fact, to find the Higgs, scientists accelerated protons and antiprotons, the anti-particles of protons.)

The construction and testing of the LHC was no easy task, and it wasn't accomplished without incident. Nevertheless, in 2010, the collider began its systematic search for the Higgs boson, and in 2012, it found the particle, with a mass of about 125 GeV — another stunning confirmation of the Standard Model.

Although the discovery was a great success for theoreticians and experimentalists, some physicists were almost let down: Sometimes it's more interesting when an experiment reveals a new puzzle than when it confirms something we already believed! For the discovery of the Higgs boson, François Engelert and Peter Higgs were awarded the 2013 Nobel Prize. (Robert Brout had died in 2011.)

Although the Higgs particle was a big part of the motivation for constructing the LHC, the particle collider can do much more. In fact, it can reach much higher energies, up to 13 TeV (100 times more than the mass of the Higgs!), which makes it the ideal instrument to look for new particles, such as the ones predicted by supersymmetry and string theory.

TIP

For an in-depth discussion of the Higgs mechanism, we recommend Lisa Randall's *Higgs Discovery: The Power of Empty Space* or Sean Carroll's *The Particle at the End of the Universe: How the Hunt for the Higgs Boson Leads Us to the Edge of a New World*, both of which are devoted entirely to the topic of the discovery of the Higgs boson.

From Big to Small: The Hierarchy Problem in Physics

The Standard Model of particle physics is an astounding success, but it hasn't answered every question physics hands it. One of the major questions that remains is the *hierarchy problem*, which seeks an explanation for the observed values of the masses that appear in the Standard Model. Many physicists feel that string theory will ultimately be successful at resolving the hierarchy problem.

For example, if you count the Higgs boson (and both types of W bosons), the Standard Model of particle physics has 18 elementary particles. The masses of these particles aren't predicted by the Standard Model. Physicists had to find them by experiment and plug them into the equations to get everything to work out right.

However, in quantum field theory everything gets corrected by quantum effects. This means that physicists must put some value in their model, which is different from the one measured. What they put in is the "bare mass," which is the actual mass, minus the quantum effects. In this way, once the quantum effects are taken into account, the "actual mass" will match the experiment. (This is one of the many quirks of the process of renormalization in quantum field theory.)

In most cases, these corrections change the mass just by a little bit. Something very bizarre happens for the mass of the Higgs (and for another quantity, we'll get to that later). Studying in detail how corrections appear in that case, it becomes clear that they are of the order of the "largest allowed mass" in the theory.

This means that if there is no particle to be discovered beyond the standard model, the quantum corrections to the Higgs mass will be of the order of the Planck mass (beyond which we expect some sort of quantum gravity to kick in).

It is possible that something new will happen before that, due to some yet undiscovered contribution. For instance, some believe that new physics will appear at the "Grand unification scale" of 10^{16} GeV, where the strong force may unify with the electroweak one.

In any case, the corrections will be much larger, billions of billions of times larger, than the actual mass of the Higgs. Strictly speaking, that is not a problem because physicists are free to choose the "bare" massess as they want. On the other hand, it is highly suspicious that the bare mass of the Higgs has to be chosen to be a very specific enormous number, just so that it will cancel another enormous number, leaving a reasonable result in the end. In Figure 8-5, you get a feeling for the huge gap in energy between the Planck scale and the electroweak scale (where the actual Higgs mass naturally lives).

At the bottom of this scale is the *vacuum energy*, which is the energy generated by all the strange quantum behavior in empty space — virtual particles exploding into existence and quantum fields fluctuating wildly due to the uncertainty principle. These fluctuations are related to the cosmological constant that we will describe in Chapter 9.

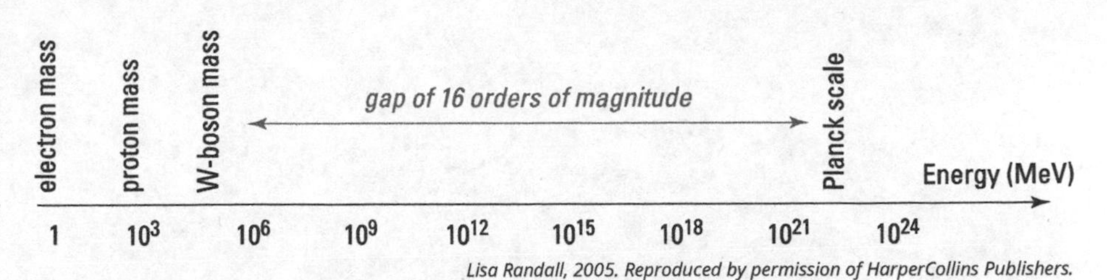

FIGURE 8-5: The hierarchy problem in physics relates to the large gap between the weak scale and Planck scale of length and energy.

The vacuum energy is actually another example of the hierarchy problem. Quantum field theory would suggest that it too should be corrected, and that again the corrections should be of the order of the Planck mass — enormously bigger than the actual vacuum energy! Again, to make things work, a delicate cancellation of huge numbers would be needed.

Of course, one could just dismiss all these issues: After all, if we can specify one number, no matter how big, and obtain sensible results, why should we worry? On the other hand, most physicists are very uneasy with the "fine tuning" required to make things work out.

Supersymmetric theories as well as string theory can provide a way out of the hierarchy problem, at least in principle, because they provide a natural mechanism to cancel large quantum fluctuations.

IN THIS CHAPTER

» Putting things where they belong in the universe

» Going back to the universe's teeny-tiny beginnings

» Getting enlightened about dark matter and dark energy

» Seeing how black holes bend and break space

Chapter 9
Physics in Space: Considering Cosmology and Astrophysics

One of humankind's first scientific acts was probably to look into the heavens and ask questions about the nature of that vast universe. Today, scientists are still fascinated by these questions, and with good reason. Though we know much more about what makes up the heavens than our cave-dwelling ancestors knew, the black space between the stars still holds many mysteries — and string theory contributes to the search for answers to many of those mysteries.

In this chapter, you find out what physicists, astronomers, astrophysicists, and cosmologists have uncovered about the workings of the universe independent of string theory. As scientists have discovered how the universe works, their findings have led to more difficult questions, which string theorists hope to answer. We cover some of the more complex points about the universe in Chapter 16. This chapter gives you the background that will help you understand the ties between cosmology, astrophysics, and string theory.

In the following pages, we explore the consequences of Einstein's relativity, where scientists find that the universe seems to have had a beginning. At this point, scientists were able to determine where the particles in our universe come from. The theory of the universe's origin grows more complex with the introduction of a rapidly expanding early universe. We also introduce you to two of cosmology's biggest mysteries: the presence of unseen dark matter and repulsive gravity in the form of dark energy. Finally, we provide a glimpse into black holes, objects that later become important to string theory.

SO MANY SCIENTISTS, SO MANY NAMES

The names for different types of space scientists can get rather confusing. Here are some guidelines:

- **Astronomer:** This is the classical term for a scientist who studies the heavens. Since Galileo, optical telescopes have been the primary tool used to examine celestial bodies. Today, telescopes can be radio, X-ray, or gamma ray telescopes, which see light in the nonvisual spectrum. Traditionally, astronomers have devoted more time to classifying and describing bodies in space than to attempting to explain the phenomena.
- **Astrologer:** In ancient times, the terms *astrologer* and *astronomer* were essentially synonyms. Since Copernicus, those terms have become more distinct; today, they represent radically different disciplines, with astrology well outside the bounds of science. Calling any scientist who studies the universe an astrologer is a serious faux pas. An astrologer tries to find connections between human behaviors and the motion of celestial bodies, generally introducing a vague or supernatural mechanism as the basis for these connections. See *Astrology For Dummies,* 2nd Edition, by Rea Orion (Wiley) for more information on the field.
- **Astrophysicist:** This term applies to someone who studies the physics of interactions within and between stellar bodies. Astrophysicists seek to apply the principles of physics to create general laws governing the behavior of stellar interactions.
- **Cosmologist:** This term is used for a type of astrophysicist who focuses on the evolution of the universe — the processes of how the universe changes over time. A cosmologist rarely cares about a specific stellar body or solar system, and galaxies are frequently too parochial for these explorers of space. Cosmologists often focus their attention on theories that use unimaginably large scales of time, space, and energy. The study of the big bang, or of the universe's end, are two examples of the cosmologist's domain.

The Enlightened Universe and the Birth of Modern Astrophysics

Modern astronomy and modern physics both started taking shape in the 1500s. The work of astronomers Nicolaus Copernicus and Galileo Galilei was key to this revolution.

The most revolutionary idea of the time was that models of the cosmos should be based on facts and data rather than on what Aristotle or Ptolemy had written centuries before. In astronomy, this thinking resulted in the *heliocentric model* (with the sun, rather than our relatively insignificant planet, at the center of the solar system), which dislodged Earth from its special place in the universe.

That assumption, known as the *Copernican principle*, says that there's really nothing inherently special about Earth's place in the cosmos (aside from the convenience of oxygen and other properties that allow life to exist, of course). In other words, Copernicus corrected our beliefs about what's where in the universe.

Everything doesn't revolve around Earth

Before Polish astronomer Nicolaus Copernicus's time, the most fashionable model in astronomy, called the *Ptolemaic model*, was based on the idea that all the celestial objects — planets, moons, stars, and so on — were on concentric spheres, each of which was centered around Earth. Over the centuries (from about 150 BCE to 1500 CE), though, scientific observations made it clear that this wasn't the case.

To preserve the Ptolemaic model, astronomers made many modifications over the years. Celestial objects were mounted on spheres, which were then mounted on other spheres. The very elegance that made the Ptolemaic model so appealing was gone, replaced with a mishmash of geometric nonsense that only partially conformed with scientific observations — which were growing more and more precise because of new technologies. It was a prime time for a scientific revolution.

In his book *On the Revolutions of the Celestial Spheres*, Copernicus explained his heliocentric model, making it clear that the sun, not Earth, sat at center stage in the universe. He still placed celestial bodies on spheres and made other assumptions that haven't borne the test of time, but his model was a major improvement over the Ptolemaic model.

Copernicus's work detailing his heliocentric model was published upon his death in 1543 because he feared retribution from the Catholic Church if he published it earlier (although he did hand out versions of the theory to friends about 30 years earlier). Some Indian writers had made the same heliocentric claim as far back as the seventh century CE, and some Islamic astronomers and mathematicians had studied the idea as well, but it's unclear to what degree Copernicus was aware of their work.

Copernicus was a theorist, not an observational astronomer. His key insight was the idea that Earth doesn't have a distinct position within the universe, a concept that was named the *Copernican principle* in the mid-20th century.

Beholding the movements of heavenly bodies

One of the greatest observational astronomers of this revolutionary age was Tycho Brahe, a Danish nobleman who lived from 1546 to 1601. Brahe made an astounding number of detailed astronomical observations. He used his family's wealth to found an observatory that corrected nearly every astronomical record of the time, including those in Ptolemy's *Almagest.*

Using Brahe's measurements, his assistant Johannes Kepler was able to create rules governing the motion of the planets in our solar system. In his three laws of planetary motion, Kepler realized that the planetary orbits are elliptical rather than circular.

More important, Kepler discovered that the motion of the planets isn't uniform. A planet's speed changes as it moves along its elliptical path. Kepler showed that the heavens are a dynamic system, a detail that later helped Isaac Newton show that the sun constantly influences the planets' motion.

Galileo, by using the telescope, later realized that other planets have moons and determined that the heavens aren't static. He also publicly supported Copernicus's theory, which was a rather risky stance at the time. Cutting a long and fascinating piece of history short, the Catholic Church charged him with heresy. To get away with only house arrest, Galileo was forced to recant his observations about the movements of heavenly bodies, and of Earth in particular. Reportedly, after his recantation, he still uttered the famous words "And yet it moves!" (There are conflicting versions of this utterance, and it may well be a myth . . . but we cannot resist the temptation to mention it.)

Galileo's work, together with Kepler's, laid the foundation for Newton's law of gravity. With gravity introduced, the final nail had been placed in the scientific consensus behind the geosynchronous (Earth-centric) view. Astronomers and physicists now knew that Earth circles the sun, as the heliocentric model described. Newton's theory became the universally accepted description of gravity until

Einstein came along. (The Catholic Church officially endorsed the heliocentric view in the 19th century. In 1992, Pope John Paul II formally apologized for Galileo's treatment.)

Introducing the Idea of an Expanding Universe

Even two centuries after Newton, Albert Einstein was strongly influenced by the concept of an unchanging universe. His general theory of relativity predicted a dynamic universe — one that changed substantially over time — so he introduced an idea he called the *cosmological constant* into the theory to make the universe static and eternal. This would prove to be a mistake when, several years later, astronomer Edwin Hubble discovered that the universe was expanding! Even today, the consequence of the cosmological constant in general relativity has enormous impact upon physics, causing string theorists to rethink their whole approach.

The equations of general relativity Einstein developed showed that the very fabric of space was expanding or contracting. This made no sense to him, so in 1917 he added the cosmological constant to the equations. That term represented a form of repulsive gravity that exactly balanced out the attractive pull of gravity.

When Hubble showed that the universe was indeed expanding, Einstein allegedly called the introduction of the cosmological constant his "biggest blunder" and removed it from the equations. The concept of a cosmological constant would return over the years, however, as you see in the "Dark Energy: Pushing the Universe Apart" section later in this chapter. With the discovery of dark energy, Einstein's "blunder" was found to be a necessary parameter in the theory (even though physicists for most of a century assumed the cosmological constant's value was zero).

Discovering that energy and pressure have gravity

In Newton's gravity, bodies with mass were attracted to each other. Einstein's relativity showed that mass and energy were related. Therefore, mass and energy both exerted gravitational influence. Not only that, but it was possible that space itself could exert a pressure that warped space. Several models were constructed to show how this energy and pressure affected the expansion and contraction of space.

When Einstein created his first model based on the general theory of relativity, he realized that it implied an expanding or collapsing universe. At the time, no one had any particular reason to think the universe was not static and Einstein assumed that this was a flaw in his theory.

Einstein's general relativity equations allowed for the addition of an extra term while remaining mathematically viable. Einstein found that the term cosmological constant could represent a positive energy (or negative pressure) uniformly distributed throughout the fabric of space-time itself, which would act as an *anti-gravity*, or repulsive form of gravity. This was an ingredient in trying to formulate a model of a static universe.

In 1917, the same year Einstein published his equations containing the cosmological constant, Dutch physicist Willem de Sitter applied them to a universe without matter. (As we explain in Chapter 4, this is a frequent step in scientific analysis — you strip a scientific theory of all the complications and consider it in the simplest cases.)

In this *de Sitter space*, the only thing that exists is the energy of the vacuum — the cosmological constant itself. Even in a universe containing no matter at all, this means that space will expand. A de Sitter space has a positive value for the cosmological constant, which can also be described as a positive curvature of space-time. A similar model with a negative cosmological constant (or a negative curvature, in which expansion is slowing) is called an *anti-de Sitter space*. (More on the curvature of space-time in a bit.)

In 1922, Russian physicist Aleksandr Friedmann turned his hand to solving the elaborate equations of general relativity, but decided to do so in the most general case by applying the *cosmological principle*, which can be seen as a more general case of the Copernican principle and consists of two assumptions:

- The universe looks the same in all directions (it's *isotropic*).
- The universe is uniform no matter where you go (it's *homogenous*).

With these assumptions, the equations become much simpler. Einstein's original model and de Sitter's model both ended up being special cases of this more general analysis. Friedmann was able to define the solution depending on just three parameters:

- Hubble's constant (the rate of expansion of the universe)
- Lambda (the cosmological constant)
- Omega_m and Omega_r (average matter and radiation density in the universe)

REMEMBER

To this day, scientists are trying to determine these values as precisely as they can, but even without real values, they can define three possible solutions. Each solution matches a certain "geometry" of space, which can be represented in a simplified way by the way space naturally curves in the universe, as shown in Figure 9-1.

- **Closed universe:** There is enough matter in the universe that gravity will eventually overcome the expansion of space. The geometry of such a universe is a positive curvature, like the sphere in the leftmost image in Figure 9-1. (This matched Einstein's original model without a cosmological constant.)
- **Open universe:** There isn't enough matter to stop expansion, so the universe will continue to expand forever at the same rate. This space-time has a negative curvature, like the saddle shape shown in the middle image in Figure 9-1.
- **Flat universe:** The expansion of the universe and the density of matter perfectly balance out, so the universe's expansion slows down over time but never quite stops completely. This space has no overall curvature, as shown in the rightmost image of Figure 9-1. (Friedmann himself didn't discover this solution; it was found years later.)

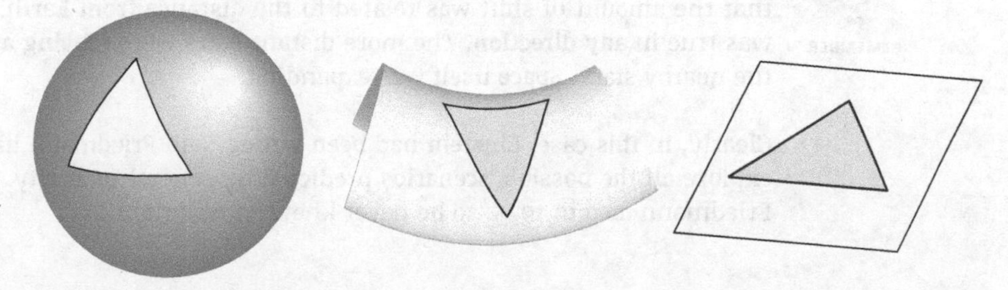

FIGURE 9-1: Three types of universes: closed, open, and flat.

These models are highly simplified, but they needed to be because Einstein's equations got very complex in cases where the universe was populated with all sorts of scattered galaxies and celestial bodies, and supercomputers didn't yet exist to perform all the math (and even physicists want to go on dates every once in a while).

Hubble drives it home

In 1927, astronomer Edwin Hubble proposed, based on his observation of distant stars, that the universe is expanding. With this new evidence, Einstein removed the cosmological constant from his equations.

Hubble had shown in 1925 that there were galaxies outside our own. Until that time, astronomers had observed white blobs of stars in the sky, which they called *nebulae*, but they disagreed about how far away the nebulae were. In his work at the Mount Wilson Observatory in California, Hubble proved that the nebulae were, in fact, distant galaxies.

While studying these distant galaxies, he noticed that the light from distant stars had a wavelength that was shifted slightly toward the red end of the electromagnetic spectrum, compared to what he expected.

This is a consequence of the wave nature of light — an object that's moving (with respect to the observer) emits light with a slightly different wavelength. This phenomenon is based on the *Doppler effect*, which is what happens to the wavelength of sound waves from a moving source. If you've ever heard a siren's pitch change as it approaches and passes you, you've experienced the Doppler effect.

In a similar way, when a light source is moving, the wavelength of the light changes. A *redshift* in light from a star means the star is moving away from the observer.

REMEMBER

Hubble saw this redshift in the stars he observed, caused not only by the motion of the stars but also by the expansion of space-time itself, and in 1929 determined that the amount of shift was related to the distance from Earth. Moreover, this was true in any direction. The more distant stars were moving away faster than the nearby stars. Space itself was expanding.

Clearly, in this case, Einstein had been wrong, and Friedmann had been right to explore all the possible scenarios predicted by general relativity. (Unfortunately, Friedmann died in 1925, so he never knew he was right.)

Finding a Beginning: The Big Bang Theory

It soon became evident that our expanding universe was once significantly smaller — so small, in fact, that it was compressed down to a single point (or, at least, a very small area). The theory that the universe started from a primordial point and has expanded ever since is known as the *big bang theory*.

The theory was first proposed in 1927, but it was controversial until the mid-sixties. In 1964 an accidental discovery at the Bell Laboratories in New Jersey provided a smoking gun for the big bang. Around the same time, on the other side of the Atlantic, Stephen Hawking and Roger Penrose showed mathematically that the sort of singularities appearing in the big bang (and in black holes) are a natural consequence of the structure of general relativity, further cementing the big bang model.

Today, the most advanced astronomical observations show that the big bang theory is likely true. String theory will hopefully help physicists understand more precisely what happened in those early moments of the universe, so understanding the big bang theory is a key component of string theory's cosmological work.

The man originally responsible for the big bang theory was Georges Lemaître, a Belgian priest and physicist who independently worked on theories similar to Aleksandr Friedmann's. Like Friedmann, Lemaître realized that a universe defined by general relativity would either expand or contract.

In 1927, Lemaître learned of Hubble's finding about distant galaxies moving away from Earth. He realized that this meant space was expanding, and he published a theory that came to be called the big bang theory. (See the nearby sidebar "What's in a name?")

TIP

Because you know that space is expanding, you can run the video of the universe backward in time in your head (rewind it, so to speak). When you do this, you realize that the universe had to be much smaller than it is now. The laws of thermodynamics (which govern the flow of heat) tell you that as the matter in the universe was compressed into a smaller and smaller amount of space, it had to be incredibly hot and dense.

The big bang theory reveals that the universe came from a state of dense, hot matter, but it says nothing about how the matter got there or whether anything else existed before the big bang (or even if the word "before" has any meaning when you're talking about the beginning of time). We explore these speculative topics in Chapter 16.

Going to bat for the big bang: Cosmic microwave background radiation

One of the major converts to the big bang theory was physicist George Gamow, who realized that if the theory were true, a residual trace of background radiation would be spread throughout the universe. Attempts to find this radiation failed for many years, until an unexpected problem in 1965 accidentally detected it.

Gamow is known to many as the author of a number of popular books on science, but he was also a theoretician and experimentalist who liked to throw out ideas right and left, seemingly not caring whether they bore fruit.

WHAT'S IN A NAME?

The name "big bang" was given to the theory by Fred Hoyle, one of the theory's biggest critics, who argued in favor of an alternative "steady state theory" to explain the evidence. Instead of an expanding universe, in Hoyle's framework new particles were created, and that's what generated the evidence that was being interpreted as an expansion. In a 1949 series of BBC radio broadcasts, Hoyle spoke dismissively of the idea that everything in the universe was created in one sudden "big bang" in the distant past.

The name stuck, much to big bang theorists' dismay. Strictly speaking, the big bang theory doesn't include a bang. Rather, the theory states that the universe started from a tiny blob and rapidly expanded, eventually creating the galaxies, stars, and planets we now observe. There's neither big nor bang in the theory.

Turning his attention to cosmology and the big bang, Gamow noted in 1948 that the dense ball of matter (probably neutrons, he hypothesized) would emit black body radiation, which had been worked out in 1900 by Max Planck. A black body emits radiation at a definable wavelength based on the temperature.

Gamow's two students, Ralph Alpher and Robert Herman, published a paper in 1948 that contained the calculation for the temperature, and therefore the radiation, of this original ball of matter. The men calculated the temperature to be about 5 degrees above absolute zero, although it took nearly a year for Gamow to agree with their calculation. This radiation is in the microwave range of the electromagnetic spectrum, so it's called the *cosmic microwave background radiation (CMBR)*.

Although the CMBR was a successful theoretical breakthrough, it went largely unnoticed at the time. Nobody conducted a serious experiment to look for this radiation, even while Gamow, Alpher, and Herman tried to gain support.

In 1965, a Princeton University team led by Robert Dicke had independently developed the theory and was attempting to test it. Dicke's team failed to discover the CMBR, however, because while they were putting the finishing touches on their equipment, someone else beat them to it.

A few miles away, at New Jersey's Bell Laboratory, Arno Penzias and Robert Wilson were having trouble of their own. Their Holmdel Horn Radio Telescope — which was more sophisticated than Princeton University's telescope — was picking up a horrible static when they attempted to detect radio signals in space. No matter where they pointed the silly thing, they kept getting the same static. The two men even cleaned bird droppings off the telescope, but to no avail. In fact, the static got worse on the unobstructed telescope. Unbeknownst to them, the static that was troubling them so much was precisely the cosmic radiation that Dicke was looking for!

Fortunately, Penzias and Dicke had a mutual friend in astronomer Bernard Burke, and upon discovering the problems the two men were having, he introduced them. Penzias and Wilson ended up earning the 1978 Nobel Prize in physics for accidentally discovering the CMBR (at a temperature of 2.7 degrees above absolute zero — Gamow's calculation had been slightly high).

Decades of research has only confirmed the big bang theory, most recently in the picture of the CMBR obtained by the Plank satellite of the European Space Agency (ESA) in 2018. The picture obtained by this satellite, shown in Figure 9-2, is like a baby picture of the universe when it was just 380,000 years old (13.7 billion years ago). Before this, the universe was dense enough to be opaque, so no light can be used to look further back than that.

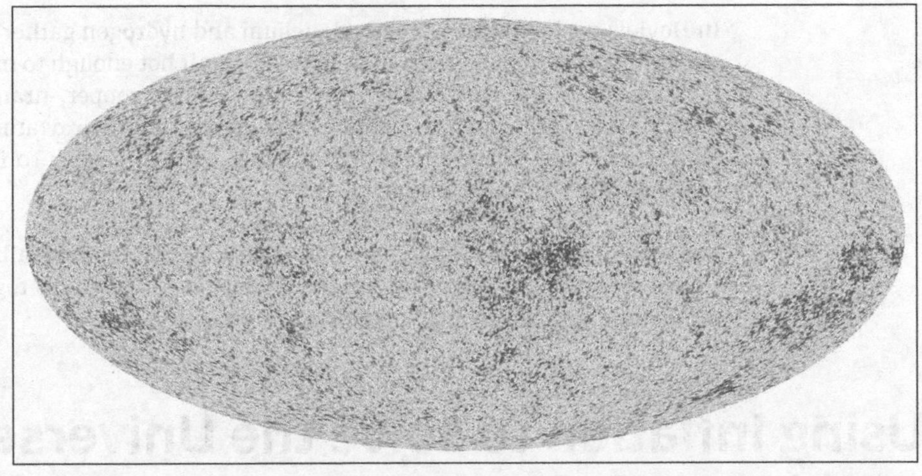

FIGURE 9-2: ESA's Planck satellite image shows an almost perfectly uniform cosmic microwave background radiation.

Courtesy of ESA

TIP

For more information on the Planck satellite, check out ESA's official Planck satellite website (`https://www.esa.int/Science_Exploration/Space_Science/Planck`).

Understanding where the chemical elements came from

George Gamow and Fred Hoyle, while differing strongly on the big bang theory, were the key figures in determining the process of *stellar nucleosynthesis*, in which atoms are made inside stars. Gamow theorized that elements were created by the heat of the big bang. Hoyle showed that the heavier elements were actually created by the intense heat of stars and supernovas.

Gamow's original theory was that as the intense heat of the expanding universe cooled, the lightest element, hydrogen, was formed. The energy at this time was still high enough to cause hydrogen molecules to interact, perhaps fusing into helium atoms. Estimates show that nearly 75 percent of the visible universe is made up of hydrogen and 25 percent is helium, with the rest of the elements on the periodic table making up only trace amounts on the scale of the entire universe.

That proved fortunate because Gamow couldn't figure out how to cook up many of the heavier elements in the big bang. Hoyle tackled the problem, assuming that if he could show all the elements were made in stars, then the big bang theory would fail. Hoyle's work on stellar nucleosynthesis was published in 1957.

In Hoyle's nucleosynthesis method, helium and hydrogen gather inside stars and undergo nuclear fusion. Even that, however, isn't hot enough to make atoms more massive than iron. The heavier elements — zinc, copper, uranium, and many others — are created when massive stars go through their deaths and explode in giant *supernovas.* These supernovas produce enough energy to fuse the protons together into the heavy atomic nucleus.

The elements are then blown out into space by the supernova blast, drifting as clouds of stellar dust. Some of this stellar dust eventually falls together under the influence of gravity to form planets, such as Earth.

Using Inflation to Solve the Universe's Problems of Flatness and Horizon

In trying to understand the universe, scientists faced two major problems: the *flatness problem* and the *horizon problem.* To solve these two problems, they modified the big bang theory with the *inflation theory,* which states that the universe expanded *very, very rapidly* shortly after it was created. Today, the principles at the heart of inflation theory have a profound impact on the way string theory is viewed by many physicists, which becomes clear in Chapter 16.

REMEMBER

These two problems can be stated simply as:

- **Horizon problem:** The CMBR is essentially the same temperature in all directions.
- **Flatness problem:** The universe appears to have a flat geometry.

The universe's issues: Too far and too flat

The horizon problem (sometimes called the *homogeneity problem*) is that no matter which direction you look in the universe, you see basically the same thing (refer to Figure 9-2). The CMBR temperatures throughout the universe are, to a *very* high level of measurement, almost exactly the same in every direction. While this initially sounds nice, in view of our cosmological principle, it really shouldn't be the case if you think about it more carefully.

If you look in one direction in space, you're actually looking back in time. The light that hits your eye (or telescope) travels at the speed of light, so it was emitted years ago. This means there's a boundary of 14 billion (or so) light-years in all directions. (The boundary is actually farther because space itself is expanding, but you can ignore that for the purposes of this example.) If there's anything farther away than that, there's no way for it to have ever communicated with us. So you look out with your powerful telescope and see the CMBR from 14 billion light-years away (call this Point A).

If you now look 14 billion light-years in the opposite direction (call this Point B), you see exactly the same sort of CMBR in that direction. Normally, you'd take this to mean that all the CMBR in the universe has somehow diffused throughout the universe, like heat in an oven. Somehow, the thermal information is communicated between Points A and B.

REMEMBER

But Points A and B are 28 billion light-years apart, which means, because no signal can go faster than the speed of light, *there's no way they could have communicated with each other in the entire age of the universe*. How did they become the same temperature if there's no way for heat to transfer between them? This is the horizon problem.

The flatness problem has to do with the geometry of our universe, which appears to be very close to a flat geometry (refer to Figure 9-1). The matter density and expansion rate of the universe appear to be nearly perfectly balanced, even 14 billion years later when minor variations should have grown drastically. Because this hasn't happened, physicists need an explanation for why the minor variations haven't increased dramatically. Did the variations not exist? Did they not grow into large-scale variations? Did something happen to smooth them out? The flatness problem seeks a reason why the universe has such a seemingly perfectly flat geometry.

Rapid expansion early on holds the solutions

In 1980, astrophysicist Alan Guth proposed the inflation theory to solve the horizon and flatness problems (although later refinements by Andrei Linde, Andreas Albrecht, Paul Steinhardt, and others were required to get it to work). In this model, the early universal expansion accelerated at a rate much faster than we see today.

It turns out that the inflation theory solves both the flatness problem and the horizon problem (at least to the satisfaction of most cosmologists and astrophysicists). The horizon problem is solved because the different regions we see used to be close enough to communicate, but during inflation, space expanded so rapidly that these close regions were spread out to cover all of the visible universe.

TIP

The flatness problem is resolved because the act of inflation actually flattened the universe. Picture an uninflated balloon, which can have all kinds of wrinkles and other abnormalities. As the balloon expands, though, the surface smoothes out. According to inflation theory, this happened to the fabric of the universe as well.

In addition to solving the horizon and flatness problems, the inflation theory provides the seeds for the structure that we see in our universe today. Tiny energy variations during inflation, due simply to quantum uncertainty, became the sources for matter to clump together, eventually forming galaxies and clusters of galaxies.

One issue with the inflation theory is that the exact mechanism that would cause — and then turn off — the inflationary period isn't known. Many technical aspects of inflation theory remain unanswered, though the models include a bosonic field of spin zero called an *inflaton field* and a corresponding theoretical particle called an *inflaton*. Most cosmologists today believe that some form of inflation likely took place in the early universe.

Some variations and alternatives to this model are posed by string theorists and other physicists. Two creators of inflation theory, Andreas Albrecht and Paul J. Steinhart, have worked on alternative theories as well; see Chapter 16 for Steinhart's ekpyrotic theory and Chapter 20 for Albrecht's variable speed of light cosmology.

Dark Matter: The Source of Extra Gravity

Astronomers have discovered that the gravitational effects observed in our universe don't match the amount of matter seen. To account for these differences, the universe appears to contain a mysterious form of matter that we can't observe, called *dark matter*. Throughout the universe, there's approximately six times as much dark matter as normal visible matter — and string theory may explain where it comes from!

In the 1930s, Swiss astronomer Fritz Zwicky first observed that some galaxies were spinning so fast that the stars in them should fly away from each other. Unfortunately, Zwicky had personality clashes with many in the astronomy community, so his views weren't taken very seriously.

In 1962, American astronomer Vera Rubin made the same discoveries and had nearly the same outcome. Though Rubin didn't have the same issues of temperament that Zwicky did, many disregarded her work because of her gender.

Rubin maintained her focus on the problem and, by 1978, had studied 11 spiral galaxies, all of which (including our own Milky Way) were spinning so fast that the laws of physics said they should fly apart. Together with work by others, this was enough to convince the astronomy community that something strange was happening.

Whatever is holding these galaxies together, observations now indicate that there has to be far more of it than there is of the visible matter that makes up the *baryonic matter* that we're used to — the matter that comprises you, this book, this planet, and the stars.

Physicists have made several suggestions about what could make up this dark matter, but so far no one knows for sure. String theorists have some ideas, which you can read about in Chapter 16.

Dark Energy: Pushing the Universe Apart

Einstein's cosmological constant allowed for a uniform repulsive energy throughout the universe. Since Hubble discovered the expansion of the universe, most scientists have believed that the cosmological constant was zero (or possibly slightly negative). Recent findings have indicated that the expansion rate of the universe is actually increasing, meaning that the cosmological constant has a positive value. This repulsive gravity — or *dark energy* — is actually pushing the universe apart. This is one major feature of the universe that we hope string theory may be able to explain.

In 1998, two teams of astronomers announced the same results: Studies of distant *supernovas* (exploding stars) showed that the stars looked dimmer than expected. The only way to rationalize this was if the stars were somehow farther away than expected, but physicists had already accounted for the expansion of the universe. The eventual explanation was startling: The rate of expansion of the universe was accelerating.

To explain this, physicists realized that there had to be some sort of repulsive effect that worked on large scales (see Figure 9-3). On small scales, the usual attractive force of gravity rules, but on larger scales, a repulsive effect of dark energy seemed to take over. (This doesn't contradict the idea that the universe is flat — but it makes the fact that it is flat, while still expanding, a very unusual and unexpected set of circumstances, which required very narrow parameters on the early conditions of our universe.)

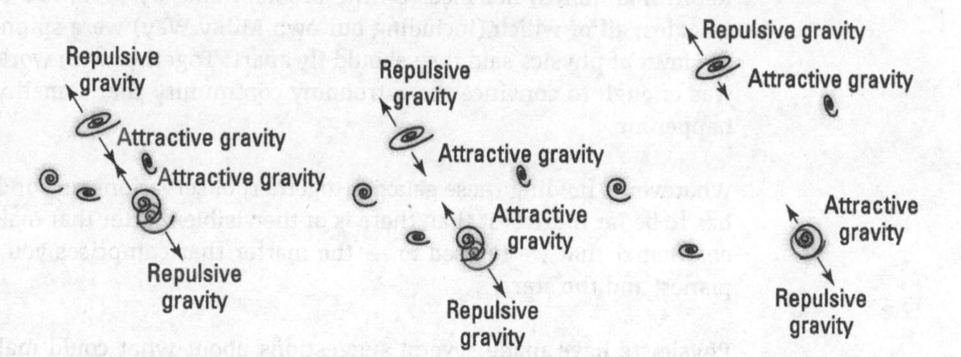

FIGURE 9-3: A repulsive effect pushes galaxies apart, while the familiar gravitational attraction tries to pull them together.

REMEMBER

A similar sort of repulsion effect is also theorized by inflation, but that's a rapid hyper-expansion in the early phases of the universe. Today's expansion due to dark energy may be remnants of the inflation dynamics, or it may be an entirely distinct phenomenon.

The finding of dark energy (or a positive cosmological constant, which it's roughly similar to) creates major theoretical hurdles, especially considering how weak dark energy is. For years, quantum field theory predicted a huge cosmological constant, but most physicists assumed that some property (such as supersymmetry, which does reduce the cosmological constant value) canceled it out to zero. Instead, the value is nonzero, but it differs from theoretical predictions by nearly

120 decimal places! (You can find a more detailed explanation of this discrepancy in Chapter 16.)

In fact, results from the Planck probe show that the vast majority of material in our current universe — about 68 percent — is made up of dark energy (remember from relativity that matter and energy are different forms of the same thing: $E = mc^2$, after all). The Planck satellite data from 2018, shown in Figure 9-4, also allows you to compare the composition of today's universe with the material present in the universe 13.7 billion years ago. The dark energy was a vanishingly small slice of the pie 13.7 billion years ago, but today it eclipses matter and drives the universe's expansion.

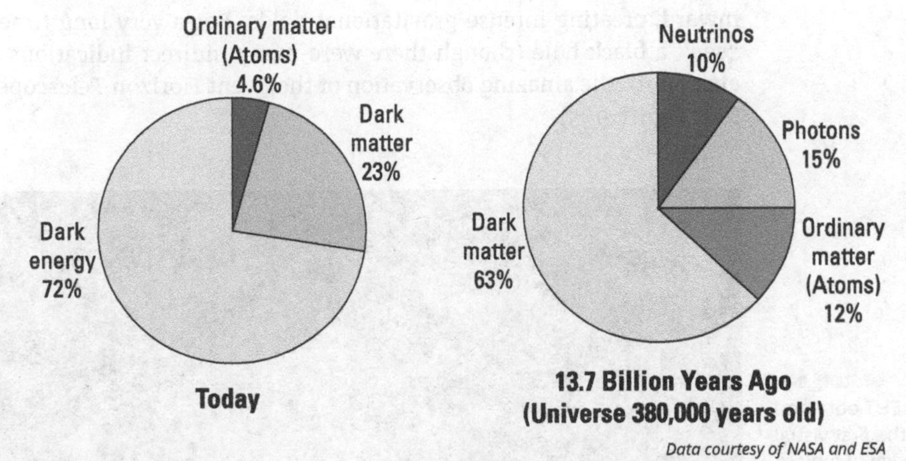

FIGURE 9-4: The Planck data allows you to compare today's universe with the distant past.

The history of the universe is a fascinating topic for study, and trying to understand the meaning of dark energy is one of the key aspects of modern cosmology. It's also one of the key challenges to modern variations of string theory, as you see in Chapter 11.

Today, many string theorists devote attention to these cosmological mysteries of the universe's origins and evolution because they provide a universal playground on which the ideas of string theory can be explored, potentially at energy levels where string behavior may manifest itself. In Chapters 14 and 16, you discover what behaviors string theorists may be looking for and what the implications are for the universe.

Stretching the Fabric of Space-Time into a Black Hole

One of the consequences of Einstein's general theory of relativity was a solution in which space-time curved so much that even a beam of light became trapped. These solutions became known as *black holes*, and the study of them is one of the most intriguing fields of astrophysics and physics as a whole. The application of string theory to study black holes is one of the most significant pieces of evidence in favor of string theory.

Black holes are believed to form when stars die and their massive bulk collapses inward, creating intense gravitational fields. For a very long time, no one could "see" a black hole (though there were many indirect indications of their existence) until the amazing observation of the Event Horizon Telescope (EHT) in 2019 (see Figure 9-5).

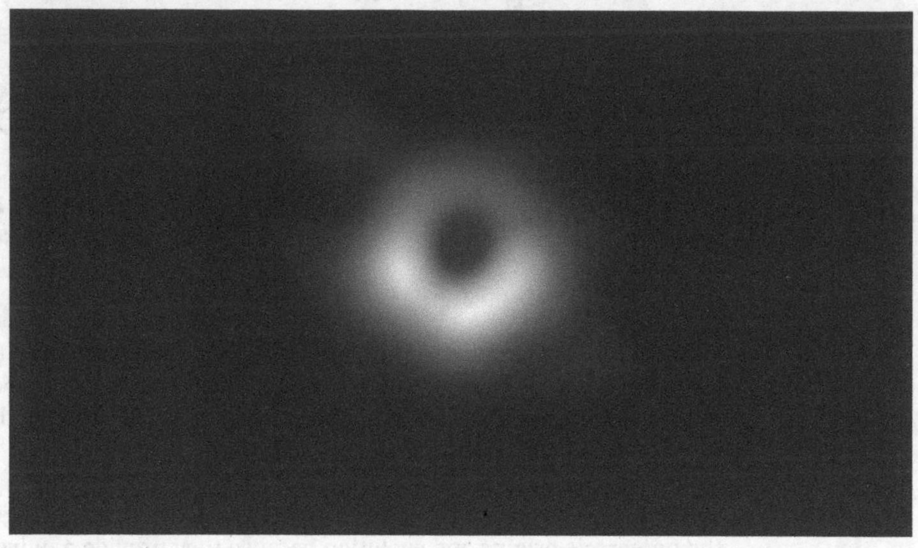

FIGURE 9-5: The EHT obtained the first visual image of a black hole in 2019. The luminous halo you see is not the black hole itself, but its accretion disk (the matter being sucked inside the black hole).

Courtesy of Event Horizon Telescope Collaboration

What goes on inside a black hole?

According to the general theory of relativity, it's possible that the very fabric of space-time bends an infinite amount. A point where curvature becomes infinite is called a space-time *singularity*. If you follow space-time back to the big bang, you'd reach a singularity. Singularities also exist inside black holes, as shown in Figure 9-6.

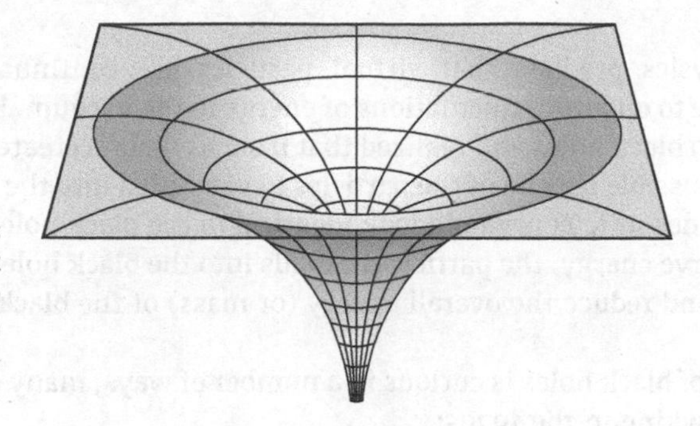

FIGURE 9-6: Inside a black hole, space-time stretches to an infinite singularity.

Because general relativity says that the curvature of space-time is equivalent to the force of gravity, the singularity of a black hole has infinite gravity. Any matter going into a black hole would be ripped apart by this intense gravitational field as it neared the singularity.

For this reason, black holes provide an excellent theoretical testing ground for string theory. Gravity is normally so weak that quantum effects aren't observed, but inside a black hole, gravity becomes the dominant force at work. A theory of quantum gravity, such as string theory, would explain exactly what happens inside a black hole.

What goes on at the edge of a black hole?

The edge of a black hole is called the *event horizon*, and it represents a barrier that even light can't escape. If you were to go near the edge of a black hole, relativistic effects would take place, including *time dilation.* To an outside observer, it would look like time was slowing down for you, eventually coming to a stop. (You, on the other hand, would notice nothing — until you got close enough to the singularity and the black hole's intense gravitational forces ripped you apart, of course.)

It was previously believed that things only get sucked into a black hole, but American physicist Stephen Hawking famously showed that black holes emit radiation, which we now call *Hawking radiation.* (This was proposed in 1974, a year after the equally groundbreaking realization by Israeli physicist Jacob Bekenstein that black holes possess *entropy* — a thermodynamic measure of disorder in a system. The entropy measures the number of different ways to arrange things in a system.) Hawking radiation is actually tiny, and it has nothing to do with the radiation emitted by matter falling into the black hole, which can be seen in the EHT picture (refer to Figure 9-5).

Quantum physics predicts that virtual particles are continually created and destroyed, due to quantum fluctuations of energy in the vacuum. Hawking applied this concept to black holes and realized that if such a pair is created near the event horizon, it's possible for one of the particles to get pulled into the black hole while the other one doesn't. This would look identical to the black hole emitting radiation. To preserve energy, the particle that falls into the black hole must have negative energy and reduce the overall energy (or mass) of the black hole.

The behavior of black holes is curious in a number of ways, many of them demonstrated by Hawking in the 1970s:

- » A black hole's entropy is proportional to the surface area of the black hole (the area of the event horizon), unlike conventional systems where entropy is proportional to volume. This was Bekenstein's discovery.
- » If you put more matter into a black hole, it cools down.
- » As a black hole emits Hawking radiation, the energy comes from the black hole, so it loses mass. This means the black hole heats up, losing energy (and therefore mass) more quickly.

In other words, Stephen Hawking showed in the mid-1970s that a black hole will evaporate (unless it is "fed" more mass than it loses in energy). He did this by applying principles of quantum physics to a problem of gravity. Hawking's calculation involved some approximations that are only valid when the black hole is sufficiently big. After the black hole evaporates down to the size of the Planck length, a quantum theory of gravity is needed to explain what happens to it.

Hawking's solution is that the black hole evaporates at that point, emitting a final burst of random energy. This solution results in the so-called *black hole information paradox*, because quantum mechanics doesn't allow information to be lost, but the radiation from the evaporation doesn't seem to carry the information about the matter that originally went into the black hole. This is one of the most puzzling questions in quantum gravity since Hawking's discovery. We discuss the black hole information paradox and its potential resolutions in greater detail in Chapter 16.

3 Building String Theory: A Theory of Everything

IN THIS PART . . .

Build on the fundamentals of string theory.

Unify the various theories into one theory.

Consider strings interacting with strings.

Get clarity from the holographic principle.

Test string theory.

IN THIS CHAPTER

- » Knowing how string theory got its start
- » Focusing on the theory's basic concepts
- » Saving the science with superstrings and supergravity
- » Cheering on the first superstring revolution

Chapter 10
Early Strings and Superstrings: Unearthing the Theory's Beginnings

A year before astronauts set foot on the moon, no one had ever heard of string theory. The concepts at the core of the theory were being neither discussed nor debated. Physicists struggled to complete the Standard Model of particle physics but had abandoned their hopes for a theory of everything (if they ever had any such hope in the first place).

In other words, no one was looking for strings when physicists found them.

In this chapter, we tell you about the early beginnings of string theory, which quickly failed to do anything the creators expected (or wanted) it to do. Then we explain how, from these humble beginnings, several elements of string theory began to spring up, which drew more and more scientists to pursue it.

Bosonic String Theory: The First String Theory

The first string theory, which became known as *bosonic string theory*, said that all the particles physicists have observed are actually the vibration of multidimensional "strings." But the theory had consequences that made it unrealistic to use to describe our reality.

A dedicated group of physicists worked on bosonic string theory between 1968 and the early 1970s, when the development of superstring theory (which said the same thing as bosonic string theory but fit reality better) supplanted it. (We explain this superior theory in the later section "Supersymmetry Saves the Day: Superstring Theory.")

Even though bosonic string theory was incomplete, it's a building block of more general stringy setups, and a very convenient place to test new methods and theories before moving on to the more modern superstring models.

Explaining the scattering of particles with early dual resonance models

String theory was born in 1968 as an attempt to explain the scattering of particles (specifically hadrons, like protons and neutrons) within a particle accelerator. Originally, it had nothing to do with strings. The early predecessors of string theory were known as *dual resonance models*.

The initial and final state of particle interactions can be recorded in a table called an *S-matrix*. At the time, finding a mathematical structure for this S-matrix was considered to be a significant step toward creating a coherent model of particle physics.

Gabriele Veneziano, a physicist at the particle accelerator laboratory at the European Organization for Nuclear Research in Geneva (CERN), realized that an existing mathematical formula seemed to capture the mathematical features of the S-matrix (see the sidebar "Applications of pure mathematics to physics" for more on this formula.) (Physicist Michio Kaku has stated that Mahiko Suzuki, also at CERN, made the same discovery at the same time but was persuaded by a mentor not to publish it.)

APPLICATIONS OF PURE MATHEMATICS TO PHYSICS

Physicists frequently find the math they need was created long before it was needed. For example, the equation physicist Gabriele Veneziano used to explain particle scattering was the Euler beta function, which was discovered in the 1700s by Swiss mathematician Leonhard Euler. Also, when Einstein began to extend special relativity into general relativity, he soon realized that traditional Euclidean geometry wouldn't work. His space had to curve, and Euclid's geometry only described flat surfaces.

Fortunately for Einstein, in the mid-1800s the German mathematician Bernhard Riemann had worked on a form of non-Euclidean geometry (named *Riemannian geometry*). The mathematics Einstein needed for the general theory of relativity had been created a half century earlier as an intellectual exercise, well before any application to gravity could be imagined. (As fascinating as revolutionizing the foundations of geometry may be, it was hardly practical.)

This happened several times in the history of string theory. Calabi-Yau manifolds, discussed at the end of this chapter, are one example. Another example is when string theorists were attempting to determine the appropriate number of dimensions to make their theories stable and consistent. One way to mathematically frame this problem came from the journals of Indian mathematical genius Srinivasa Ramanujan (referenced in the film *Good Will Hunting*), who died in 1920. The specific mathematics in this case was a function called the Ramanujan function.

Veneziano's explanation has been called the *dual resonance model*, the *Veneziano amplitude*, or just the *Veneziano model*. The dual resonance model was close to the correct result for how hadrons interact, but not quite correct. At the time Veneziano developed the model, particle accelerators weren't precise enough to detect the differences between model and reality. (Eventually, it would be shown that the alternative theory of quantum chromodynamics is the correct explanation of hadron behavior, as discussed in Chapter 8.)

After the dual resonance model was formed, hundreds of theoretical papers were published in attempts to modify the parameters a bit. This is the way theories are approached in physics; after all, an initial guess at a theory is rarely precisely correct and typically requires subtle tweaks — to see how the theory reacts, how much it can be bent and modified, and so on — so that ultimately it fits with the experimental results.

The dual resonance model would have nothing to do with that sort of tinkering — it simply didn't allow for any changes that would still enable it to be valid. The mathematical parameters of the theory were too precisely fixed. Attempts to modify the theory in any way quickly led to a collapse of the entire theory. Any slight disturbance would send it toppling over like a dagger balanced on its tip. Mathematically, it was locked into a certain set of values. In fact, it has been said by some that the theory had absolutely no adjustable parameters — at least in the form first found by Veneziano.

This isn't the way theories are supposed to behave. If you have a theory and modify it so the particle mass, for example, changes a bit, the theory shouldn't collapse — it should just give you a different result.

When a theory can't be modified, there are only two possible reasons: Either it's completely wrong, or it's completely right! For several years, dual resonance models looked like they might be completely right, so physicists continued to ponder what they might mean.

Exploring the first physical model: Particles as strings

The first basic physical interpretation of string theory was particles as vibrating strings. As the strings, each representing a particle, collided with each other, the S-matrix described the result.

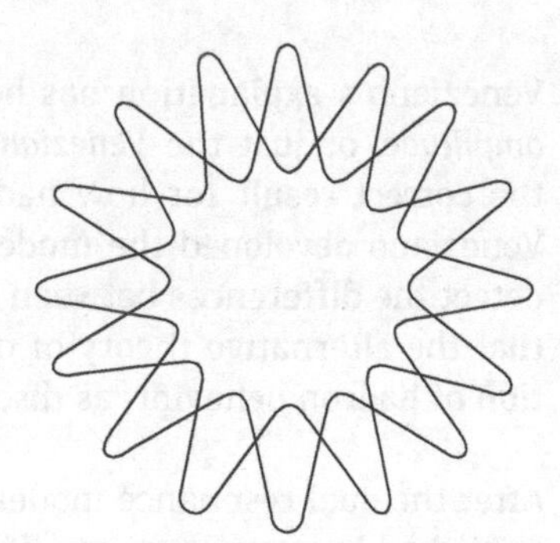

FIGURE 10-1: Most people think of particles as balls or points. In string theory, scientists view them as vibrating strings instead.

What are these strings like? The strings that we are talking about are almost like rubber bands, at least classically. There's a certain "springiness" to them. A phrase that may explain them well is "filaments of energy" (as string theorist Brian Greene and others have called them). Though most people think of particles as balls of matter, physicists have long thought of them as little bundles of waves (called *wave packets*), which is in line with describing them as strings. (Strictly speaking, physicists treat particles as having no size whatsoever, at least in quantum field theory.)

This interpretation was put forth independently by Yoichiro Nambu, Holger Nielsen, and Leonard Susskind in 1970, earning all three men positions as founders of string theory.

According to Einstein's work, mass is a form of energy, an insight demonstrated dramatically by the creation of the atomic bomb. Quantum theory showed physicists that matter is represented by the mathematics of wave mechanics, so even a particle has a wavelength associated with it.

In string theory, matter again takes on a new form. Particles of different types are different vibrational modes of these fundamental entities: energetic rubber bands, or *strings*. (Classical vibrations and strings are discussed in Chapter 5.) In essence, the more the string vibrates, the more energy (and therefore mass) it possesses.

Through all the transformations string theory has undergone in the years since its discovery, this central concept remains (fairly) constant, although in recent years new objects in addition to strings have been introduced (which we explain in Chapter 11 in our discussion of branes).

REMEMBER

The basic physical model couldn't have been simpler: The particles and forces in nature are really interactions between vibrating strings of energy.

Bosonic string theory loses out to the Standard Model

The dual resonance model was created for the express purpose of explaining the S-matrix particle scattering, which was now explained in terms of the Standard Model of particle physics — gauge fields and quantum chromodynamics. (See Chapter 8 for more on these concepts.) There was no point to string theory in light of the success of the Standard Model.

Also it was clear that dual resonance models were only approximately correct. In 1969, physicists showed that Veneziano had discovered only the first term in an infinite series of terms. Although this term was the most important, it was not the end of the story. More terms were needed to fully account for the dynamics of the strings, namely their interactions.

Terms could be added (which Michio Kaku did in 1972), correcting for the different ways the strings could collide, but that made the theory less elegant. There were growing indications that string theory might not work the way everyone had thought it would and that, indeed, quantum chromodynamics explained the behavior of the particle collisions better.

The early string theorists had therefore spent a lot of time giving meaning to a theory that seemed to (almost) accurately predict the S-matrix, only to find that the majority of particle physicists weren't interested in it. It had to be very frustrating to have such an elegant model that was quickly falling into obscurity.

But a few string theorists weren't about to give up on it quite yet.

Why Bosonic String Theory Doesn't Describe Our Universe

By 1974, bosonic string theory was quickly becoming a mathematical mess, and attempts to make the theory mathematically consistent caused more trouble for the model than it already had. Playing with the math introduced four conditions that should have, by all rights, spelled the end of the early string theory:

- The existence of massless particles, which hadn't been observed experimentally
- The existence of tachyons, which move faster than the speed of light
- The fact that fermions, such as electrons, can't exist
- The existence of 25 spatial dimensions
- The fact that certain configurations of interacting strings gave rise to infinities

The cause of these problems was a reasonable constraint built into string theory. No matter what else string theory did, it needed to be consistent with existing physics — namely, special relativity and quantum theory.

The Standard Model of particle physics was consistent with both theories (though it still had trouble reconciling with general relativity), so string theory also had to be consistent with both. If it violated a half century of established physics, there was no way it could be a viable theory.

Physicists eventually found ways to modify the theory to be consistent with these existing physical laws. Unfortunately, the modifications resulted in the four problematic features outlined in the bulleted list. It wasn't just that these features were possible, but they were now seemingly essential components of the theory.

Massless particles

One side effect of creating a consistent string theory is that it has to contain certain objects that can never be brought to rest. Because mass is a measure of an object while it's at rest, these sorts of particles are called *massless particles*. It would be a major problem for string theory if the massless particles it predicted didn't really exist.

Overall, though, this wasn't a terribly disturbing problem because scientists know for certain that at least one particle exists only in a state of motion: the photon. (The gluons, whose existence was directly observed only in 1979, are also massless particles.)

In the original formulation of the Standard Model of particle physics, it was assumed that a particle called the *neutrino* might have a mass of zero. However, in 1998 it was discovered that neutrinos (which come in three flavours) have a mass, though it's really tiny even for an elementary particle. Still, the neutrino is a fermion, not a boson.

There was one other possible massless particle: the graviton. The *graviton* is the theoretical gauge boson that may be responsible for the force of gravity under quantum field theory.

The existence of massless particles in string theory was unfortunate, but it was a surmountable problem. String theorists needed to uncover the properties of massless particles and prove that their properties were consistent with the known universe.

Tachyons

A bigger problem than massless particles was the *tachyon*, a particle predicted by bosonic string theory that travels faster than the speed of light (which creates all sort of puzzles in relativity). Under a consistent bosonic string theory, the mathematical formulas demanded that tachyons exist, but the presence of tachyons in the theory represents a fundamental instability in the theory.

At best, it means that we are describing a theory that would make sense in the wrong way — for instance, by taking the wrong reference state for the empty vacuum of the theory. Physicists would respond that this false vacuum would decay to something else, and that process is hard to pin down in general. At worst, the presence of tachyons could mean the whole theory is nonsense. Even if you believed that the tachyon of bosonic string theory could be removed by some decay, it would be much better to do without it from the get-go.

Strictly speaking, Einstein's theory of relativity doesn't absolutely forbid an object from traveling faster than the speed of light. What it says is that it would require an infinite amount of energy for an object to *accelerate* to the speed of light. Therefore, in a sense, the tachyon would still be consistent with relativity because it would *always* be moving faster than the speed of light (and wouldn't ever have to accelerate to that speed).

Mathematically, when you calculate a tachyon's mass and energy using relativity, it would contain imaginary numbers. (An *imaginary number* is the square root of a negative number.)

This was exactly how string theory equations predicted the tachyon: They were consistent only if particles with imaginary mass existed. But what is imaginary mass? What is an imaginary energy? These physical impossibilities give rise to the problems with tachyons.

The presence of tachyons is in no way unique to bosonic string theory. For example, the Standard Model contains a certain vacuum in which the Higgs boson is actually a type of tachyon as well. In this case, the theory isn't inconsistent; it just means that the solution that was applied wasn't a stable solution. It's like trying to place a ball at the top of a hill — any slight movement will cause the ball to roll into a nearby valley. Similarly, this tachyon solution decays into a stable solution without the tachyons.

Unfortunately, in the case of bosonic string theory, there was no clear way to figure out what happened during the decay, or even if the solution ended up in a stable solution after decaying into a lower energy state.

With all these problems, physicists don't view tachyons as actual particles that exist, but rather as mathematical artifacts that fall out of the theory as a sign of certain types of inherent instabilities. Any solution that contains tachyons quickly decays because of these instabilities.

Some physicists (and science fiction authors) have explored notions of how to treat tachyons as actual particles, a speculative concept that comes up briefly in Chapter 17. But for now, just know that tachyons were one of the things that made physicists decide, at the time, that bosonic string theory was a failure.

No electrons allowed

The real issue in bosonic string theory was the one that it's named after. The theory predicted only the existence of bosons, not fermions. Photons could exist, but not quarks or electrons. That isn't good because fermions make up all the matter we see around us.

Every elementary particle observed in nature has a property called a *spin*, which is either an integer value (−1, 0, 1, 2, and so on) or a half-integer value (−½, ½, and so on). Particles with integer spins are *bosons*, and particles with half-integer spins are *fermions*. One key finding of particle physics is that all particles fall into one of these two categories.

To apply to the real world, string theory had to include both types of particles, and the original formulation didn't. The only particles allowed under the first model of string theory were bosons. This is why it would come to be known to physicists as the *bosonic string theory*.

25 space dimensions, plus 1 of time

Dimensions are the pieces of information needed to determine a precise point in space. (Dimensions are generally thought of in terms of up/down, left/right, forward/backward.) In 1974, Claude Lovelace discovered that bosonic string theory could be physically consistent only if it were formulated in 25 spatial dimensions (Chapter 15 delves into the idea of the additional dimensions in more depth), but so far as anyone knows, we have only three spatial dimensions!

Relativity treats space and time as a continuum of coordinates, so that means the universe has a total of 26 dimensions in string theory, as opposed to the four dimensions that we perceive.

It's unusual that this requirement would be implicit in the theory. Einstein's relativity has three spatial dimensions and one time dimension because those are the conditions he used to create the theory. He didn't begin working on relativity and just happen to stumble upon three spatial dimensions, but rather intentionally built them into the theory from the beginning. If he'd wanted a 2-dimensional or 5-dimensional relativity, he could have built the theory to work in those dimensions.

With bosonic string theory, the equations actually demanded a certain number of dimensions to be mathematically consistent. The theory falls apart in any other number of dimensions!

The reason for extra dimensions

TIP

The reason for extra dimensions comes from requiring string theory to have two crucial features. First of all, it has to be a quantum theory. Secondly, but equally importantly, it must obey the laws of Einstein's special relativity (which we encountered in Chapter 6).

Before we take into consideration quantum effects, string theory is literally a model describing a string, which moves around following the rules of special relativity. (This is how the theory was constructed in the first place!) These rules dictate that the description of the string must be invariant not only under transformations such as rotations and translations, but also under the transformations introduced by Hendrik Lorentz, that corresponds to redefining, or *boosting*, the overall speed of the string.

The trouble comes when we want to create a quantum theory out of this classical model. In fact, as we discussed in Chapter 8, the fact that a theory is invariant under some symmetry like rotations or Lorentz boosts does not guarantee that this invariance is preserved at the quantum level. In other words, quantum fluctuations could spoil the invariance. This is precisely what happens if we try to construct a quantum theory of a string in four space-time dimensions: the invariance demanded by Einstein's relativity is lost!

The same problem shows up in five, six, seven space-time dimensions . . . but miraculously, the invariance is safe in 26 spacetime dimensions. This means that it is only possible to build a quantum theory of a relativistic (bosonic) string in 26 spacetime dimensions — no more, no less. This is called the *critical dimension* for the string.

Of course, you may shrug and just give up the demand for relativistic invariance. If that's how you feel, you can consider strings in four dimensions with no problem (and some physicists do study these string theories, which are called *non-critical)*. But we have seen that Einstein's special relativity is the building block on which the last century of fundamental physics has been built, and most physicists have been rather squeamish about giving it up.

For this reason, most string theorists accept that dealing with strings means dealing with extra dimensions. The question then becomes: What to do with them?

Dealing with the extra dimensions

The physical conception of these extra dimensions was (and still is) the hardest part of the theory to comprehend. Everyone can understand three spatial dimensions and a time dimension. Give us a latitude, longitude, altitude, and time, and we can meet you anywhere on the planet. You can measure height, width, and length, and you experience the passage of time, so you have a regular familiarity with what those dimensions represent.

What about the other 22 spatial dimensions? It was clear that these dimensions had to be hidden somehow. The Kaluza-Klein theory assumed that extra dimensions were rolled up, but rolling them up in precisely the right way to achieve results that made sense was difficult. This was achieved for string theory in the mid-1980s through the use of Calabi-Yau manifolds, as we discuss later in this chapter.

No one has any direct experience with these strange other dimensions. That the idea came out of the symmetry relationships associated with a relatively obscure new theoretical physics conjecture certainly didn't offer much motivation for physicists to accept it. And for more than a decade, most physicists didn't.

Supersymmetry Saves the Day: Superstring Theory

Despite bosonic string theory's apparent failures, some brave physicists stayed committed to their work. Why? Well, physicists can be a passionate bunch (nearly obsessive, some might say). Another reason was that by the time these problems were fully realized, many string theorists had already moved on from bosonic string theory anyway.

With the development in 1971 of *supersymmetry*, which allows for bosons and fermions to coexist, string theorists were able to develop *supersymmetric string theory*, or, for short, *superstring theory*, which took care of the major problems that destroyed bosonic string theory. This work opened up whole new possibilities for string theory.

Almost every time you hear or read the phrase "string theory," the person probably really means "superstring theory." Since its discovery, supersymmetry has been applied to virtually all forms of string theory. The only string theory that really has nothing to do with supersymmetry is bosonic string theory, which was created before supersymmetry. For all practical discussion purposes (with anyone who isn't a theoretical physicist), "string theory" and "superstring theory" are the same term.

Fermions and bosons coexist . . . sort of

Symmetries exist throughout physics. A *symmetry* in physics is basically any situation where two properties can be swapped throughout the system and the results are precisely the same.

The notion of symmetry was picked up by Pierre Ramond in 1970, followed by the work of John Schwarz and Andre Neveu in 1971, to give hope to string theorists. Using two different techniques, they showed that bosonic string theory could be generalized in another way to obtain non-integer spins. Not only were the spins non-integer, but they were precisely half-integer spins, which characterize the fermion. No spin 1/4 particles showed up in the theory, which is a check that the theory is compatible with special relativity. Physicists have never observed spin 1/4 particles, so the fact that they don't show up in the theory also means we don't have to explain an inconsistency with the natural world, something which is always awkward in science.

Including fermions in the model meant introducing a powerful new symmetry between fermions and bosons, called supersymmetry. *Supersymmetry* can be summarized as a symmetry that allows you to simultaneously make the following two changes without changing the outcome of any experiment. (Remember, symmetries allow you to make changes without affecting the results of experiments!)

- Every boson can be switched with a corresponding fermion.
- Every fermion can be switched with a corresponding boson.

In Chapter 11, we discuss the reasons to believe that supersymmetry is true, as well as ways it can be experimentally observed. For now, it's enough to know that it's needed to make string theory work.

Of course, as you'll anticipate if you're looking for trends in the story of string theory, things didn't quite fall out right. Fermions and bosons have very different properties, so getting them to switch places without affecting the possible outcomes of an experiment isn't easy.

Physicists know about a number of bosons and fermions, but when they began looking at the properties of the theory, they found that the correspondence didn't exist between known particles. A photon (which is a boson) doesn't appear to be linked by supersymmetry with any of the known fermions.

Fortunately for theoretical physicists, this messy experimental fact was seen as only a minor obstacle. They turned to a method that has worked for theorists since the dawn of time. If you can't find evidence of your theory, hypothesize it!

Double your particle fun: Supersymmetry hypothesizes superpartners

Under supersymmetry, the corresponding bosons and fermions are called *superpartners*. The superpartner of a standard particle is called a *sparticle*.

WHO DISCOVERED SUPERSYMMETRY?

The origins of supersymmetry are a bit confusing because it was discovered around the same time by four separate groups of scientists.

In 1971, Russians Evgeny Likhtman and Yuri Golfand created a consistent theory containing supersymmetry. A year later, two more Russians, Vladimir Akulov and Dmitri Volkov, put forth their supersymmetry theory. These theories were in only two dimensions, however.

Because of the Cold War, communication between Russia and the noncommunist world wasn't very good, so many physicists didn't hear about the Russian scientists' work. European physicists Julius Wess and Bruno Zumino were able to create a 4-dimensional supersymmetric quantum theory in 1973, probably aware of the Russian work. Theirs was noticed by the Western physics community at large.

Then, of course, we have Pierre Ramond, John Schwarz, and Andre Neveu, who developed supersymmetry in 1970 and 1971, in the context of their superstring theories. It was only on later analysis that physicists realized their work and the later work hypothesized the same relationships.

Many physicists consider this repeated discovery as a good indication that there's probably something to the idea of supersymmetry in nature, even if string theory itself doesn't prove to be correct.

Because none of the existing particles are superpartners of any other particle, if supersymmetry is true, there are twice as many particles as we currently know about. For every standard particle, a sparticle that has never been detected experimentally must exist. Sparticles are one of the key pieces of evidence the Large Hadron Collider is looking for.

REMEMBER

If we mention a strangely named particle that you've never run into, it's probably a sparticle. Because supersymmetry introduces so many new particles, it's important to keep them straight. Physicists have introduced a Dr. Seuss-like naming convention to identify the hypothetical new particles:

- The superpartner of a fermion begins with an "s" before the standard particle name; so the superpartner of an "electron" is a "selectron," and the superpartner of a "quark" is a "squark."
- The superpartner of a boson ends in an "–ino"; so the superpartner of a "photon" is a "photino," and the superpartner of the "graviton" is the "gravitino."

Table 10-1 shows the names of standard particles and their corresponding superpartners.

TABLE 10-1 **Some Superpartner Names**

Standard Particle	Superpartner
Lepton	Slepton
Muon	Smuon
Neutrino	Sneutrino (or neutralino)
Top Quark	Stop Squark
Gluon	Gluino
Higgs boson	Higgsino
W boson	Wino
Z boson	Zino

Even though there is an elementary superpartner called a "sneutrino," there exists no elementary particle or superpartner called a "sneutron" (because the neutron is not an elementary particle to begin with).

Some problems get fixed, but the dimension problem remains

The introduction of supersymmetry into string theory helped with some of the major problems of bosonic string theory. Fermions now existed within the theory, which had been the biggest problem. Tachyons vanished from superstring theory. Massless particles were still present in the theory, but weren't seen as a major issue. Even the dimensional problem improved, dropping from 26 space-time dimensions down to a mere ten.

The supersymmetry solution was elegant. Bosons — the photon, graviton, and Z and W bosons — are carriers of force. Fermions — the electron, quarks, and neutrinos — are units of matter. Supersymmetry created a new symmetry, one between matter and forces.

In 1972, Andre Neveu and Joel Scherk resolved the massless particle issue by showing that string vibrational states could correspond to the gauge bosons, such as the massless photon.

The dimensional problem remained, although it was better than it had been. Instead of 25 spatial dimensions, superstring theory became consistent with a "mere" nine spatial dimensions (plus one time dimension, for a total of ten dimensions). Many string theorists of the day believed this was still too many dimensions to work with, so they abandoned the theory for other lines of research.

One physicist who turned his back on string theory was Michio Kaku, one of today's most vocal advocates of string theory. Kaku's PhD thesis involved completing all the terms in the Veneziano model's infinite series. He'd created a field theory of strings, so he was working in the thick of string theory. Still, he abandoned his work on superstring theory, believing there was no way it could be a valid theory. That's how serious the dimensional problem was.

The handful of people who remained dedicated to string theory after 1974 faced serious questions about how to proceed. With the exception of the dimensional problem, they had resolved nearly all the issues with bosonic string theory by transforming it into superstring theory.

The only question was what to do with it.

Supersymmetry and Quantum Gravity in the Disco Era

By 1974, the Standard Model had become the theoretical explanation of particle physics and was being confirmed in experiment after experiment. With a stable foundation, theoretical physicists now looked for new worlds to conquer, and many decided to tackle the same problem that had vexed Albert Einstein for the last decades of his life: quantum gravity.

As a consequence of the Standard Model's success, string theory wasn't needed to explain particle physics. Instead, almost by accident, string theorists began to realize that string theory might just be the very theory that would solve the problem of quantum gravity.

The graviton is found hiding in string theory

The graviton is a particle that, under predictions from unified field theory, would mediate the gravitational force (see Chapter 2 for more on the graviton). In a very real sense, the graviton *is* the force of gravity. One major feature of string theory is that it not only includes the graviton, but requires its existence as one of the massless particles discussed earlier in this chapter.

In 1974, Joel Scherk and John Schwarz demonstrated that a spin-2 massless particle in superstring theory could actually be the graviton. This particle was represented by a closed string (which forms a loop), as opposed to an open string, where the ends are loose. Figure 10-2 shows both types of strings.

Open strings

Closed strings

FIGURE 10-2: String theory allows for open and closed strings. Open strings are optional, but closed strings have to exist.

REMEMBER

String theory demands that closed strings must exist, though open strings may or may not exist. Some versions of string theory are perfectly mathematically consistent but contain *only* closed strings. No theory contains only open strings because if you have open strings, you can construct a situation where the ends of the strings meet each other and, voilà, a closed string exists. (Cutting closed strings to get open strings isn't always allowed.)

From a theoretical standpoint, this was astounding (in a good way). Instead of trying to shoehorn gravity into the theory, the graviton fell out as a natural consequence. If superstring theory was the fundamental law of nature, then it required the existence of gravity in a way that no other proposed theory had ever done!

Immediately, it became clear to Schwarz and Scherk that they had a potential candidate for quantum gravity on their hands.

Even while everyone else was fleeing from the multiple dimensions their theory predicted, Scherk and Schwarz became more convinced than ever that they were on the right track.

The other supersymmetric gravity theory: Supergravity

Supergravity is the name for theories that attempt to apply supersymmetry directly to the theory of gravity without the use of string theory. Throughout the late 1970s, this work proceeded at a faster pace than string theory, mainly because it was popular while the string theory camp had become a ghost town. Supergravity theories prove important in the later development of M-theory, which we cover in Chapter 11.

In 1976, Daniel Freedman, Sergio Ferrara, and Peter van Nieuwenhuizen applied supersymmetry to Einstein's theory of gravity, resulting in a theory of supergravity. They did this by introducing the superpartner of the graviton, the gravitino, into the theory of general relativity. That immediately made the problem of infinities that plagues quantum gravity a little more manageable, which got physicists very excited (though it wasn't clear if all infinite results would drop out of supergravity).

Building on this work, Eugene Cremmer, Joel Scherk, and Bernard Julia were able to show in 1978 that supergravity could be written, in its most general form, as an 11-dimensional theory. Supergravity theories with more than 11 dimensions just don't exist.

Most versions of supergravity (the jury is still out on some) ultimately fell prey to the mathematical inconsistencies that plagued most quantum gravity theories: They work fine as a classical theory as long as you kept it away from the quantum realm. This left room for superstring theory to rise again in the mid-1980s, but supergravity didn't go away completely. We return to the idea of the 11-dimensional supergravity theory in Chapter 11.

String theorists don't get no respect

During the late 1970s, string theorists were finding it hard to be taken seriously, let alone find secure academic work. The name of the game was quantum field theory, gauge theories, and, to some extent, supergravity.

There had been earlier issues in getting recognition for string theory work. The journal *Physics Review Letters* didn't consider Leonard Susskind's 1970 work — interpreting the dual resonance model as vibrating strings — significant enough to publish. Susskind himself recounts how physics giant Murray Gell-Mann laughed at him for mentioning string theory in 1970. (The story ends well, with Gell-Mann expressing interest in the theory in 1972.)

As the decade progressed, two of the major forces behind string theory would run into hurdle after hurdle in getting a secure professorship. John Schwarz had been denied tenure at Princeton in 1972 and spent the next 12 years at CalTech in a temporary position, never sure if the funding for his job would be renewed. Pierre Ramond, who had discovered supersymmetry and helped rescue string theory from oblivion, was denied tenure at Yale in 1976.

Against the backdrop of professional uncertainty, the few string theorists continued their work through the late 1970s and early 1980s, helping deal with some of the extra dimensional hurdles in supergravity and other theories, until the day came when the tables turned and they were able to lay claim to the high ground of theoretical physics.

A Theory of Everything: The First Superstring Revolution

The year 1984 is marked by many as the start of "the first superstring revolution." The major finding that sparked the revolution was the proof that string theory contained no mathematical inconsistencies, unlike many of the quantum gravity theories, including supergravity, studied during the 1970s.

For nearly a decade, John Schwarz had been working on showing that superstring theory could be a quantum theory of gravity. His major partner in this work, Joel Scherk, had died in 1980, a tragic blow to the cause. By 1983, Schwarz was working with Michael Green, one of the few individuals who had been persuaded to work on string theory during that time.

Typically, two major problems arose in theories of quantum gravity: anomalies and infinities. Neither is a good sign for a scientific theory.

- **Infinities** occur when a computation of some physical quantity (for instance, the probability of producing a certain particle in a scattering experiment) gives an arbitrarily large result that cannot be "renormalized" away (see Chapter 8 for a discussion of the infinities and the renormalization process).
- **Anomalies** are cases where quantum mechanical processes can violate a symmetry that is supposed to be preserved. This can be just a quirk of the quantum world if the symmetry is optional to begin with, or it can spell disaster if the symmetry is a crucial building block of the model.

Superstring theory was actually pretty good at avoiding infinities.

TIP

One simplification that allows you to understand, in very general terms, how superstring theory avoids infinities is that the distance value never quite reaches zero. Dividing by zero (or a value that can get arbitrarily close to zero) is the mathematical operation that results in an infinity. Because the strings have a tiny bit of length (call it L), the distance never gets smaller than L, so the gravitational force is obtained by dividing by a number that never gets smaller than L^2. This means the gravitational force will never explode up to infinity, as happens when the distance approaches zero without a limit.

String theory also had no anomalies (at least under certain conditions), as Schwarz and Green proved in 1984. They showed that certain 10-dimensional versions of superstring theory had exactly the constraints needed to cancel out all anomalies.

This changed the whole landscape of theoretical physics. For a decade, superstring theory had been ignored while every other method of creating a quantum theory of gravity collapsed in on itself under infinities and anomalies. Now, this discarded theory had risen from the ashes like a mathematical phoenix — both finite and anomaly-free.

Theorists began to think that superstring theory had the potential to unify all the forces of nature under one simple set of physical laws with an elegant model in which everything consisted of different energy levels of vibrating strings. Not only would string theory predict the existence of quantum gravity, but you would also get a plethora of particles and their superpartners — hopefully exactly those existing in nature — along with all of their various physical properties, like masses and interaction strengths. It was the ideal that had eluded Einstein: a fundamental theory of all natural law that explained all observed phenomena.

But We've Got Five Theories!

In the wake of 1984's superstring revolution, work on string theory reached a fever pitch. If anything, it proved a little too successful. It turned out that instead of one superstring theory to explain the universe, there were five, given the colorful names

- Type I
- Type IIA
- Type IIB
- Type HO
- Type HE

And, once again, each one *almost* matched our world . . . but not quite.

By the time the decade ended, physicists had developed and dismissed many variants of string theory in hopes of finding the one true formulation of the theory.

Instead of one formulation, though, five distinct versions of string theory proved to be self-consistent. Each had some properties that made physicists think it would reflect the physical reality of our world — and some properties that are clearly not true in our universe.

The distinctions between these theories are mathematically sophisticated. We introduce their names and basic definitions mainly because of the key role they play in M-theory, which we discuss in Chapter 11.

Type I string theory

Type I string theory involves both open and closed strings. It contains a form of symmetry that's mathematically designated as a symmetry group called O(32). This happens to be the group of rotations and reflections in 32 dimensions. (We'll try to make that the most mathematics you need to know related to symmetry groups.)

Type IIA string theory

Type IIA string theory involves closed strings where the vibrational patterns are symmetrical, regardless of whether they travel left or right along the closed string. Type IIA open strings are attached to structures called D-branes (which we discuss in greater detail in Chapter 11) with an odd number of dimensions.

Type IIB string theory

Type IIB string theory involves closed strings where the vibrational patterns are asymmetrical, depending on whether they travel left or right along the closed string. Type IIB open strings are attached to D-branes (discovered in 1995 and covered in Chapter 11) with an even number of dimensions.

Two strings in one: Heterotic strings

A new form of string theory, called *heterotic string theory*, was discovered in 1985 by the Princeton team of David Gross, Jeff Harvey, Emil Martinec, and Ryan Rohm. This version of string theory sometimes combines some features of bosonic string theory with some of superstring theory.

A distinction of the heterotic string is that the string vibrations in different directions resulted in different behaviors. "Left-moving" vibrations resembled the old bosonic string, while "right-moving" vibrations resembled the Type II strings. The heterotic string seemed to contain exactly the properties that Green and Schwarz needed to cancel out anomalies within the theory.

TECHNICAL STUFF

It was ultimately shown that only two mathematical symmetry groups could be applied to heterotic string theory, which resulted in stable theories in ten dimensions: O(32) symmetry and $E_8 \times E_8$ symmetry. These two groups gave rise to the names Type HO and Type HE string theory.

Type HO string theory

Type HO is a form of heterotic string theory. The name comes from the longer name Heterotic O(32) string theory, which describes the symmetry group of the theory. It contains only closed strings whose right-moving vibrations resemble the Type II strings and whose left-moving vibrations resemble the bosonic strings. The similar theory, Type HE, has subtle but important mathematical differences regarding the symmetry group.

Type HE string theory

Type HE is another form of heterotic string theory, based on a different symmetry group from the Type HO theory. The name comes from the longer name Heterotic $E_8 \times E_8$ string theory, which describes the symmetry group of the theory. It also contains only closed strings whose right-moving vibrations resemble the Type II strings and whose left-moving vibrations resemble the bosonic strings.

How to Fold Space: Introducing Calabi-Yau Manifolds

The problems of extra dimensions continued to plague string theory, but these were solved by introducing the idea of *compactification*, in which the extra dimensions curl up around each other, growing so tiny that they're extremely hard to detect. The mathematics that explain how this might be achieved had already been developed in the form of complex *Calabi-Yau manifolds*, an example of which is shown in Figure 10-3. The challenge is that string theory offers no real way to determine exactly which of the many Calabi-Yau manifolds (or a similar type of folded structure) is right!

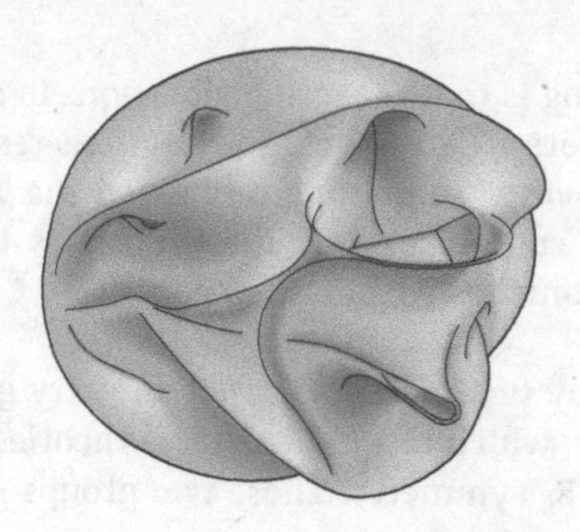

FIGURE 10-3: According to string theory, the universe has extra dimensions that are curled up in tiny compact manifolds, like Calabi-Yau manifolds and their generalizations.

When the extra dimensions were first discovered in the 1970s, it was clear that they must be hidden in some way. After all, we certainly don't see more than three spatial dimensions.

One suggestion was the solution that had been proposed by Theodor Kaluza and Oskar Klein a half century earlier: The dimensions could be curled up into a very small size.

Early attempts to curl up these extra dimensions ran into problems because they tended to retain the symmetry between left- and right-handed particles (called *parity* by physicists), which isn't always retained in nature. This parity violation in the Standard Model is crucial in understanding the operation of the weak nuclear force.

For string theory to work, there had to be a way to compactify the extra six dimensions while still retaining a distinction between the left-handed and right-handed particles.

In 1985, the Calabi-Yau manifolds (created for other purposes years earlier by mathematicians Eugenio Calabi and Shing-Tung Yau) were used by Edward Witten, Philip Candelas, Gary Horowitz, and Andrew Strominger to compactify the extra six space dimensions in just the right way. These manifolds preserved supersymmetry just enough to replicate certain aspects of the Standard Model.

One benefit of the Calabi-Yau manifolds is that the geometry of the folded dimensions gives rise to different types of observable particles in our universe. If the Calabi-Yau shape has three holes (or rather higher-dimensional analogs of holes), three families of particles will be predicted by the Standard Model of particle physics. (Obviously, by extension, a shape with five holes will have five families, but physicists are concerned about only the three families of particles they know exist in this universe.)

Moreover, one can tweak the Calabi-Yau manifolds to reproduce the parity violation, which has been established experimentally in the Standard Model.

Unfortunately, there are tens of thousands of possible Calabi-Yau manifolds for six dimensions, and string theory offers no reasonable means of determining which is the right one. For that matter, even if physicists could determine which one is the right one, they'd still want to answer the question of why the universe folded up the extra six dimensions in that particular configuration.

When Calabi-Yau manifolds were introduced as a possible solution, some vocal members of the string theory community hoped that one specific manifold would fall out as the right one. That hasn't proved to be the case, which is what many string theorists would have expected in the first place: that the specific Calabi-Yau manifold is a quantity that has to be determined by experiment. In fact, it's now known that some other geometries for folded spaces can also maintain the needed properties. We talk about the implications of this folded space — what it could really mean — in Chapters 15 and 16.

String Theory Loses Steam

The rising tide of string theory research couldn't last forever, and by the early 1990s, some physicists were giving up any hope of finding one single theory. Just as the earlier introduction of multiple dimensions had warded off new physicists, the rise of so many distinct yet consistent versions of string theory gave many physicists pause. Physicists who were motivated purely by the drive to find a quick and easy "theory of everything" began turning away from string theory when it became clear that there was nothing quick and easy about it. As the easier problems got solved and the harder ones remained, only the truly dedicated scientists retained the motivation to work through the complications.

In 1995, a second string theory revolution would come along, with the rise of new insights that would help convince even many of the skeptics that work on string theory would ultimately bear significant fruit. That second revolution is the topic of Chapter 11.

IN THIS CHAPTER

» Dualities present a puzzle

» Re-energizing the movement with M-theory

» Thinking about branes

» Using multidimensional branes in string theory

Chapter 11
M-Theory and Beyond: Bringing String Theory Together

Chapter 10 ends with five versions of string theory. Theorists continued their work but were uncertain how to take these findings. A new insight was needed to generate further progress in the field.

In this chapter, we explain how that insight came about in the form of M-theory, which unified the five string theories into one theory. We discuss how string theory was expanded to include objects with more than one dimension called branes. We introduce some possible insights that may help explain what M-theory is trying to describe.

Introducing the Unifying Theory: M-Theory

At a conference in 1995, physicist Edward Witten proposed a bold resolution to the problem of five distinct string theories. In his theory, based on newly discovered dualities, each of the existing theories was a special case of one overarching string theory, which he enigmatically called *M-theory*. One of the key concepts required for M-theory was the introduction of branes (short for membranes) into string theory. *Branes* are fundamental objects in string theory that have more than one spatial dimension.

Witten didn't thoroughly explain the true meaning of the name M-theory, leaving it as something that each person can define for themself. There are several possibilities for what the "M" could stand for: membrane, magic, mother, mystery, or matrix. Witten probably took the "M" from membrane because membranes featured so prominently in the theory, but he didn't want to commit himself to requiring them so early in the development of the new theory.

Although Witten didn't propose a complete version of M-theory (in fact, we're still waiting on one), he did outline certain defining traits that M-theory would have:

- 11 dimensions (10 space dimensions plus 1 time dimension)
- Dualities that result in the five existing string theories all being different explanations of the same physical reality
- Branes — like strings, but with more than one spatial dimension

Translating one string theory into another: Duality

The core of M-theory is the idea that each of the five string theories introduced in Chapter 10 is actually a variation on one theory. This new theory — M-theory — is an 11-dimensional theory that allows for each of the existing theories (which are 10-dimensional) to be equivalent if you make certain assumptions about the geometry of the space involved.

The basis for this suggestion was the understanding of dualities that were being recognized among the various string theories. A *duality* occurs when you can look at the same phenomenon in two distinct ways, taking one theory and mapping it to another theory. In a sense, the two theories are equivalent. By the mid-1990s, growing evidence showed that at least two dualities existed between the various string theories: *T-duality* and *S-duality*.

These dualities were similar to earlier dualities conjectured in 1977 by Claus Montonen and David Olive in quantum field theory. In the early 1990s, Indian physicist Ashoke Sen and Israeli-born physicist Nathan Seiberg did work that expanded on the notions of these dualities. Witten drew upon their work, more recent work by Chris Hull and Paul Townsend, and his own work to present M-theory.

Topological duality: T-duality

One of the dualities discovered at the time was called *T-duality,* which refers to either *topological duality* or *toroidal duality*, depending on whom you ask. (*Toroidal* is a reference to the simplest space, which is a *torus*, or donut shape. *Topological* is a precise way of defining the structure of that space, as explained in the nearby sidebar "Topology: The mathematics of folding space." In some cases, the T-duality has nothing to do with a torus, and in other cases, it's not topological.) The T-duality related the Type II string theories to each other and the heterotic string theories to each other, indicating that they were different manifestations of the same fundamental theory.

In the T-duality, you have a dimension that is compactified into a circle (of radius R), so the space becomes something like a cylinder. It's possible for a closed string to wind around the cylinder, like thread on a spindle. (This means that both the dimension and the string have radius R.) The number of times the closed string winds around the cylinder is called the *winding number*. A second number represents the momentum of the closed string.

TOPOLOGY: THE MATHEMATICS OF FOLDING SPACE

TECHNICAL STUFF

Topology allows you to study mathematical spaces by eliminating all details from the space except certain sets of properties that you care about. Two spaces are topologically equivalent if they can be continuously deformed into one another, even if they differ in other details. Certain actions may be more easily performed on one of the spaces than the other. You can perform actions on that space and work backward to find the resulting effect on the topologically equivalent space. It can be far easier than trying to perform these actions directly on the original space.

One of the key components of topology is the study of how different topological spaces relate to each other. If there is some manipulation relating the two spaces, and if this manipulation can be performed without breaking or reconnecting each space in a new way, the two spaces are topologically equivalent.

To picture this, imagine a donut (or torus) of clay that you slowly and meticulously recraft into the shape of a coffee mug. The hole in the center of the donut never has to be broken in order to be turned into the handle of the coffee mug. On the other hand, there's no way to turn a donut into a pretzel without introducing breaks into the space — a donut and a pretzel are topologically distinct.

Here's where things get interesting. For certain types of string theory, if you wrap one string around a cylindrical space of radius R and the other around a cylindrical space of radius $1/R$, the winding number of one theory seems to match the momentum number of the other theory. (Momentum, like just about everything else, is quantized.)

In other words, T-duality can relate a string theory with a large compactified radius to a different string theory with a small compactified radius (or, alternately, wide cylinders with narrow cylinders). Specifically, for closed strings, T-duality relates to the following types of string theories:

- Type IIA and Type IIB superstring theories
- Type HO and Type HE superstring theories

The case for open strings is a bit less clear. When a dimension of superstring space-time is compactified into a circle, an open string doesn't wind around that dimension, so its winding number is 0. This means that it corresponds to a string with momentum 0 — a stationary string — in the dual superstring theory.

The end result of T-duality is an implication that Type IIA and IIB superstring theories are really two manifestations of the same theory, and Type HO and HE superstring theories are really two manifestations of the same theory.

Strong-weak duality: S-duality

Another duality that was known in 1995 is called *S-duality,* which stands for *strong-weak duality.* S-duality is connected to the concept of a *coupling constant,* which is the value that tells the interaction strength of the string by describing how probable it is that the string will break apart or join with other strings.

The coupling constant g in string theory describes the interaction strength due to a quantity known as the *dilation field,* ϕ. If you had a high positive dilation field ϕ, the coupling constant $g = e^{\phi}$ becomes very large (or the theory becomes strongly coupled). If you instead had a dilation field $-\phi$, the coupling constant $g = e^{-\phi}$ becomes very small (or the theory becomes weakly coupled).

Because of the mathematical methods that string theorists have to use to approximate the solutions to string theory problems (see nearby sidebar "Perturbation theory: String theory's method of approximation"), it was very hard to determine what would happen to string theories that were strongly coupled.

In S-duality, a strong coupling in one theory relates to a weak coupling in another theory in certain conditions. In one theory, the strings break apart and join other strings easily, while in the other theory, they hardly ever do so. In the theory where the strings break and join easily, you end up with a chaotic sea of strings constantly interacting.

TIP

Trying to follow the behavior of individual strings is similar to trying to follow the behavior of individual water molecules in the ocean — you just can't do it. So what do you do instead? You look at the big picture. Instead of looking at the smallest particles, you average them out and look at the unbroken surface of the ocean, which, in this analogy, is the same as looking at the strong strings that virtually never break.

PERTURBATION THEORY: STRING THEORY'S METHOD OF APPROXIMATION

TECHNICAL STUFF

The equations of string theory are incredibly complex, so they often can only be solved through a mathematical method of approximation called *perturbation theory*. This method, frequently used in quantum mechanics and quantum field theory, is a well-established mathematical process.

In perturbation theory, physicists arrive at a first-order approximation, which is then expanded with other terms that refine the approximation. The goal is that the subsequent terms will become so small so quickly that they'll cease to matter. Adding even an infinite number of terms will result in converging onto a given value. In mathematical speak, *converging* means that you keep getting closer to the number without ever passing it.

Consider the following example of convergence: If you add a series of fractions, starting with ½ and doubling the denominator each time (½ + ¼ + ⅛ + . . . well, you get the idea), you'll always get closer to a value of 1, but you'll never quite reach 1. The reason for this is that the numbers in the series get small very quickly and stay so small that you're always just a little bit short of reaching 1. In fact, for many practical applications, the first few terms in the sum give a good enough approximation of the final result (so that we need not bother with all the other terms).

However, if you add numbers that double (2 + 4 + 8 + . . . well, you get the idea), the series doesn't converge at all. The solution keeps getting bigger as you add more terms. In this situation, the solution is said to *diverge*, or become infinite.

The dual resonance model that Gabriele Veneziano proposed — which sparked all of string theory — was found to be only a first-order approximation of what later came to be known as string theory. Work over the last 50 years has largely been focused on trying to find situations in which the theory built around this original first-order approximation can be absolutely proved to be safe from infinities and also matches the physical details observed in our own universe.

REMEMBER

S-duality introduces Type I string theory to the set of dual theories that T-duality started. Specifically, it shows that the following dualities are related to each other:

- Type I and Type HO superstring theories are S-dual to each other.
- Type IIB is S-dual to itself.

If you have a Type I superstring theory with a very strong coupling constant, it's theoretically identical to a Type HO superstring theory with a very weak coupling constant. So these two types of theories, under these conditions, yield the exact same predictions for masses and charges.

Using two dualities to unite five superstring theories

Both T-duality and S-duality relate different string theories together. Here's a review of the existing string theory relationships:

- Type I and Type HO superstring theories are related by S-duality.
- Type HO and Type HE superstring theories are related by T-duality.
- Type IIA and Type IIB superstring theories are related by T-duality.

With these dualities (and other, more subtle ones that relate Type IIA and IIB together with the heterotic string theories), relationships exist to transform one version of string theory into another one — at least for some specially selected string theory conditions.

REMEMBER

To solve these equations of duality, certain assumptions have to be made, and not all of them are necessarily valid in a string theory that would describe our own universe. For example, some of these dualities are understood only in cases of perfect supersymmetry, while our own universe exhibits (at best) broken supersymmetry.

String theory skeptics aren't convinced that these dualities in some specific states of the theories relate to a more fundamental duality of the theories at all levels. Physicist (and string theory skeptic) Lee Smolin calls this the pessimistic view and dubs the string theory belief in the fundamental nature of these dualities the optimistic view.

Still, in 1995, it was hard not to be in the optimistic camp (and, in fact, many theorists had never stopped being optimistic about string theory). The very fact that these dualities existed at all was startling to string theorists. The discovery wasn't planned, but came out of the mathematical analysis of the theory. This was seen as powerful evidence that string theory was on the right track. Instead of falling apart into a bunch of different theories, superstring theory was actually pulling back together into one single theory — Edward Witten's M-theory — which manifested itself in a variety of ways.

The second superstring revolution begins: Connecting to the 11-dimensional theory

The period immediately following the proposal of M-theory has been called the "second superstring revolution" because it once again inspired a flurry of research into superstring theory. The research this time focused on understanding the connections between the existing superstring theories and the 11-dimensional theory Witten had proposed.

Witten wasn't the first one to put forth this sort of a connection. The idea of uniting the different string theories into one theory by adding an 11th dimension had been proposed by Mike Duff of Texas A&M University, but it never caught on among string theorists. Witten's work on the subject, however, resulted in a picture where the extra dimension could emerge from the unifications inherent in M-theory — which prompted the string theory community to look at his work more seriously.

In 1994, Witten and his colleague Paul Townsend had discovered a link between the 10-dimensional superstring theory and an 11-dimensional theory, which had been proposed back in the 1970s: supergravity.

Supergravity resulted when you applied supersymmetry to the equations of general relativity (in other words, when you introduced a particle called the gravitino — the superpartner to the graviton — to the theory). In the 1970s this was pretty much the dominant approach to trying to get a theory of quantum gravity.

What Witten and Townsend did in 1994 was take the 11-dimensional supergravity theory from the 1970s and curl up one of the dimensions. They then showed that a membrane in 11 dimensions that has one dimension curled up behaves like a string in 10 dimensions.

Again, this is a recurrence of the old Kaluza-Klein idea, which comes up again and again in the history of string theory. By taking Kaluza's idea of adding an extra dimension (and Klein's idea of rolling it up very small) and assuming certain symmetry conditions, Witten showed that it was possible to demonstrate that dualities existed between the string theories.

There were still issues with an 11-dimensional universe. Physicists had shown supergravity didn't work because it allowed infinities. In fact, every theory except string theory allowed infinities. But Witten wasn't concerned about that because supergravity was only an approximation of M-theory, and M-theory would, by necessity, have to be finite.

It's important to realize that neither Witten nor anyone else proved that all five string theories could be transformed into each other in our universe. In fact, Witten didn't even propose what M-theory actually was.

What Witten did in 1995 was provide a theoretical argument to support the idea that there *could be* a theory — which he called M-theory — that united the existing string theories. Each known string theory was just an approximation of this hypothetical M-theory, which was not yet known. He also believed that at low energy levels, M-theory was approximated by the 11-dimensional supergravity theory.

Branes: Stretching Out a String

In a sense, the introduction of M-theory marks the end of "string theory" because it ceases to be a theory that contains only fundamental strings. M-theory also contains multidimensional membranes called *branes*. Strings are merely 1-dimensional objects, and therefore only one of the types of fundamental objects that make up the universe, according to the new M-theory.

Branes have at least three key traits:

- Branes exist in a certain number of dimensions, from zero to nine.
- Branes can contain an electrical and magnetic charge.
- Branes have a tension, indicating how resistant they are to influence or interaction.

String theory became more complex with the introduction of multidimensional branes. The first branes, called *D-branes*, entered string theory in 1989. Another type of brane, called a *p-brane* (not to be confused with the term you used to tease your younger sibling), was later introduced. Subsequent work showed that these two types of branes were in fact the same thing.

If you haven't realized it by now, different things ending up being related (or the same) is a common theme in string theory.

Branes are objects of multiple dimensions that exist within the full 10-dimensional space required by string theory. In the language of string theorists, this full space is called the *bulk*.

A major reason string theorists didn't originally embrace branes was that introducing more elaborate physical objects went against the goal of string theory. Instead of simplifying the theory and making it more fundamental, branes made the theory more complicated and introduced more types of objects that didn't appear to be necessary. These were the exact features of the Standard Model that string theorists hoped to avoid.

In 1995, though, Joe Polchinski proved that it wasn't possible to avoid them. Any consistent version of M-theory had to include higher-dimensional branes.

The discovery of D-branes: Giving open strings something to hold on to

The motivation for D-branes came from work by Joe Polchinski, Jin Dai, and Rob Leigh of the University of Texas, and independent work performed at the same time by Czech physicist Petr Hořava. While analyzing the equations of string theory, these physicists realized that the ends of open strings didn't just hover out in empty space. Instead, it was as if the end of the open string was attached to an object, but string theory at the time didn't have objects (other than strings) for it to attach to.

To solve this problem, the physicists introduced the *D-brane*, a surface that exists within the 10-dimensional superstring theory so open strings can attach to it. These branes, and the strings attached to them, are shown in Figure 11-1. (The "D" in D-brane comes from Johann Peter Gustav Lejeune Dirichlet, a German mathematician whose relationship to the D-brane stems from a special type of boundary condition, called the *Dirichlet boundary condition*, which the D-branes exhibit.)

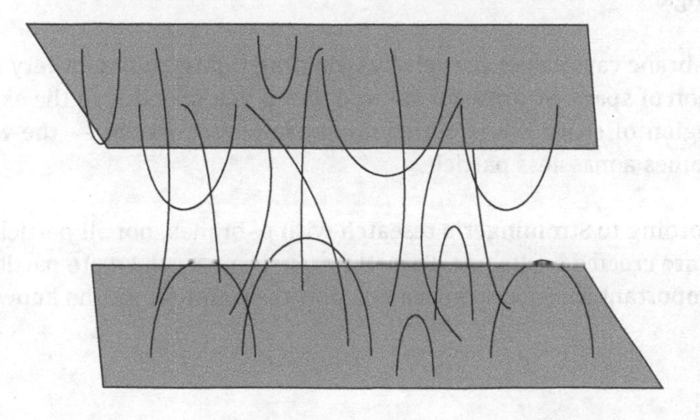

FIGURE 11-1: Open strings attach to the brane at each end. The ends can attach to the same brane or to different branes.

TIP

It's easiest to visualize these branes as flat planes, but D-branes can exist in any number of dimensions from zero to nine, depending on the theory. A 5-dimensional D-brane would be called a D5-brane.

D-branes can multiply quickly. You could have a D5-brane intersecting a D3-brane, which has a D1-brane extending off it. Open superstrings could have one end on the D1-brane and the other end on the D5-brane, or on some other D5-brane in another position, and D9-branes (extended in all ten dimensions of space-time) could be in the background of all of them. At this point, it becomes quite difficult to picture this 10-dimensional space or keep all the possible configurations straight in any meaningful way.

In addition, D-branes can be either finite or infinite in size. Scientists honestly don't know the real limitations of how these branes behave. Prior to 1995, few people paid much attention to them.

Creating particles from p-branes

In the mid-1990s, Andrew Strominger performed work on another type of brane, called *p-branes*, which were solutions to Einstein's general relativity field equations. The *p* represents the number of dimensions, which again can go from zero to nine. (A 4-dimensional *p*-brane is called a 4-brane.)

The *p*-branes expanded infinitely far in certain directions but finitely far in others. In those finite dimensions, they actually seemed to trap anything that came near them, similar to the gravitational influence of a black hole. This work has provided one of the most amazing results of string theory: a way to describe some aspects of a black hole (see the section "Using branes to explain black holes").

In addition, the *p*-branes solved one problem in string theory: Not all of the existing particles could be explained in terms of string interactions. With the *p*-branes, Strominger showed that it was possible to create new particles without the use of strings.

A *p*-brane can make a particle by wrapping tightly around a very small, curled-up region of space. Strominger showed that if you take this to the extreme — picture a region of space that's curled up as small as possible — the wrapped *p*-brane becomes a massless particle.

REMEMBER

According to Strominger's research with *p*-branes, not all particles in string theory are created by strings. Sometimes, *p*-branes can create particles as well. This is important because strings alone don't account for all the known particles.

Deducing that branes are required by M-theory

Strongly motivated by Edward Witten's proposal of M-theory, Joe Polchinski began working intently on D-branes. His work proved that D-branes weren't just a hypothetical construct allowed by string theory, but they were essential to any version of M-theory. Furthermore, he proved that D-branes and *p*-branes were describing the same objects.

In a flurry of activity that would characterize the second superstring revolution, Polchinski showed that the dualities needed for M-theory only worked consistently in cases where the theory also contained higher-dimensional objects. An M-theory that contained *only* 1-dimensional strings would be an inconsistent M-theory.

Polchinski defined the types of D-branes string theory allows and some of their properties. His D-branes carried charge, which meant they interacted with each other through something similar to the electromagnetic force.

A second property of D-branes is tension. The tension in a D-brane indicates how easily an interaction influences the D-brane, like ripples moving across a pool of water. A low tension means a slight disturbance results in large effects on the D-brane. A high tension means it's harder to influence (or change the shape of) the D-brane.

If a D-brane had a tension of zero, then a minor interaction would have a major result — like someone blowing on the surface of the ocean and parting it like Moses parts the Red Sea in *The Ten Commandments*. An infinite tension would mean the exact opposite: No amount of work would cause changes to the D-brane.

TIP

If you picture a D-brane as the surface of a trampoline, you can more easily visualize the situation. When the weight of your body lands on a trampoline, the tension in the trampoline is weak enough that it gives a bit, but strong enough that it eventually bounces back, hurling you into the air. If the tension in its surface were significantly weaker or stronger, a trampoline would be no fun whatsoever; you'd either sink until you hit the ground, or you'd hit a flat, immovable surface that doesn't sink (or bounce) at all.

REMEMBER

Together, these two features of D-branes — charge and tension — mean they aren't just mathematical constructs but are tangible objects in their own right. If M-theory is true, D-branes have the capacity to interact with other objects and move from place to place.

Uniting D-branes and p-branes into one type of brane

Though Polchinski was aware of Strominger's work on *p*-branes — they regularly discussed their projects over lunch — both scientists thought the two types of branes were distinct. Part of Polchinski's 1995 work on branes included the realization that D-branes and *p*-branes were actually one and the same object. At energy levels where predictions from string theory and general relativity match up, the two are equivalent.

It may seem odd that this hadn't occurred to either of the men before 1995, but there was no reason to expect that the two types of branes would be related to each other. To a layperson, they sound basically the same — multidimensional surfaces existing in a 10-dimensional space-time. Why *wouldn't* you at least consider that they're the same things?

Well, part of the reason may be based on the specific nature of scientific research. When you're working in a scientific field, you're quite specific about the questions you're asking and the ways in which you're asking them. Polchinski and Strominger were asking different questions in different ways, so it never occurred to either of them that the answers to their questions might be the same. Their knowledge blinded them from seeing the commonalities. This sort of tunnel vision is fairly common, and it's part of the reason sharing research is so encouraged within the scientific community.

Similarly, for a layperson, the dramatic differences between these two types of branes are less clear. Just as someone who doesn't study much religion may be confused by the difference between Episcopalian and Catholic theological doctrines, to a priest of either religion, the differences are well-known, and the two are seen as extremely distinct.

In the case of branes, though, the layperson would have had clearer insight on the issue than either of the experts. The very details that made D-branes and *p*-branes so intriguing to Polchinski and Strominger hindered their ability to see past the details to the commonalities — at least until 1995, when Polchinski finally saw the connection.

Because of equivalence, both D-branes and *p*-branes are typically just referred to as branes. When referencing their dimensionality, the *p*-brane notation is usually the one used. Some physicists still use the D-brane notation because there are other types of branes that physicists talk about. (In other chapters of this book, we mainly refer to them as branes, thus saving wear and tear on our keyboard's D key.)

Using branes to explain black holes

One of the major theoretical insights that string theory has offered is the ability to understand some black hole physics. These revelations are directly related to work on p-branes, which, in certain configurations, can act somewhat like black holes.

The connection between branes and black holes was discovered by Andrew Strominger and Cumrun Vafa in 1996. This is one of the few aspects of string theory that can be cited as actively confirming the theory in a testable way, so it's rather important.

The starting point is similar to Strominger's work on p-branes to create particles: Consider a tightly curled region of a space dimension that has a brane wrapped around it. In this case, though, you're considering a situation in which gravity doesn't exist, which means you can wrap multiple branes around the space.

The brane's mass limits the amount of electromagnetic charge the brane can contain. A similar phenomenon happens with electromagnetically charged black holes. These charges create an energy density, which contributes to the mass of the black hole. This places a limit on the amount of electromagnetic charge a stable black hole can contain.

In the case where the brane has the maximum amount of charge — called an *extremal configuration* — and the case where the black hole has the maximum amount of charge — called an *extremal black hole* — the two objects share some properties. This allows scientists to use a thermodynamic model of an extremal configuration brane wrapped around extra dimensions to extract the thermodynamic properties they would expect to obtain from an extremal black hole. Also, you can use these models to relate near-extremal configurations with near-extremal black holes. This resulted in a spectacular confirmation of the black-hole thermodynamic properties found by Jacob Bekenstein and Stephen Hawking by completely different arguments (see Chapter 9).

Black holes are one of the mysteries of the universe that physicists would most like to have a clear explanation for. To find more details on how string theory relates to black holes, skip over to Chapter 16.

REMEMBER

String theory wasn't built with the intention of designing this relationship between wrapped branes and black holes. The fact that an artifact extracted purely from the mathematics of string theory would correlate so precisely with a known scientific object like a black hole, and one that scientists specifically want to study in new ways, was seen by everyone as a major step in support of string theory. It's just too perfect, many think, to be mere coincidence.

Getting stuck on a brane: Brane worlds

With the discovery of all these new objects, string theorists have begun exploring what they mean. One major step is the introduction of *brane world* scenarios, where our 3-dimensional universe is actually a 3-brane.

Ever since the inception of string theory, one of the major conceptual hurdles has been the addition of extra dimensions. These extra dimensions are required so the theory is consistent, but we certainly don't seem to experience more than three space dimensions. The typical explanation has been to compactify the extra six dimensions into a tightly wound object roughly the size of the Planck length.

In the brane world scenarios, the reason we perceive only three spatial dimensions is that we live inside a 3-brane. There's a fundamental difference between the space dimensions on the brane and those off the brane.

The brane world scenarios are a fascinating addition to the possibilities of string theory, in part because they may offer some ways in which we can have consistent string theories without resorting to elaborate compactification scenarios. Not everyone is convinced, however, that compactifications can be eliminated from the theory, and even some brane world theories include compactification as well.

See Chapter 12 for a discussion of some proposed brane world scenarios that offer intriguing explanations for certain aspects of our universe, such as how to resolve the hierarchy problem (from Chapter 8).

Matrix Theory as a Potential M-Theory

A year after the proposal of M-theory, Leonard Susskind introduced a suggestion for what the "M" could stand for. *Matrix theory* proposes that the fundamental units of the universe are 0-dimensional point particles, which Susskind calls *partons* (or D0-branes). (No, these particles have nothing to do with legendary singer-songwriter Dolly Parton.) Partons can assemble into all kinds of objects, creating the strings and branes required for M-theory. In fact, most string theorists believe that matrix theory is equivalent to M-theory.

Matrix theory was developed by Susskind, Tom Banks, Willy Fischler, and Steve Shenker in the year after Witten proposed M-theory. (The paper on the topic wasn't published until 1997, but Susskind presented the concept at a 1996 string theory conference prior to publication.) The theory is also approximated by

11-dimensional supergravity, which is one of the reasons string theorists think it's appropriate to consider it equal to M-theory.

The name "parton," which Susskind uses in his book *The Cosmic Landscape* (and we've used here) to describe these 0-dimensional branes, comes from a term used by the Nobel Prize-winning quantum physicist (and string theory skeptic) Richard P. Feynman. Both Feynman and his colleague and rival Murray Gell-Mann were working to figure out what made up hadrons. Though Gell-Mann proposed the quark model, Feynman described a more vague theory where hadrons were made up of smaller pieces that he just called partons.

YET ANOTHER STRING THEORY: F-THEORY

Another theory that sometimes gets discussed is called F-theory (the name is a joking reference to the idea that the M in M-theory stands for mother). However, more than a fundamental theory in and of itself, F theory should be regarded as a powerful tool within string theory to construct realistic solutions, hopefully reproducing our own universe.

Cumrun Vafa proposed F-theory in 1996 after noticing that certain complicated solutions of Type IIB string theory could be described in terms of a simpler solution of a different theory with 12 dimensions, up from the 10 dimensions of superstrings or the 11 dimensions of M-theory. Unlike M-theory, where all the dimensions of space-time are treated on equal footing, two of the dimensions of F-theory are fundamentally different from the rest: They *always* have to be curled up. So now to get to three space dimensions, we have eight small dimensions instead of six!

This makes it seem as though the theory is getting more complicated, but in fact the F-theory description is often simpler. The eight dimensions include not only all the information from the previous six but also information about the branes that exist in the solution (those setups could get complicated). This is an example of a common theme in the development of string theory: More and more of the theory's details, such as which particles exist and how they interact, or which branes live where can be described simply in terms of the geometry of the extra dimensions. This geometry is often easier to understand and analyze.

F-theory has received more attention in the past few years because its rich structure allows solutions that reproduce many of the phenomena of the Standard Model and GUT theories (see Chapter 14 for more on those).

One intriguing aspect of partons, noted by Witten, is that as they get close to each other, it becomes impossible to tell where they actually are. This may be reminiscent of the uncertainty principle in quantum mechanics, in which the position of a particle can't be determined with absolute precision, even mathematically (let alone experimentally). It's impossible to test this the same way scientists can test the uncertainty principle because there's no way to isolate and observe an individual parton. Even light itself would be made up of a vast number of partons, so "looking" at a parton is impossible.

Unfortunately, the mathematics involved in analyzing matrix theory is difficult, even by the standards string theorists use. For now, research continues, and string theorists are hopeful that new insights may show more clearly how matrix theory can help shed light on the underlying structure of M-theory.

IN THIS CHAPTER

» Understanding how strings can split and join

» Feynman-like diagrams for string theory

» Dark energy poses a challenge

» Explaining how extra dimensions may uncurl

» Navigating the landscape of string theory

Chapter 12
Exploring Strings and Their Landscape

We know that if a string is oscillating, we can interpret its oscillations as particles. But how do strings interact with each other? In this chapter, we look at how strings can join and merge, producing diagrams that are actually quite similar to the diagrams Richard Feynman introduced for quantum field theory. We consider how the physics of string theory may be reflected in the universe around us. Finally, we examine how the discovery of dark energy has complicated string theorists' efforts to match experimental data and how physicists are struggling to find "the right version" of string theory in a vast landscape.

Strings and Fields: String Field Theory

Since the time of Newton, physicists have been trying to describe how things interact with each other. In Chapters 7 and 8, we see that the classical concept of force is replaced by the exchange of particles that "mediate" that force. For instance, for Charles-Augustin de Coulomb, the force between two electrical charges propagated instantaneously from one to the other; that's what we would call the electric force, or Coulomb force. For Richard Feynman and friends, the same force was the result of photons being exchanged by the two charges, see also Figure 8-3 in Chapter 8. How does string theory describe interactions?

It turns out that the most natural way to introduce interactions in string theory is to allow strings to merge together and split apart. You may worry that this will add more specimens to the zoo of string theories we've already encountered. Fortunately, that isn't the case.

Given a free string theory, there's one — and only one — possible way, mathematically, for the strings to interact. That's excellent news for string theorists! First of all, the fact that there's any way at all for strings to interact wasn't obvious. Second, the fact that there's only one way means we shouldn't second-guess ourselves.

Much like in quantum field theory, we need a parameter to measure the strength of interactions between strings. This is called the *string coupling constant*, denoted with the variable g.

REMEMBER

There are two important parameters that describe the behavior of a string. One is the *string tension*, called ' ("alpha-prime"), which measures how stiff or elastic the string is. If the tension is small, the string is big and floppy; as the tension becomes larger, the string becomes smaller and stiffer. Hence, the string tension is inversely related to the string size. Just like tuning a guitar string, changing the tension results in different types of oscillations (different particles, in string theory). The second parameter is the string interaction constant g, which measures how likely the string is to break apart or merge.

Splitting and joining of strings and how to avoid infinities

The best way to visualize the splitting and joining of strings is to think about soap bubbles. This example isn't completely accurate because a soap bubble looks like a sphere, whereas a string should create a loop that looks more like a circle. (That is to say, the soap bubble has one dimension more than the string.) However, with a little imagination, it's probably close enough.

When blowing soap bubbles into the wind, you may see two bubbles touch each other and merge into a single, slightly bigger bubble. The converse is also possible: A big bubble may separate into two smaller bubbles. (Of course, soap bubbles can also pop during or after the process of merging or separating. Popping isn't something we expect from strings — sometimes an analogy can only take you so far.)

What's helpful about our analogy is that the process of splitting and merging seems quite random, and depends in a complicated way on the shape of the bubbles and how they evolve with time. This is also the case for strings: The splitting and merging happens dynamically rather than at some predefined instant or place.

The right side of Figure 12-1 depicts a "movie" of the merging of two closed strings into one. Recall that a closed string looks like a circle at each given instant in time. As time flows upward, we go from having two disjointed circles to having only one. You can see that the merging strings look quite similar to the annihilation of an electron with a positron, shown on the left side of Figure 12-1, only the lines have now been "fattened out."

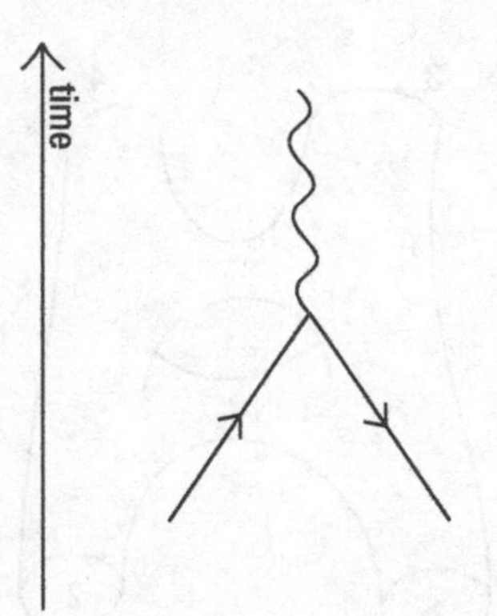

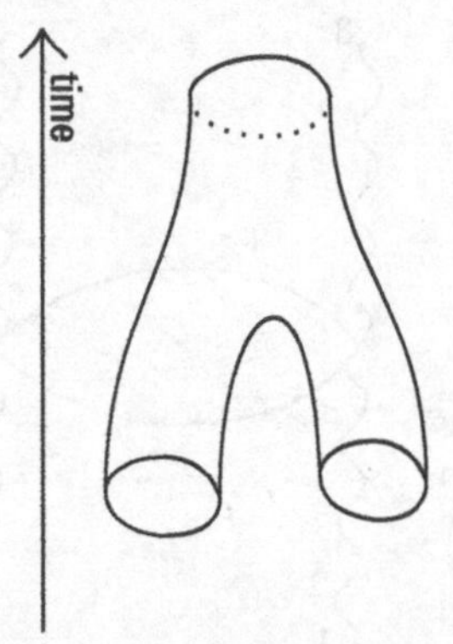

FIGURE 12-1: (Left) The interaction of an electron and a positron releases a photon in quantum electrodynamics. (Right) Two strings merge into a single one.

There's a big advantage in this new type of interaction between strings. When we consider interacting particles, we understand that they are always *meeting at a point*. That was an issue because as quantum field theory had the bad habit of giving infinite results when particles got infinitely close to each other. (For more on these infinities and how physicists deal with them, have a look at Chapter 8.) What's most important is that infinities are especially dangerous when dealing with quantum gravity. In fact, they're the reason it wasn't possible to define a quantum version of general relativity through the recipe used for electromagnetism and the other forces.

REMEMBER

Strings don't interact at a point. The interaction point is smeared over a region whose size is, roughly speaking, the size of the string. The precise size of the string depends on its tension, but it should be roughly the size of the Planck length. Therefore, in a very concrete sense, string theory *regularizes* the interactions between particles, and this naturally happens at the Planck scale.

However, if we zoom out and look at distances much larger than the Planck length, the strings are so tiny that they *almost* look like ordinary particles — in other words, like points. Therefore, at large enough distances (which means low enough energies), we go back to the quantum field theory that we know and love.

This nice structure of interactions was one of the features that got string theorists very excited from the early days.

Trying to visualize how strings create loops

Once we let strings merge and split, it's no surprise that they can create loops similar to the loops we encountered in Chapter 8 for elementary particles. Figure 12-2 shows a simple string loop compared to a loop in quantum electrodynamics.

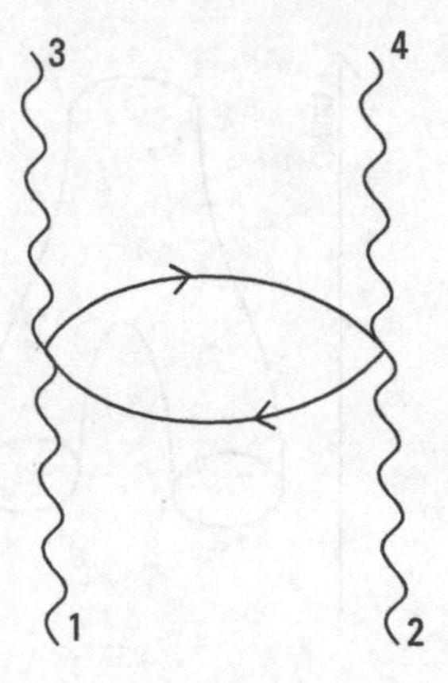

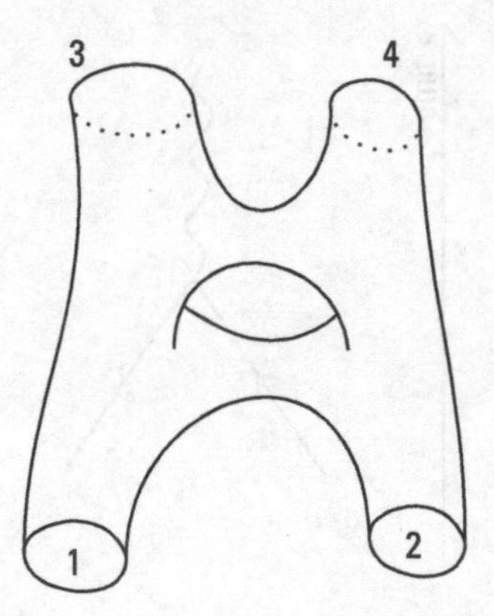

FIGURE 12-2: (Left) A loop of virtual particles in quantum electrodynamics. (Right) A loop in string theory.

There are two big problems with loops. The first one is that you can make them as complex as you want — in fact, not only can you consider more and more general loops, but *you have to* consider them. The second problem is that, especially when you're dealing with complicated loops, there are many different ways to give them shape. Figure 12-3 shows two seemingly different loops that are really two different shapes of the same loop. (See Chapter 11 for an explanation of how this makes the two loops topologically equivalent.)

Building on our experience with quantum field theory, the most natural way to account for all these possibilities is, add them all up. Of course, this sum will depend on the string coupling constant *g*. If *g* is small, it's unlikely that the strings will split and merge (so we can consider only loops that involve very few interactions). If g is big enough, we have to consider many — in principle infinitely many — interactions. (We discuss this logic in our explanation of perturbation theory in Chapter 8, see the sidebar "Perturbation theory: String theory's method of approximation".)

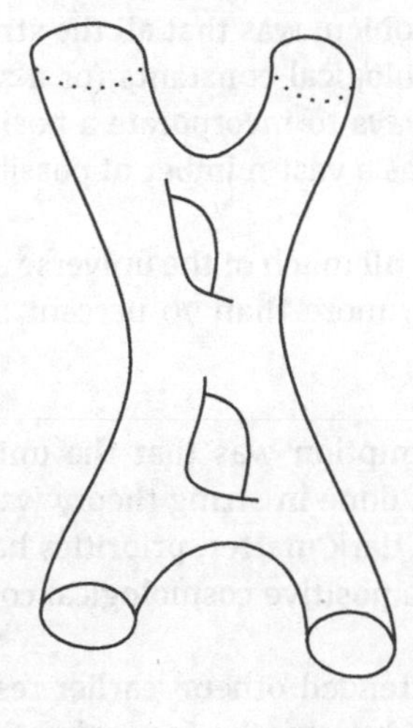

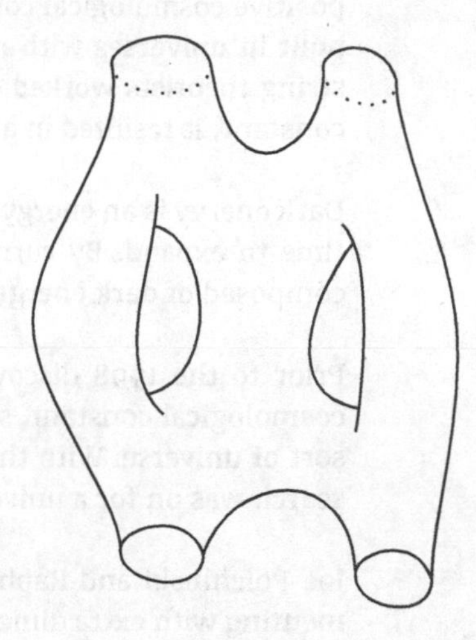

FIGURE 12-3: Two apparently different string loops that can deform into each other.

Because the challenges we encounter in dealing with string interactions are very similar to those of quantum field theory, it makes sense to try to use the quantum field theory machinery to solve them. Therefore, physicists have been trying to develop *string field theory* — something that looks like quantum field theory, but for strings.

String field theory is a monumental endeavor: Ordinary quantum field theory *for particles* is already very complicated. Here, we have a theory like quantum field theory, but *for strings*, each of which accounts for infinitely many particles!

Perhaps unsurprisingly, string field theory is still a work in progress. However, understanding string interactions is a crucial part of understanding string theory.

Even if we don't deal with all the complications of string field theory, and you just want to study a simple diagram (like the one on the right side of Figure 12-2), you'll run into a problem. You have to add up all the shapes of the string loop. The result looks like a donut, but it could be a circular donut, an elongated one, or anything in between. Even this relatively simple sum is quite challenging technically, to the extent that it has been worked out only in the simplest cases.

String Theory Gets Surprised by Dark Energy

The discovery of dark energy in 1998 meant that our universe needed to have a positive cosmological constant. The problem was that all the string theories were built in universes with negative cosmological constants (or a zero value). When string theorists worked out possible ways to incorporate a positive cosmological constant, it resulted in a theory that has a vast number of possibilities!

Dark energy is an energy that seems to fill much of the universe and causes space-time to expand. By current estimates, more than 70 percent of the universe is composed of dark energy.

Prior to the 1998 discovery, the assumption was that the universe had a zero cosmological constant, so all the work done in string theory was focused on that sort of universe. With the discovery of dark matter, priorities had to change. The search was on for a universe that had a positive cosmological constant.

Joe Polchinski and Raphael Bousso extended others' earlier research by experimenting with extra dimensions that had *electric flux* (a number that represents the intensity of an electric field through a surface) wrapped around them. Branes carried charge, so they could also have flux. This construction had the potential to limit some parameters of the theory in a way that couldn't vary continuously.

In 2003, a Stanford group including Renata Kallosh, Shamit Kachru, Andrei Linde, and Sandip Trivedi (KKLT) released a paper that showed ways to extend the Polchinski-Bousso thinking to construct string theories with a positive cosmological constant. The trick was to create a universe and then wrap it with branes and anti-branes (anti-branes are the "anti-particles" of branes) to contain the electric and magnetic flux. This introduced the potential for two effects:

- It allowed a small positive cosmological constant.
- It stabilized the extra dimensions in string theory.

Whether the KKLT proposal, or some modification thereof, succeeds in achieving these two goals, it is still a topic of debate among string theorists – there are many subtle technical issues with it that are still being explored.

But even if we might know how to create a universe of positive curvature in string theory, there is still the question of how to wrap the extra dimensions onto themselves. And for this, a new problem arises: There far too many solutions!

Considering Proposals for Why Dimensions Sometimes Uncurl

Most string theory proposals have been based on the concept that the extra dimensions required by the theory are curled up so small that they can't be observed. With M-theory and brane worlds, it may be possible to overcome this restriction.

A few scenarios have been proposed to try to describe a mathematically coherent version of M-theory that would allow the extra dimensions to be extended. If any of these scenarios hold true, they have profound implications for how (and where) physicists should be looking for the extra dimensions of string theory.

Measurable dimensions

One model that has gotten quite a bit of attention was proposed in 1998 by Savas Dimopoulos, Nima Arkani-Hamed, and Gia Dvali. In this theory, some of the extra dimensions could be as large as a millimeter without contradicting known experiments, which means it may be possible to observe their effects in experiments conducted at the European Organization for Nuclear Reasearch's Large Hadron Collider (LHC). (The proposal has no unique name, but we'll call it *MDM* for *millimeter dimension model.*)

When Dimopoulos introduced MDM at a 1998 supersymmetry conference, it was a somewhat subversive act. He was making a bold statement: Extra dimensions were as important, if not more so, than supersymmetry.

Many physicists believe that supersymmetry is the key physical principle that will prove to be the foundation of M-theory. Dimopoulos proposed that the extra dimensions — previously viewed as an unfortunate mathematical complication to be ignored as much as possible — may be the fundamental physical principle M-theory was looking for.

In MDM, a pair of extra dimensions could extend as far as a millimeter away from the 3-dimensional brane we reside on. If they extend much more than a millimeter, someone would have noticed by now, but at a millimeter, the deviation from Newton's law of gravity would be so slight that no one would be any the wiser. So, gravity radiating out into extra dimensions would explain why it's so much weaker than the brane-bound forces.

The way it works is, everything in our universe is trapped on our 3-dimensional brane *except gravity*, which can extend off our brane to affect the other dimensions.

Unlike in string theory, the extra dimensions wouldn't be noticeable in experiments except for gravity probes, and in 1998, gravity hadn't been tested at distances shorter than a millimeter.

Now, don't get too excited yet. Experiments to look for these extra millimeter-sized dimensions have been done, including at the LHC, and it turns out, they probably don't exist. Of course, this depends on how big, precisely, the extra dimensions are. When you want to explore smaller and smaller dimensions, you need higher and higher energies. So far, the LHC has operated with energies of up to 13 TeV (tera-electronvolts, a unit of energy, see Chapter 8) without finding any indication of the extra dimensions. It's still possible that they're there, quite a bit smaller than a millimeter, but larger than the typical scale of string compact dimensions, which would show up only at much higher energy scales.

Infinite dimensions: Randall-Sundrum models

If the idea of a millimeter-sized dimension turned heads, the 1999 proposal by Lisa Randall and Raman Sundrum was even more spectacular. In the *Randall-Sundrum models*, gravity behaves differently in different dimensions, depending on the geometry of the branes.

In the original Randall-Sundrum model, called *RS1*, a brane sets the strength of gravity. In this *gravitybrane*, the strength of gravity is extremely large. As you move in a fifth dimension away from the gravitybrane, the strength of gravity drops exponentially.

An important aspect of the RS1 model is that the strength of gravity depends only on your position within the fifth dimension. Because our entire 3-brane (this is a brane world scenario where we're trapped on a 3-brane of space) is at the same fifth-dimensional position, gravity is consistent everywhere in the 3-brane.

In a second scenario, called *RS2*, Randall and Sundrum realized that the 3-brane we're stuck in may have its own gravitational influence. Though gravitons can drift away from the 3-brane into other dimensions, they can't get very far because of the pull of our 3-brane. Even with large dimensions, the effects of gravity leaking into other dimensions would be incredibly small. Randall and Sundrum called the RS2 model *localized gravity*.

In both models, the key feature is that gravity on our own 3-brane is essentially always the same. If that weren't the case, we'd have noticed the extra dimensions before now.

In 2000, Lisa Randall proposed another model with Andreas Karch called *locally localized gravity*. In this model, the extra dimension contains some negative vacuum energy. Locally localized gravity goes beyond the earlier models because it allows gravity to be localized in different ways in different regions. Our local area looks 4-dimensional and has 4-dimensional gravity, but other regions of the universe may follow different laws.

Understanding the Current Landscape: A Multitude of Theories

As far back as 1986, Andrew Strominger found that there were a vast number of consistent string theory solutions and observed that all predictive power may have been lost. Actually, when considering a negative cosmological constant (or zero), you apparently end up with an infinite family of possible theories.

REMEMBER

With a positive cosmological constant — as needed in our universe, thanks to dark energy — things get better, but not by much. There are now a finite number of ways to roll up the branes and anti-branes to obtain a positive cosmological constant. How many ways? Some estimates have indicated as many as 10^{500} possible ways to construct such a string theory, and that might be a low estimate!

This is an enormous problem if the goal of string theory is to develop a single unified theory. The vision of both the first and second superstring revolutions (or at least the vision guiding some bandwagoners who jumped on board) was a theory that would describe our universe with no experimental observations required.

In 2003, Leonard Susskind published "The Anthropic Landscape of String Theory," in which he very publicly gave up the idea that a unique solution to string theory, reproducing our universe, would be discovered. In the paper, Susskind introduced the concept of "the landscape" of string theories: a vast number of mathematically consistent possible universes, some of which actually exist. Susskind's *string theory landscape* was his solution to the unfathomable number of possible string theories.

But with so many possibilities, does the theory have any predictive power? Can we use a theory if we don't know what the theory is?

The anthropic principle requires observers

Susskind's proposed solution involves relying on something known as the *anthropic principle*. This principle indicates that the universe has the properties it does because we're here to observe them. If it had vastly different properties, we wouldn't exist. Other areas of the *multiverse* (the vast collection of existing universes) may have different properties, but they're too far away for us to see.

REMEMBER

The anthropic principle was coined by Cambridge astrophysicist Brandon Carter in 1974. It exists in two basic versions:

» **Weak anthropic principle:** Our location (or region) of space-time possesses laws such that we exist in it as observers.

» **Strong anthropic principle:** The universe is such that there has to exist a region of space-time within it that allows observers.

If you're reading these two variations of the anthropic principle and scratching your head, you're in good company. Even string theorists who later embraced the anthropic principle — such as Susskind and Joe Polchinski — once despised it as totally unscientific. This is in part because the anthropic principle (in its strong form) is sometimes invoked to require a supernatural designer of the universe, something that most scientists (even religious ones) try to avoid in their scientific work. (Conversely, it's also often used, in the weak form, as an argument *against* a supernatural designer, as Susskind does in his book *The Cosmic Landscape*.)

For the anthropic principle to make sense, you have to consider an array of possible universes. Figure 12-4 shows a picture of the energy levels of possible universes, where each valley represents a particular set of string theory parameters.

TIP

According to the weak anthropic principle, the only portions of the multiverse we can ever observe are the ones where the parameters allow us to exist.

In this sense, the weak anthropic principle is almost a given — it's just always going to be true. That's part of the point of it. Because we're here, we can use the fact that we're here to explain the properties the universe has. In the string theory landscape, so many possibilities are out there, ours is just one that happened to come into being, and we're lucky enough to be here.

REMEMBER

If the string theory landscape represents all the universes that are possible, the multiverse represents all the universes that actually exist. Distant regions of the multiverse may have radically different physical properties than those we observe in our own section.

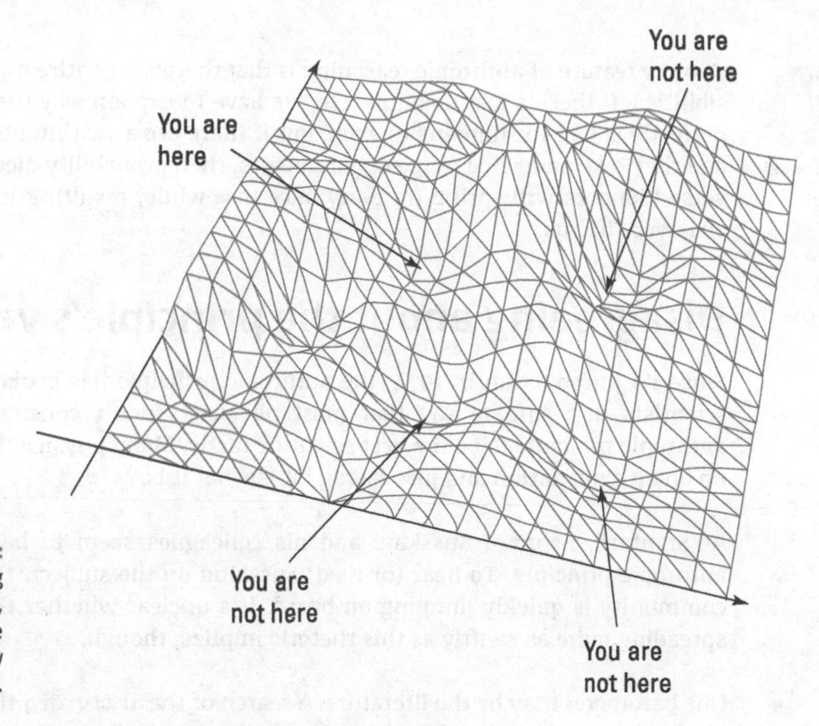

FIGURE 12-4: In the string theory landscape, only some possibilities allow life to exist.

This concept is similar to Lisa Randall's locally localized gravity (see the "Infinite dimensions: Randall-Sundrum models" section earlier in this chapter), where only our local region exhibits the gravity that we know and love in three space dimensions. Other regions may have five or six space dimensions, but that doesn't matter to us because they're so far away that we can't see them. Those other regions are different parts of the multiverse.

In 1987, Nobel Prize winner Steven Weinberg added a bit of credibility to the field. Using reasoning based on the anthropic principle, he analyzed the cosmological constant required to create a universe like ours. His prediction was a very small positive cosmological constant, only about one order of magnitude off from the value found more than a decade later.

This is a frequently cited case of when the anthropic principle led to a testable prediction, but it should be taken with some caution. Clearly, our universe is one in which galaxies formed the way they have — not too fast or too slow. Using that fact is totally uncontroversial as a means of determining the cosmological constant, but the anthropic principle goes further. It doesn't just determine the cosmological constant; it supposedly explains why the cosmological constant has that value.

The key feature of anthropic reasoning is that there's an entire multiverse of possibilities. If there's just one universe, we have to explain *why* that universe is so perfectly suited for humans to exist. But if there are a vast number of universes, and they take on a wide range of parameters, then probability dictates that a universe like ours will spring up every once in a while, resulting in life forms and observers like us.

Disagreeing about the principle's value

Since its introduction in 1974, the anthropic principle has evoked passion from scientists. It's safe to say that most physicists don't consider invoking the anthropic principle to be the best scientific tactic. Many physicists see it as giving up on an explanation and just saying "it is what it is."

At Stanford, Leonard Susskind and his colleagues seem to be embracing the anthropic principle. To hear (or read) Susskind on the subject, the string theory community is quickly jumping on board. It's unclear whether the movement is spreading quite as swiftly as this rhetoric implies, though.

One barometer may be the literature. A search of the `arXiv.org` theoretical physics database garners 450 hits for the term "anthropic." Searching for "anthropic principle" obtains 155 hits, and adding words like "string" and "brane" only causes the number of hits to drop from there. For comparison, searching on "string theory," "cosmological constant," or even the far less popular "loop quantum gravity" results in so many hits that the search cuts off at 1,000 papers.

Similarly, coverage of the anthropic principle in string theory textbooks and popular science books is quite mixed: Many don't mention it at any great length if at all, a few endorse it enthusiastically, and a few criticize it. So the jury is certainly still out on how widely the string theory community has adopted the anthropic principle.

Some string theorists, such as David Gross, appear to be strongly opposed to anything that even hints at the anthropic principle. A large number of string theorists bought into it based on the idea — championed by Edward Witten's promise of M-theory in 1995 — that there would be a single string theory solution at the end of the rainbow.

String theorists seem to be turning to the anthropic principle mostly out of a lack of other options. This certainly seems to be the case for Witten, who has made public statements indicating he may be unenthusiastically turning toward anthropic thinking.

We end this chapter in many ways worse off than we began. Instead of five distinct string theory solutions, we have one theory, but with 10^{500} or so choices of parameters. It's unclear what the fundamental physical properties of string theory are in a field of so many options. The only hope is that new observations or experiments will provide some sort of clue about which aspect of the string theory landscape to explore next.

EXPLORING THE LANDSCAPE AND AVOIDING THE SWAMPLAND

String theory seems to predict a very large landscape of possible configurations, one of which may describe our universe. In principle, it's now a matter of rolling up our sleeves and looking for this needle in the haystack. That seems like a monumental task, even if we're smart enough to rule out a bunch of them outright. For all candidate configurations, we would need to carefully compute the string theory predictions and compare them with the Standard Model's predictions. Ideally, from studying the correct string solution, we should get a picture that looks approximately like the Standard Model, some new particles, and gravity. (And this is assuming that we are at all able to carry out such a computation from start to finish! This is not the case unless there is some supersymmetry to help us out.)

Because of that, it's tempting to skip a time-consuming exploration of the string landscape and just try to guess the correct approximate theory. After all, we know that the Standard Model should be there, gravity should appear in some way, and some additional particles should show up too. How hard can it be to make an educated guess?

The swampland idea originally put forward by Harvard physicist Cumrun Vafa is that guessing would be very hard indeed. In fact, the conjecture is that most of the reasonable-looking approximate theories you may write down will have a tiny chance of being related to a complete, consistent string theory. They will live in a swampland of bad theories, not in the string landscape. And, as anyone who has watched *The NeverEnding Story* knows all too well, the swampland is much more vast than the landscape, and not a nice place to be. Therefore, the game becomes to try to come up with rules to avoid being in the swampland, and try to exclude as many "bad" theories as possible. String theorists have formulated and are still formulating many swampland conjectures, that make different proposals of what properties a "bad" theory might have.

EXPLORING THE LANDSCAPE AND AVOIDING THE SWAMPLAND

IN THIS CHAPTER

» Looking at optical holograms

» Using holograms to peer inside black holes

» The holographic principle and the boundary of the universe

Chapter 13
Gaining Insights from the Holographic Principle

Another key insight into string theory comes from the *holographic principle*, which relates a theory in space to a theory defined only on the boundary of that space. Many physicists believe that the holographic principle should hold in any sensible quantum gravity theory, and indeed, it was first formulated without any reference to string theory. But it is in string theory that holography can be really made precise.

In this chapter, we look at what makes holograms so special and how they can be used to encode information in fewer dimensions. We see how the holographic principle can be applied to the study of black holes. Finally, we consider how the idea of holograms has been applied to string theory.

What's a Hologram?

Most people have experienced *holograms* as images (often on stickers and sometimes on posters) that seemingly morph into different images depending on the viewing angle. The image subtly shifts when you view it from different perspectives so that it (kind of) looks like a 3-dimensional object. But clearly the image is on a 2-dimensional surface, so the fact that you're able to see 3-dimensional information that includes depth (i.e., it looks 3D) is astounding when you first encounter a hologram. Some holograms are so detailed that you can even look "around" elements within the image.

REMEMBER

An *optical hologram* is a 2-dimensional image that contains the 3-dimensional information of an object. When viewing an optical hologram, you can tilt the image and watch the shape move, as if you're looking at it from slightly different perspectives. The process of making a hologram is called *holography*, which means "writing of the whole."

Creating optical holograms

Optical holograms are created by using light waves to produce interference patterns. The interference patterns are recorded on a certain type of film (if you're doing it the old-school way — these days, various types of sensors can do the trick), infusing the image with more than just the 2-dimensional information.

Traditional optical holography involves using lasers in a darkroom to illuminate a 3-dimensional object. For example, say the object is a small plastic horse (which is what one of your authors used to create an optical hologram in his undergraduate optics course). The film is placed near the horse, and the laser light reflected from the horse lands on the film. The curves and contours of the figure cause the light to reflect off it at different angles, and the film records the various interference patterns from the laser light.

The film must be developed at that point, but unlike ordinary film on which an image is visible, the image on the holographic film cannot be seen. If you were to look at the film, you'd see an unimpressive blank sheet of plastic that certainly doesn't seem to contain information about anything, let alone about a small plastic horse.

Because you used a laser of a specific frequency to create the original light, you have to illuminate the film with a laser using the same frequency. At that point, the image of the horse suddenly appears on the clear film, and you can see contours, movement, and perspective shifting in a 3-dimensional way on the 2-dimensional surface, indicating that you're viewing a hologram.

TIP

The hologram images that most people see — on stickers or movie posters, for example — work in normal lighting. These "white light holograms" don't require you to use lasers set at a particular wavelength to view the images; they're designed specifically to work in ordinary light. But at their core they involve the same idea: looking at 3-dimensional information on a 2-dimensional surface.

More bang for your buck: Encoding information in fewer dimensions

Holograms are surprising in many ways, especially in terms of the information they contain. Consider our toy horse. If you use your camera (or your phone) to snap a picture of it, you get a digital file in which a 2-dimensional table of pixels is encoded. There's a clear correspondence between each pixel in the picture and a little square in the 2-dimensional view that was in front of the camera, but this captures only a small part of that information. This view from the camera is of 3-dimensional objects (a toy horse, in this example), with a depth that cannot be fully captured in the resulting 2-dimensional image. If the picture is taken from the side, you don't really know how wide the toy horse is. You would have to take several pictures from several angles to really get a feel for that.

In an optical hologram, you see a bit more. An optical hologram is like a window that allows you to look at an object as if it's still in front of the camera. That's why, when you look at a hologram of a toy horse from different angles, you see different features of the horse. A consequence of creating a hologram is that information is encoded in a different way than an ordinary picture records information. If you drew a little grid on the hologram, you wouldn't be able to assign a color to each "pixel" in a way that corresponds to a "pixel" of the image! This is actually impossible because each pixel of the hologram shows something different depending on the angle from which you view it.

Another way to understand this concept is to think about a photograph of a couple that's cut in half. Each image in the cut photo shows only two people. If you were to cut a hologram of a couple in half, you would have two smaller holograms, both of which would still depict both of the people (see Figure 13-1). It would be like looking through a window that's half as wide. All this goes to show that a crucial aspect of holograms is how they encode information.

So far, we've been talking about optical holograms, which depict 3-dimensional-looking objects but still describe a 2-dimensional surface (the surface of a toy horse, for instance). What if we consider a hologram that's a bit less, shall we say, superficial? Instead of an optical hologram that contains the surface information about an object, it would be a hologram that tells us everything about the information within the object. This *true hologram* would need to depict all the 3-dimensional features of an object, including its interior.

TIP

The true hologram we're talking about is a thought experiment (or *Gedankenexperiment*, if we use the German term Einstein made popular among physicists). Although it would be cool to make a hologram that describes a 3-dimensional object, it isn't possible to do that with an optical hologram. Not only are we devoid of the technology, but there are good reasons to believe it's impossible, as we will see in a moment. Why talk about true holograms then? It turns out that they do seem to be possible *not as optical holograms*, but in quantum gravity!

FIGURE 13-1: Cutting an optical hologram in half is very different than cutting a photograph in half because holographic information is encoded in a different way.

Even thinking about a true hologram presents two major issues, which can be seen by taking the 3-dimensional object and dividing it in many small 3-dimensional boxes. Each of the boxes contains some basic information about the object, similar to how each pixel contains some basic bit of information in a digitized image. We can refine this description by making the boxes smaller and smaller — much like increasing the resolution of a picture. At the same time, we repeat this description for the 2-dimensional hologram, dividing it in a large number of square pixels. Increasing the number of 2-dimensional pixels also refines our picture.

The first problem we encounter is that the map between the 3-dimensional pixels of the object and the 2-dimensional pixels of the hologram cannot match up one-to-one. This is a little odd, but it isn't a dealbreaker by itself. After all, a similar thing happens with optical holograms (refer to Figure 13-1).

The second issue is more crucial. Imagine we have an object that sits in a cube with edges measuring 1 centimeter. We divide it into small boxes of 0.1 millimeter, which means that in total we have 100^3 (that's one million) boxes, each containing a bit of information. So we need our hologram to be big enough to store one million bits of information. If we want to increase the resolution of the object by a factor of ten, using boxes with an edge of 0.01 millimeter, we would need $1{,}000^3$ (or one billion) boxes — that's one billion bits of information. A ten-fold increase in resolution results in a thousand-fold increase in the number of bits we need.

Now suppose that we have a true hologram that perfectly describes our object. Clearly, as we refine our 3-dimensional description by making the boxes smaller and smaller, we need to improve the resolution of the hologram too. Remember, even though the pixels describe all the 3-dimensional elements inside the object, the hologram we're imagining is actually 2-dimensional. A 2-dimensional object has an area instead of a volume, so the thousand-fold increase in information means each side of the cube's surface would need to contain the square root of that. The square root of 1,000 is roughly 32, so each 10-fold increase in resolution requires a 32-fold increase in the information stored on the surface, to compensate for a thousand-fold increase in information about the 3-dimensional object.

For these reasons, it seems impossible to create a true hologram. The amount of information in two and three dimensions just seems to scale too differently to do it systematically. But gravity turns this reasoning upside down: Precisely because of how information scales, physicists argue that there *must* be true "gravitational" holograms.

Using Holograms to Understand Black Holes

Physicists gained incredible insight into gravity from studying black holes, which eventually led to the formulation of the holographic principle.

Black holes were first understood as mathematical objects: formal solutions to Einstein's equations of general relativity, the simplest of which Karl Schwarzschild had already found in 1915, the same year Einstein published his general relativity work. Now we have extensive evidence that black holes are astronomical objects that truly exist in our universe and in our galaxy (you can read more about them in Chapter 9).

Going down a black hole

A black hole is a region of space where gravity distorts the geometry of space-time in a very severe way — so severe that nothing can escape the gravitational pull of the black hole. If you stand on the surface of Earth and launch a projectile sufficiently fast, you know it will escape the planet's gravitational pull and shoot off to infinity (this makes for an excellent exam question in your freshman physics course, or a great idea for a novel about space exploration if you're Jules Verne). With a black hole, this is impossible. Even if you shoot a projectile as fast as possible (with the speed of light, which means your projectile is actually a photon), it will be never able to escape the inside of the black hole.

The black hole, therefore, has an interior out of which nothing can escape. If we look toward a black hole, we can see only up to its *event horizon* — the last boundary from where light can still barely escape.

Many scientists have studied the bizarre properties of black holes. Among them, Jacob Bekenstein and Stephen Hawking achieved a particularly surprising set of results (see Chapter 9 for more on their work). Bekenstein and Hawking found that black holes behave like thermodynamic objects — in other words, like hot stoves. They have a temperature (which is typically extremely low compared to stars or even planets) and are radiating away a little bit of their mass. Moreover, they have an *entropy*, or a degree of disorder or randomness, which is the key concept that allows us to turn them into holograms.

In thermodynamics, entropy is something that measures the amount of disorder of a system. When you consider some thermodynamic object, like a balloon filled with helium, you normally describe it with a few macroscopic properties: its temperature, pressure, and density, for example. But the helium gas is made up of an enormous number of tiny atoms, each with its own position and velocity. The concepts of temperature, pressure, and density describe only the average state of the gas: There are many possible configurations of the individual atoms that give rise to the same temperature, pressure, and density. Entropy measures how many of those configurations there are. In this sense, it's a measure of information, because we can use different configurations to encode a message.

Black holes and entropy

Bekenstein's surprising insight was that the maximum possible amount of entropy — that is, information — that a black hole may contain is proportional to the area of its event horizon. In other words, if we have a spherical black hole of radius R, the amount of information inside it is proportional to R^2, not to R^3.

This is very confusing at first because we've been thinking of information about an object as emerging from little boxes that encode some basic bit of that information (for instance, "Is there a particle inside the box or not?"). The number of small boxes inside a solid sphere clearly scales like R^3. The catch is that in a black hole, we can't overlook the gravitational interaction.

To understand the interplay of entropy and gravity, you can try to imagine creating a black hole by squeezing the whole mass of a planet or a star into a tiny space. (Don't try this at home!) Simplifying it quite a bit and using the sun as our star, we would need to squeeze all the sun's mass, which is normally contained in about a 700,000-kilometer radius, into a small ball with, say, a 3-kilometer radius. That's how insanely densely packed a black hole is! Once we do this, we'd have a black hole with a 3-kilometer radius.

Now imagine that we want to throw something else, like your phone, into the black hole. The mass of the black hole wouldn't change by much — not with respect to the mass of the sun! It would increase only a tiny bit. Because of this, the radius (and therefore the area) of the black hole would grow by only a tiny amount. And, by applying the equations of general relativity, Bekenstein and Hawking were able to show that the extra surface is exactly what's needed to account for the information contained in your phone.

If you take Bekenstein's intuition about black holes very seriously, you could show that all the information about the content of a black hole is encoded on its surface. Right there, on the event horizon, we have a number of pixels that grow like R^2. Bekenstein told us how big these pixels should be: They should have the size of twice the Planck length, to be precise. That's incredibly small, and as a result, even the smallest black hole could contain an enormous amount of information. For instance, a black hole whose radius is a billionth of a billionth of a millimeter would have enough of these two-Planck-length–sized "bits" on its surface to encode all the data ever produced in the history of humanity many times over.

As with optical holograms, the precise map between the "bits" on the event horizon and the data inside the black hole would be very complicated, and in fact, understanding this map in detail is an open question.

REMEMBER

In the "More bang for your buck: Encoding information in fewer dimensions" section of this chapter, we argued that we need a sufficient number of pixels in the hologram (or, in this case, on the surface of the black hole) to be able to encode all the information inside. If it weren't for gravity (and the equations of general relativity that describe it), this would be impossible. But according to Bekenstein, the maximum information inside a black hole is precisely given by the number of Plank-size pixels on the surface of the black hole! In other words, we have precisely enough pixels to fit all the information that might be in the black hole (though the map from the inside to the pixels on the surface may be tricky to nail down).

TIP

You know that black holes also emit energy in the form of radiation (see Chapter 9). This radiation carries away some of the information that was inside the black hole. Because the map between the surface and the interior of the black hole is very scrambled, black holes act as information scramblers — when you throw in some information, they garble it up and then spit it out.

If it works for black holes, it works for me

Building on the findings of Bekenstein and Hawking, Dutch physicist (and soon-to-be Nobel laureate) Gerard 't Hooft made a radical proposal in 1993. He took the laws of black holes and used them to conjecture a very fundamental property of gravity.

From Bekenstein's work, you know that the amount of information within a black hole increases in proportion to the area of the event horizon, rather than the volume. Moreover, you know that black holes are the "densest" possible objects. It stands to reason, then, that any volume of space contains an amount of information that's *at most* the amount that would be inside a black hole of the same size.

In other words, thanks to quantum gravity, the maximum amount of information in any region of the universe is given, at most, by the area of the surface of that region (or, more precisely, by the number of Planck-size pixels that may fit in that area). This is a very special feature of gravity, which we certainly don't expect from an "ordinary" quantum theory (without gravity).

You might think this sounds far-fetched, and you're right. Basically, we're talking about changing our whole understanding of the universe! However, although the conclusion is shocking, the components that 't Hooft relied on are well-established: the work of Bekenstein and Hawking, and the fact that black holes are the most tightly packed objects in the universe (certainly more tightly packed than stars, planets, and galaxies). 't Hooft's remarkable insight was to take these facts to their logical conclusions. As you'll see, it turns out that he was actually on to something — although a more detailed understanding of his ideas would come only a few years later, thanks to string theory.

To summarize:

- In any quantum theory without gravity, the information contained in a region of space grows at most like the volume of that region.
- In a theory of quantum gravity, the information contained in a region of space grows at most like the area of the boundary of that region.

That's already quite a radical statement, but if we make the "region of space" the whole universe, we can take it a step further:

> A quantum theory of gravity in a *D*-dimensional universe is equivalent to a quantum theory without gravity on the *(D-1)*-dimensional boundary of that universe.

TIP

In other words, the *holographic principle*, as 't Hooft's proposal is called, says that everything that happens in a space can be explained in terms of information that's somehow stored on the surface of that space. For example, picture a 3-dimensional space that resides inside the 2-dimensional curled surface of a cylinder, as in Figure 13-2. You reside inside this space, but perhaps some sort of shadow or reflection resides on the surface.

Once again, there's a key aspect of this situation that's missing from our example: A shadow contains only your outline, but in 't Hooft's holographic principle, *all* the information in the space is retained. That includes your shape, your bones, your cells, down to every last electron; the reduction in dimension doesn't equate to a reduction in information. In other words, these gravitational holograms are indeed the sort of *true holograms* (unlike the optical ones) we referenced earlier in this chapter as mere thought experiments.

FIGURE 13-2: The holographic principle says information about a space is contained on the surface.

The idea of a true hologram is hard to visualize, which is one of the reasons the holographic principle took physicists by surprise when it was proposed in 1993. It's absolutely mind-bending to think that the information needed to describe a space is proportional to the area, and not the volume of that space. A big reason this is so counterintuitive is that we like to describe a system by giving it coordinates, setting up some grid, and referencing what's going on in any cell of that grid. But the holographic description messes up this picture because the map between the grid and the pixels on the boundary isn't one-to-one.

TECHNICAL STUFF

Note that in the case of more than three space dimensions, "volume" isn't a precise term. A 4-dimensional "hypervolume" would be length times width times height times some other space dimension. For now, you can ignore the time dimension.

TIP

If you take seriously the holographic principle, you can think of it in two ways:

- Our universe, which includes gravity, is a 4-dimensional space that's equivalent to some 3-dimensional physical system without gravity "at the boundary of the universe" (whatever that means).
- As long as we can overlook the effects of gravity, our universe can be considered the 4-dimensional boundary of a 5-dimensional universe with gravity.

In Scenario 1, we live in the space inside the boundary; in Scenario 2, we're on the boundary, reflecting a higher order of reality that we don't perceive directly. Both pictures raise profound implications about the nature of the universe we live in.

Considering AdS/CFT Correspondence

Though the idea of holography was first introduced in 1993, even Leonard Susskind (who, with 't Hooft, was one of its main advocates) says he thought it would be decades before there would be any way to confirm it. The main issue is that the map between the boundary and the "bulk" of the space-time is complicated, and there were almost no cases where scientists could make detailed computations to verify 't Hooft's intuition.

Then, in 1997, everything changed. Argentine physicist Juan Maldacena published a paper inspired by the holographic principle destined to make quite a splash. He proposed something called the *anti-de Sitter/conformal field theory correspondence*, or *AdS/CFT correspondence*, which brought the holographic principle to center stage in string theory and offered, for the first time, the possibility to explicitly check the conjecture.

Checking the predictions

In his AdS/CFT correspondence, Maldacena proposed a new duality between strings propagating in D-dimensional anti-de Sitter space (AdS space) and quantum field theories (more specifically, conformal quantum field theories) on the $(D-1)$-dimensional flat space-time, which can be thought of as the boundary of the AdS space.

The main examples of this correspondence appeared when AdS space (or space-time) has dimensions $D=5$, 4, and 3. In particular, when $D=5$, the theory "on the boundary" lives in four space-time dimensions, like we are used to, and it's quite similar to quantum chromodynamics (it's a Yang-Mills gauge theory). As usual in string theory, all these constructions require supersymmetry. In fact, the simplest examples of AdS/CFT are precisely those with as much supersymmetry as possible.

Unfortunately, none of the theories originally proposed by Maldacena, nor any of the many generalizations that followed, precisely capture the physics of real-world gravity and the Standard Model. However, many of the theories were nice enough to allow for precise checks of the holographic correspondence, which led to spectacular agreement in very diverse setups: For different amounts of supersymmetry, and for various dimensions D, it seems that AdS/CFT is a general feature of string theory.

When we talk about checks of the AdS/CFT correspondence, we don't mean experimental verification (because most setups are quite far from the real world). Instead, we mean mathematical checks. In essence, AdS/CFT tells us that two theories (string theory on AdS and conformal field theory, or CFT) are one and the same. This means that any prediction made in either theory should precisely match. For this reason, theoretical physicists worked very hard to extract precise predictions from both theories (using very different techniques depending on whether they were in the AdS or the CFT camp) and match them. These types of tests have been extremely successful, and they've also helped us to better understand how to perform computations in string theory and in CFT.

AdS space, or living in an M. C. Escher painting

In order to understand AdS/CFT, you should first understand what AdS space, or anti-de Sitter space, is. In our discussion of the possible geometry of our universe in Chapter 9, we touch on *de Sitter space*, which is a fancy name for a space-time of constant positive curvature, much like a sphere. It isn't *exactly* a sphere because we want a *space-time* with positive curvature, not just a *space* with positive curvature (to keep things simple, we're being a bit cavalier with this distinction). AdS space is almost the same thing, but with *negative* curvature.

TECHNICAL STUFF

SPHERES, HYPERSPHERES, AND SADDLES

A mathematical way to define a space with positive curvature (like a sphere) is to require that it satisfy this equation:

$$X_1^2+X_2^2+...+X_D^2+X_{(D+1)}^2 = R^2$$

This means that we are describing a *D*-dimensional surface inside a *(D+1)*-dimensional ambient space. For example, in the case where *D*=1, we have a circle in the plane, and when *D*=2, we have a sphere in 3-dimensional space. Larger values of *D* describe generalizations of the sphere, sometimes called a *hypersphere.*

A space of negative curvature, called hyperbolic space or a "saddle" (for reasons you can see in Figure 13-3), would satisfy an equation something like

$$-X_1^2+X_2^2+...+X_D^2+X_{(D+1)}^2 = -R^2$$

A crucial ingredient of the AdS/CFT correspondence is the existence of a boundary for the gravity theory. To understand what this means, you need to try to visualize AdS space, or at least its close cousin, hyperbolic space. This is no easy task, but we'll give it a go. (We talk more about the curved geometries of string theory in Chapter 15.)

It's easiest to start with a sphere, which is the positive-curvature part of our image. A nice property of the sphere is that if you zoom in around any point, you see the same thing: a curved "cap," always with the same curvature.

In the case of hyperbolic space (which has negative curvature), if we zoom in around any point, we see something that looks like a curved "saddle." Now the trick is that this should happen *around every single point*, which isn't something we can visualize with our three-dimensional intuition. Figure 13-3 illustrates what our image should look like.

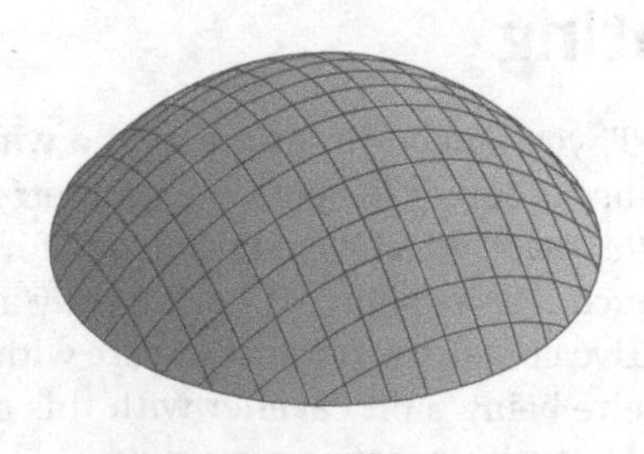

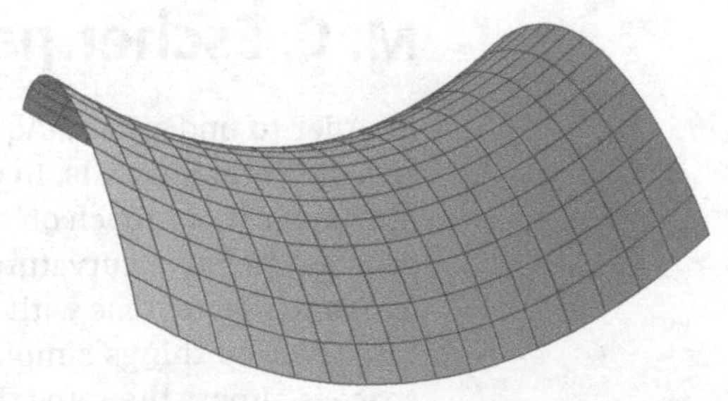

FIGURE 13-3: Left: Zooming into a sphere, we see a cap, with positive curvature. Right: Zooming into hyperbolic space, we see a saddle, with negative curvature.

Courtesy Source

A neat trick for visualizing hyperbolic space was invented by Henri Poincaré, and it's quite easy to draw in two dimensions (after all, the page of this book is 2-dimensional). The idea is to represent the space as a disk, with a special ruler for measuring the distance. When you're in the center of the disk, the ruler has a standard size, but as you progress toward the edge of the disk, it shrinks. As a result, as you look toward the edge of the circle, it seems that all measures shrink more and more.

Figure 13-4 gives you an idea of what hyperbolic geometry looks like. From the point of view of someone living in this hyperbolic universe, all the lines look straight. Moreover, the various triangles created by the intersection of the lines have edges of the same size. A person who starts walking from the center of the disk and takes a step from one tile to the next will need an infinite number of steps to get to the edge. Another interesting feature of the Poincaré disk is that the sum of the angles of the triangles is actually *smaller* than 180°. This is a crucial characteristic of hyperbolic space. (On the sphere, the sum of the angles of a triangle is *larger* than 180°.)

FIGURE 13-4: The Poincaré disk is a representation of hyperbolic space. For someone living on the disk, all the lines look straight and all the triangles have the same size.

Courtesy Source

M. C. ESCHER AND THE HYPERBOLIC PLANE

The Dutch artist Maurits Cornelius Escher (1898-1972) had a real knack for creating mind-bending visuals. You may be familiar with his famous lithograph of stairs going everywhere, and nowhere, titled *Relativity*. The title is no coincidence, as Escher was fascinated by physics and mathematics, and often discussed those subjects with world-class scientists, including mathematician George Pólya and physicist Roger Penrose. One object of his fascination was the Poincaré disk, which he depicted in his *Circle limit* series of woodcut prints highlighting its mind-bending geometry (see Figure 13-5).

FIGURE 13-5: The woodcut *Circle limit IV* by M. C. Escher depicts a tiling of the Poincaré disk where the tiles are figures of angels and devils.

The edge of the disk is the "boundary" of the universe: It's where our holographic theory would live. Of course, for most applications we would need 3-, 4-, or 5-dimensional analogs of the hyperbolic space that we're imagining. Picturing those gets more and more mind-bending, but hopefully you get the idea.

Finally, we can try to be a bit more precise about how anti-de Sitter space enters into the AdS/CFT correspondence. In Chapter 10, we note that superstring theory is very picky when it comes to dimensions: It wants to live in ten dimensions. So why are we talking about 3-, 4-, or 5-dimensional AdS space? Shouldn't we consider ten dimensional AdS space?

As it turns out, the geometries that are important for the AdS/CFT correspondence don't involve *only* AdS space. Instead, you have combinations of different geometries, like 5-dimensional AdS space and a 5-dimensional sphere. Even if the grand total gives ten dimensions, the boundary is still only 4-dimensional (in fact, a sphere has no boundary). There are also more complicated combinations, like 3-dimensional AdS space, two 3-dimensional spheres, and one 1-dimensional circle (again, a total of ten dimensions, but the boundary is only 2-dimensional).

CFTs: conformal, but nonconformist

The other side of the AdS/CFT correspondence is a *conformal field theory* (CFT), which is just a type of quantum field theory that has an additional symmetry called — you guessed it — conformal symmetry.

In our explanation of quantum field theory and the Standard Model of particle physics (see Chapters 7 and 8), we make a big deal about symmetries. For instance, in all theories of the Standard Model, there is no point or direction in space that has unique properties with respect to any other. This is what we called a symmetry, and more specifically invariance under translations and rotations. Moreover, all these theories are "relativistic invariant," which is just Einstein's way of saying that we have one more symmetry, this time about redefining the overall speed in which all the particles are moving. Conformal symmetry is yet another condition that we may impose, but it's a little more surprising. In a nutshell, it means that the result of an experiment won't depend on whether we scale the apparatus up or down. For instance, if we make everything twice as big, we would expect nothing to change.

TIP

Conformal field theory is a type of quantum field theory where lengths and distances are completely irrelevant, and only angles matter.

This sounds very strange: We know that molecules are a certain size, and atoms are much smaller, and protons are even smaller. If we were to scale up our experiments and try to have a molecule shrunk down to the size of a proton, nothing would make sense! Indeed, conformal symmetry is not at all a symmetry of the Standard Model.

Then why should we study it? First of all, conformal symmetry does appear in nature, but only in very peculiar circumstances. For instance, physicists that study things like magnets and superconductors encounter CFTs all the time. In fact, the theories that describe the transition from one phase of matter (magnetic to non-magnetic, superconducting to non-superconducting, and so on) are often CFTs.

In any case, for us, these theories aren't meant to describe the world as we see it but to provide an alternative description of quantum gravity in the funky AdS space (which also looks nothing like our world). As a result, it isn't so surprising if CFTs are a bit unusual.

Although the AdS/CFT correspondence really came into focus in string theory thanks to Maldacena's work, it's worth mentioning that it seems to be more general than that. In fact, in 1985 (well before Maldacena's revolutionary article), J. D. Brown and Marc Henneaux noticed that ordinary gravity (without any strings) in 3-dimensional anti-de Sitter space shared many qualitative features with conformal field theories in two dimensions. Their work can be seen as an independent piece of evidence pointing toward the holographic correspondence.

Understanding quantum gravity through AdS/CFT correspondence

The AdS/CFT correspondence is one of the biggest new ideas in string theory and theoretical physics. It allowed us to understand many features of quantum gravity much better.

The first challenge string theorists encountered in understanding the AdS/CFT correspondence was to make sure they grasped the correspondence in full detail. For instance, string theory in anti-de Sitter space depends on several parameters that have a string interaction constant as well as a string tension (see Chapter 11). How do these parameters appear in the dual theory?

It turns out that working out a precise map in full generality is rather complicated. Some features are universal, however. In particular, in the regime where strings behave like ordinary particles, the interactions are very strong in the dual theory (and hard to study), while when interactions are weak in the dual CFT, the strings are very "stringy" and hard to study. That makes it quite hard to compare the two pictures, but by exploiting many clever tricks, physicists managed to check the correspondence carefully. Once this was done, they had a formidable tool at their disposal.

Thanks to the AdS/CFT correspondence, it's no longer necessary to work with string theory — at least as long as you're happy to limit yourself to AdS space. You can do any computation you want in the CFT on the boundary of AdS space and find the result that way. That's a big deal because computing string interactions is very hard, and although working with a CFT also isn't easy, it does give you a lot of new tools to do the computations.

Perhaps the biggest limitation of the AdS/CFT correspondence is in its name: If we want to study real-world gravity, we have little interest in AdS space. Instead, we'd like to study strings in de Sitter space, or at least in flat space. Many string theorists have tried to formulate a holographic correspondence that could work for our universe, ideally with dark energy, dark matter, the Standard Model, and all that. Unfortunately, so far this seems very hard.

Nonetheless, AdS/CFT has been a major advance. 't Hooft's intuition for the holographic principle was compelling, but it didn't really provide us with a detailed way to map a gravitational theory to a holographic theory that lives at its boundaries. Instead, the AdS/CFT correspondence provides a detailed dictionary between the two sides of the duality, which by now has been checked quite extensively.

Before AdS/CFT, we might have believed that holography should hold for any gravity theory in some way, but we couldn't have said *precisely how*. Now, in the special case of strings in AdS space, we know *exactly* how holography works. At the very least, this hints at how precisely holography may work in the real world.

One of the major issues in quantum gravity is understanding the evolution of black holes (see Chapter 16 for more on this). Black holes form when a lot of matter is concentrated in a tiny space (as happens when certain stars collapse), and they very slowly evaporate due to Hawking radiation. As a result, all the information within the black hole also "evaporates," meaning it's carried away in the radiation. Immediately after Hawking proposed this theory, a paradox arose: It seemed that some information was lost in the evaporation process. Although it's possible that this was due to some approximation used by Hawking in his calculation, it was hard to show explicitly how to resolve the paradox. If the AdS/CFT correspondence is correct, this point is moot: In principle, you could study the whole formation and evaporation of the black hole in CFT, where the laws of quantum field theory guarantee that information is never lost.

Turning the Tables: Using Holography to Study Strongly Interacting Matter

There's another interesting use for holography: You can exploit string theory to discover more about quantum field theory. Perhaps this sounds weird because we've been going on and on about how tricky string theory is and how splendidly quantum field theory works in describing our world. There are, however, some situations where the tables are turned.

The force is strong when using AdS/CFT

In Chapter 8, we mention the idea of perturbation theory. In a nutshell, *perturbation theory* is a general approach employed by physicists (and other scientists) where, instead of computing what we want right away, we compute by a series of approximations. In quantum field theory, perturbation theory is absolutely crucial: We always start from the case where all particles are free, and then we add more and more possibilities of interactions.

Imagine we want to study the dynamics of two electrons. In the crudest picture, the electrons are free. This is, of course, a bad approximation because we know very well that electrons interact because of the electromagnetic force (which in quantum field theory is due to the exchange of photons — see Chapter 8). So, our first approximation is the case where the electrons interact, but only once (say, they exchange exactly one photon). A second order of approximation is when they interact twice; a third order is when they interact thrice, and so on.

Of course, the true answer is the sum of all these possibilities. Fortunately for us, however, the probability that we have two interactions is much smaller than the probability of one, and the probability of more and more interactions gets smaller and smaller very quickly. Therefore, for many applications you could just consider the case of one or two interactions, and you'd get a pretty precise result.

However, all this reasoning hinges on the fact that the probability of having one more interaction is small. This is true for the electromagnetic force, and it's exactly what physicists mean when they say that quantum electrodynamics (QED) is a weakly interacting theory. But our universe also has strongly interacting forces. To be more precise, it has one strongly interacting force, which is the very aptly named *strong force*.

The strong force is the interaction between quarks within the protons and neutrons (as well as within more general subatomic particles called hadrons). It's mediated by particles called *gluons*, which are analogs of the photons. The interaction between the quarks and the gluons is so strong that it's impossible to pull a single quark out of a proton or neutron. The quantum field theory that describes this dynamic is called quantum chromodynamics (QCD), and like QED, it's constructed as a gauge theory.

Because quarks and gluons interact so strongly, the usual approach of assessing them through perturbation theory isn't enough in many situations. This is where the AdS/CFT correspondence may come to the rescue. In fact, AdS/CFT is a weak-strong duality: A strongly interacting theory on one side of the duality is mapped to a weakly interacting theory on the other side. This opens up the prospect of describing strongly interacting theories, like QCD, as weakly interacting string theories (which is the regime where string theory almost becomes classical gravity and is quite easy to work with).

Understanding strongly coupled QCD, or strongly coupled theories in general, is a major challenge in physics, perhaps as important as understanding quantum gravity itself. If AdS/CFT allowed us to do that, it would be a monumental success of string theory!

Unfortunately, as is often the case, it's more complicated than that. AdS/CFT does give us a map between strongly and weakly interacting theories, but this map works only for *conformal* field theories. QCD is not a conformal field theory, and it's a good thing it isn't: Our world is not invariant when you rescale lengths! Besides, the AdS/CFT construction always has some amount of supersymmetry, which we also know doesn't exist in QCD. Therefore, we'll probably be able to get at best a qualitative insight into strong interactions of quarks from AdS/CFT.

Even with all this in mind, physicists have been trying to use AdS/CFT to better understand a strongly interacting theory like QCD with surprising success.

Cooking up a soup of quarks and gluons

When you give anything to a physicist, they will try to take it apart to see what's inside and how it works. This is pretty much what particle accelerators do, by smashing subatomic particles into each other and measuring what comes out of the process. This approach isn't as easy to employ when you're dealing with quarks, however.

Because of the enormous strength of strong interactions, it's literally impossible to pull a single quark out of a hadron and study it. The best that physicists can do is to keep smashing a bunch of hadrons together in a tiny space, so that all the quarks inside them heat up. Once the quarks attain sufficiently high energy, they aren't so tightly bound to each other by the gluons. However, this loosening is possible only if they remain within the hot soup of quarks and gluons — and we mean *hot*, like *a few trillion degrees hot* (this is one of the happy cases where it doesn't matter if you use Kelvin, Celsius, or even Fahrenheit degrees).

REMEMBER

This hot soup of quarks and gluons is called *quark-gluon plasma*. It's a bizarre state of matter, but it's of great theoretical interest because physicists believe it was the state of matter just moments after the big bang.

In other words, when the universe was still tiny and all matter was compressed in a small space, the temperature was sufficiently high to loosen up the quarks. Then, as the universe expanded, it cooled down, and all the quarks ended up confined within the hadrons.

Another remarkable fact about quark-gluon plasma is that, despite its rather extreme features, it's possible to produce in the lab. The trick is that instead of accelerating and smashing elementary particles, you can use a particle collider like the Large Hadron Collider at the European Organization for Nuclear Research in Switzerland or the Relativistic Heavy Ion Collider at Brookhaven National Laboratory in New York to accelerate *whole nuclei* of heavy atoms (like lead, which has 208 protons and neutrons) and smash them into each other. The much larger number of quarks is sufficient to create the high density and temperature needed for quark-gluon plasma.

Not only can experimental physicists create quark-gluon plasma in particle colliders, but they can also measure its properties, including its viscosity. Here, viscosity is really what you know from experience: the amount of friction within a fluid that hinders its flow. For instance, honey has a high viscosity, while water has a much lower one. Cooking oil has a higher viscosity when it's cold than it does when it's hot. In essence, viscosity is a measure of how hard it is for a liquid to move (or for another object to move through a liquid).

For the quark-gluon plasma, this kind of measurement isn't easy, and we can't expect it to be as precise as other collider data, like the mass of elementary particles. However, even after all experimental uncertainties are taken into consideration, the result turns out to be much smaller than what was expected from studying QCD.

Intriguingly, a much closer value for the viscosity of quark-gluon plasma can be found using AdS/CFT. In this case, you can try to use the correspondence that involves string theory on *5-dimensional* AdS space (as well as a 5-dimensional sphere) and a 4-dimensional CFT called "the *N*=4 supersymmetric Yang-Mills theory," which is a mouthful! Among friends, it's known as *N*=4 SYM.

Here's the lowdown on *N*=4 SYM. First of all, it's a 4-dimensional gauge theory (like electrodynamics and QCD), which is great for our purposes. Second, it's "four-fold" supersymmetric (that's the *N*=4 part of the name), which is a lot of supersymmetry — in fact, it's the maximum amount of supersymmetry for such a theory. This is bad for us. Next, for the AdS/CFT application, we need a theory with lots and lots of different types of gluons (hundreds, if not thousands). This is also bad for our purposes if you think about QCD, where there are just eight gluons. Finally, the theory is conformal, so it's invariant under rescaling (also bad for us).

All in all, there are many things in *N*=4 SYM that are at best qualitatively similar to QCD. Still, it's possible to use AdS/CFT to predict with decent accuracy the viscosity of the quark-gluon plasma, which indicates that *N*=4 SYM captures at least *some feature* of QCD in the quark-gluon plasma regime.

To be fair, there are *other* properties of the quark-gluon plasma that an experimentalist can measure, and AdS/CFT isn't as successful in predicting all of them. This is one of the many instances where string enthusiasts would point at a great success in capturing even *some qualitative features* of quark-gluon plasma, while string critics would emphasize that predictions are hit and miss. We return to this debate in Chapter 18. But for the moment, we'll count it as a victory on the side of string theory enthusiasts.

IN THIS CHAPTER

» Picking the right string theory to test

» Disproving string theory may be harder than confirming it

» Exploring two types of labs: Nature's lab and particle accelerators

Chapter 14 Putting String Theory to the Test

No matter how impressive string theory is, without experimental confirmation, it can be thought of as little more than mathematical speculation. As we discuss in Chapter 4, science is an interplay of theory and experiment. String theory attempts to structure the experimental evidence around a new theoretical framework.

One problem with string theory is that the energy required to get direct evidence for its distinct predictions is typically so high that it's very hard to reach. Newer experimental methods and tools, such as the Large Hadron Collider (described at the end of this chapter), are expanding our ability to test in higher energy ranges, possibly leading to discoveries that more strongly support string theory predictions, such as extra dimensions and supersymmetry. Probing the strings directly requires massive amounts of energy that are still far away from any experimental exploration.

In this chapter, our goal is to look at different ways that string theory can be tested, so it can be either verified or disproved. First, we explain the work that still needs to be done to complete the theory so it can make meaningful predictions. We also cover a number of experimental discoveries that would pose complications for string theory. Then we discuss ways of proving that our universe does contain supersymmetry, a key assumption required by string theory. Finally, we outline the testing apparatus — those created in deep space and particle accelerators created on Earth.

Understanding the Obstacles

As we explain in Chapter 11, string theory isn't complete. There are a vast number of different string theory solutions — literally billions of billions of billions of billions of different possible variants of string theory. So, in order to test string theory, scientists have to figure out which predictions the theory actually makes.

Before testing on string theory can take place, physicists need to filter through the massive number of possible solutions to find a manageable few that may describe our universe. Most of today's tests related to string theory are simply helping to define the current parameters of the theory. Then, after the remaining theoretical solutions are somehow assessed in a reasonable way, scientists can begin testing the unique predictions they make.

REMEMBER

Two features are common to (almost) all versions of string theory, and scientists who are looking for evidence of string theory are testing these ideas even now:

- Supersymmetry
- Extra dimensions

These are string theory's two cornerstone ideas (aside from the existence of strings themselves, of course), which have been around since the theory was reformulated into superstring theory in the 1970s. Theories that have tried to eliminate them haven't lasted very long.

Testing an incomplete theory with indistinct predictions

One of the issues in testing string theory is that this "theory" is more of a framework to talk about quantum gravity, rather than a precise physical model like quantum electrodynamics. For instance, we may use string theory to study the AdS/CFT correspondence, as we discuss in Chapter 13, with enormous success. Does this teach us anything about the predictive power of string theory in our universe? Hardly, since our universe looks nothing like anti-de Sitter space, which is the "AdS" of AdS/CFT.

Testing string theory in the real world (which, for the sake of argument, we'll assume is the universe we live in) runs into two major questions:

- How can we model our universe within the string theory framework?
- How can we check the model's claims against reality?

A truly quantitative approach to doing precision tests of string theory usually gets stuck at the first step. We never even get to wrestle with the second step! Therefore, physicists sometimes look for more qualitative tests to try to check features that should be common to "any sensible" stringy model.

A typical example is looking for supersymmetric partners of the Standard Model particles. Pretty much all stringy models have supersymmetry, so all of them should include new particles. But without knowing the precise stringy model, we don't know what these partners should look like. In all these models, supersymmetry is broken at low energies (to reproduce the fact that, in the real world, we do not observe supersymmetry at our energies scale). However, the remnants of the breaking depends from model to model. And if we make an experiment and we don't find any of these remnants, it may just be that we looked in the wrong place. You're probably starting to see why some people are getting frustrated.

But wait — there's more! Imagine that you do find a new particle, or some hint of extra dimensions. That would have all physicists jumping up and down with excitement, no doubt: It would signal that the Standard Model of particle physics isn't the end of the story, and there's some new physics to understand. But whether that new particle, or those extra dimensions, really has anything to do with string theory — well, that's another question!

Testing versus proof

There's really no way to prove, or disprove, something like string theory. You can prove that a specific prediction (such as supersymmetry, covered in the next section) is true, but that doesn't prove the theory as a whole is correct, or wrong. In a very real sense, string theory can never be proven or disproven: More than anything, it's a framework, which makes it incredibly flexible.

It could be that this framework does capture most, perhaps even all, properties of real-world quantum gravity, or perhaps it needs to be supplemented by some new ingredient that, so far, no one has discovered. This has happened before, when the ingredients of supersymmetry and branes were added to the mix (see Chapters 10 and 11).

For scientists, these nuances are known and accepted, but there's some confusion about testing versus proof among nonscientists. Most people believe that science proves things about the laws of nature beyond a shadow of a doubt, but the truth is, science dictates there's *always* a shadow of a doubt in any theory.

REMEMBER

A theory can be tested in two ways. The first is to apply the theory to explain existing data (called a *postdiction*). The second is to apply the theory to make predictions about new data that would be expected, which experiments can then look for. String theory has been quite successful at coming up with postdictions, but it hasn't been as successful at making clear predictions.

String theory has some valid criticisms that need to be addressed (see Chapter 18). Even if those criticisms are addressed, string theory will never be proven, but the longer it makes predictions that match experiments, the more support it will gain. For this to happen, of course, string theory must start making predictions that can be tested.

Analyzing Supersymmetry

One major prediction of string theory is that a fundamental symmetry exists between bosons and fermions, called *supersymmetry*. For each boson, there exists a related fermion, and for each fermion, there exists a related boson. (*Bosons* and *fermions* are types of particles with different spins; Chapter 8 has more detail about these particles.) We know that this is not an exact symmetry of nature, and therefore all realistic string models come with some some mechanism to break supersymmetry — introducing an asymmetry between the particles which we know and their (yet-to-be-found) partners

Finding the missing sparticles

Under supersymmetry, each particle has a superpartner. Every boson has a corresponding fermionic superpartner, just as every fermion has a bosonic superpartner. The names of fermionic superpartners end in "-ino," while bosonic superpartners' names start with an "s." Finding these superpartners is a major goal of modern high-energy physics. As a whole, this proposed set of superpartner particles are referred to as "sparticles."

The problem is that without a complete version of string theory, string theorists don't know which energy levels to look at. Scientists will have to keep exploring until they find superpartners and then work backward to construct a theory that contains the superpartners. This seems only slightly better than the Standard Model of particle physics, where the properties of all 18 fundamental particles have to be entered by hand.

Also, there doesn't appear to be any fundamental theoretical reason *why* scientists haven't found superpartners yet. If supersymmetry does unify the forces of physics and solve the hierarchy problem, then scientists would expect to find low-energy superpartners.

Instead, scientists have explored energy ranges into a few thousand giga-electronvolts (GeVs) but still haven't found any superpartners. So the lightest superpartner would appear to be much heavier than the 18 observed fundamental particles. Some theoretical models predict that the superpartners could be 1,000 times heavier than protons, so their absence is understandable but still frustrating.

Right now, the best proposal for a way to find supersymmetric particles outside a high-energy particle accelerator (see "LHC finds a boson, but no superpartners" at the end of this chapter) is the idea that the dark matter in our universe may actually be the missing superpartners (see "Analyzing dark matter and dark energy" later in this chapter). If that's the case, a lot of superpartners are out there waiting for us to discover them, if we can only figure out how to do it.

Testing implications of supersymmetry

If supersymmetry exists, then some physical process occurs that causes the symmetry to become spontaneously broken as the universe goes from a dense high-energy state into its current low-energy state. In other words, as the universe cooled down, the superpartners had to somehow decay into the particles we observe today. If theorists can model this spontaneous symmetry-breaking process in a way that works, it may yield some testable predictions.

The main problem is something called the *flavor problem*. In the Standard Model, there are three flavors (or generations) of particles. *Electrons*, *muons*, and *taus* are three different flavors of *leptons* (an elementary particle that can't be broken down), as we note in Chapter 8 (refer to Table 8-1).

In the Standard Model, these particles don't directly interact with each other. (They can exchange a gauge boson, so there's an indirect interaction.) Physicists assign each particle number based on its flavor, and these numbers are a conserved quantity in quantum physics. The electron number, muon number, and tau number don't change, in total, during an interaction. An electron, for example, gets a positive electron number but gets zero for both muon and tau numbers.

REMEMBER

Because of this, a muon (which has a positive muon number but an electron number of zero) can never decay into an electron (with a positive electron number but a muon number of zero), or vice versa. In the Standard Model and in supersymmetry, these numbers are conserved, and interactions between the different flavors of particles are prohibited. This is a very strong constraint on how any new particle may fit into the theory.

However, no one claims that our universe has a perfectly preserved supersymmetry — it's predicted that our universe has *broken supersymmetry.* There's no guarantee that the broken supersymmetry will conserve the muon and electron numbers, and creating a theory of spontaneous supersymmetry breaking that keeps this conservation intact is actually very hard. Succeeding at creating such a theory may provide a testable hypothesis, allowing for experimental support of string theory.

Testing Gravity from Extra Dimensions

The testing of gravity produces a number of ways to see if string theory predictions are true. When physicists test for gravity outside our three dimensions, they

- Search for a violation of the inverse square law of gravity
- Search for certain signatures of gravity waves in the cosmic microwave background radiation (CMBR)

It's possible that further research will uncover other ways to determine the behavior of string theory or related concepts.

Checking the inverse-square law

If extra dimensions are compactified in the ways that string theorists have typically treated them, then there are implications for the behavior of gravity. Since the days of Newton, we have always assumed that the gravitational force between two bodies scales like the inverse square of their distance. This is also true in general relativity, as long as we assume the universe to have three spatial dimensions, but it would not be true in a higher number of dimensions. Extra dimensions might lead to being a violation of the inverse-square law, especially if gravitational force extends into these extra dimensions at small scales. Current experiments seek to test gravity to an unprecedented level, hoping to see these sorts of differences from the established law.

The behavior of gravity has been tested down to under a millimeter, so any compactified dimensions must be smaller than that. Recent models indicate that they may be as large as that, so scientists want to know if the law of gravitation breaks down around that level.

As of this book's publication, no evidence has been found to confirm the extra dimensions at this level, but only time will tell.

Searching for gravity waves to understand inflation

General relativity predicts that gravity moves in waves through space-time. These gravitational waves arise whenever two massive objects are moving around each other, but in practice they're detectable only when we're dealing with very heavy objects that move very fast. Recently, astronomers have developed enormous detectors, called *interferometers,* that are sensitive enough to measure the passing of a gravitational wave due to an astronomical event like the merging of two black holes.

Gravitational waves are extremely interesting because they give us a window into astrophysical or cosmological events with large energies — big collisions and similar events. The biggest bang of all was, well, the big bang, and gravitational waves can teach us a lot about that too.

Broadly speaking, in the early stages of the universe, you would have seen a very hot and very dense soup of particles of all types — so dense that particles would've been constantly bumping into each other. As the universe expanded, it cooled down and got a bit diluted, to the point where particles were extremely unlikely to hit anything at all.

If you think about the universe right now, it's mostly empty space. The chance that a photon coming out of the sky will hit something on Earth is low; most of them will just travel undisturbed forever in our solar system, which is a relatively busy piece of real estate, cosmologically speaking.

So, at some point, the universe went from being a cramped place, where particles bumped into each other constantly, to a deserted one. When that happened, the photons that were bouncing around between all the charged particles just started traveling away undisturbed. Looking at them now by studying the CMBR gives us a snapshot of the early universe. (You can read more about the CMBR later in this chapter in "Using outer space rays to amplify small events" as well as in Chapter 9.)

But there's more: The *gravitons* (particles that mediate gravity — see Chapter 10) that were bouncing around in the hot soup also stopped bouncing as soon as things cooled down and particles got more separated. And they also just kept traveling on undisturbed, becoming what we call *primordial gravitational waves*. This is another snapshot of the early universe!

What's really exciting is that because gravity and electromagnetic radiation interact quite differently, these are two truly different snapshots, taken at slightly different times (though both were very early in the history of the universe). Together, they give us a two-frame movie of the early universe.

Astronomers are working diligently to measure these primordial waves, though this may require an even bigger interferometer (including one built out of satellites orbiting around the planet!). Their detection (or non-detection) will give hard constraint on the early evolution of the universe.

This is one of the key areas where string theory can be put to the test. After all, string theory is a theory of quantum gravity that's applicable even at the earliest stages of the universe, and describing what happened after the big bang (the process called *inflation*) is one of its goals.

One caveat (because in this story, there's always a caveat) is that the details of string-inspired inflation depend on the finer features of string theory, so once again it isn't clear that a sharp experimental result on the primordial gravitational waves will rule out *all* string-inspired models.

Disproving String Theory Sounds Easier Than It Is

With any theory, it's typically easier to disprove it than to prove it, although one criticism of string theory is that it may have become so versatile, it can't be disproved. We elaborate on this concern in Chapter 18, but in the following sections we assume that string theorists can pull together a specific theory. Having a working theory in hand makes it easier to see how it could be proved wrong.

Violating relativity

String theories are constructed on a background of space-time coordinates, so physicists assume relativity is part of the environment. If relativity turns out to be in error, then physicists will need to revise this assumption. However, that would be quite a blow to most of our current understanding of theoretical physics (like all of the Standard Model), which solidly rests on the bedrock of relativity.

There are theories that predict errors in relativity, most notably the *variable speed of light* (VSL) cosmology theories of John Moffat, and Andreas Albrecht and João Magueijo. Moffat went on to create a more comprehensive revision of general relativity with his *modified gravity* (MOG) theories. (These theories are addressed in Chapter 20.) The majority of work since the VSL and MOG theories were presented has shown that there isn't much support for them, but they all represent approaches where theorists question the current assumptions upon which core physical concepts — including concepts central to string theory — are built.

Even if some of these speculative concepts were shown to represent physical reality, though, there's reason to think that string theory would survive. Elias Kiritsis and Stephon Alexander have both proposed VSL theories within the context of string theory. Alexander went on to do further work in this vein with the "bad boy of cosmology," Magueijo, who is fairly critical of string theory as a whole.

Could proton decay spell disaster?

If one of the older attempts at unification of forces (called *grand unification theories*, or GUTs) proves successful, it will have profound implications for string theory. One of the most elegant GUTs was the 1974 Georgi-Glashow model, proposed by Howard Georgi and Sheldon Glashow. This theory has one flaw: It predicts that protons decay, and experiments over the last 35 years haven't shown this to be the case. Even if proton decay is detected, string theorists may be able to save their theory.

The Georgi-Glashow model allows quarks to transform into electrons and neutrinos. Because protons are made of specific configurations of quarks, if a quark inside a proton were to suddenly change into an electron, the proton itself would cease to exist as a proton. The nucleus would emit a new form of radiation as the proton decayed.

This quark transformation (and resulting proton decay) exists because the Georgi-Glashow model uses an SU(5) symmetry group. In this model, quarks, electrons, and neutrinos are the same fundamental kind of particle, manifesting in different forms. The nature of this symmetry is such that the particles can, in theory, transform from one type into another.

Of course, these decays can't happen very often, because we need protons to stick around if we're going to have a universe as we know it. The calculations showed that a proton decays at a very small rate: less than one proton every 10^{33} years.

This is a very small decay rate, but there's a way around it by having a lot of particles in the experiment. Scientists created vast tanks filled with ultrapure water and shielded from cosmic rays that may interfere with protons (and give false decay readings). They then waited to see if any of the protons decayed.

After 35 years, there has been no evidence of proton decay, and these experiments are constructed so there could be as many as a few decays a year. The results from Super-Kamiokande, a neutrino observatory in Japan, show that an average proton would take at least 10^{35} years to decay. To explain the lack of results, the Georgi-Glashow model has been modified to include longer decay rates, but most physicists don't expect to observe proton decay anytime soon (if at all).

If scientists did finally discover the decay of a proton, that would mean the Georgi-Glashow model would need to be looked at anew. String theory gained success in part because of the failure of all other previous models, so if the Georgi-Glashow predictions work, it may indicate poor prospects for string theory.

The string theory landscape remains as resilient as ever, and some predictions of string theory allow for versions that include proton decay. The decay time frame predicted is roughly 10^{35} years — exactly the lower limit allowed by the Super-Kamiokande neutrino observatory.

The renewal of GUT wouldn't disprove string theory, even though the failure of GUT is part of the reason string theory was originally adopted. String theories can now incorporate GUTs in low-energy domains. But string theory can't tell us whether we should anticipate that GUT exists or protons decay. Maybe; maybe not — string theory can deal with it either way. This is just one of the many cases where string theory shows a complete ambivalence to experimental evidence, which some critics say makes it "un-falsifiable" (as we discuss at greater length in Chapter 18).

Seeking mathematical inconsistencies

Given that string theory exists only on paper right now, one major problem would be definitive proof that the theory contains mathematical inconsistencies. This is the one area where string theory has proved most adaptable, successfully avoiding inconsistencies for more than 30 years.

Of course, scientists know that string theory isn't the whole story — the true theory is an 11-dimensional M-theory, which hasn't yet been defined. Work continues on various string theory approximations, but a serviceable definition of the fundamental theory — M-theory — may still prove a challenge.

One weakness is in the attempt to prove string theory finite. In Chapter 18, you can read about the controversy over whether this has been achieved. (It appears that even among string theorists there's a growing acknowledgment that the theory hasn't been proved finite to the degree it was once hoped it would be.)

To create his theory of gravity, Newton had to develop calculus. To develop general relativity, Einstein had to make use of differential geometry and develop (with the help of his friend Marcel Grossman) tensor calculus. Quantum physics was developed hand in hand with group representation theory by innovative mathematician Hermann Weyl. (*Group representation theory* is the mathematical study of how symmetries can act on various mathematical objects, which is at the heart of modern physics.)

THE MATHEMATICS OF STRING THEORY

It's sometimes said that string theory is a 21st-century theory that was accidentally discovered in the 20th century. One of the reasons is that much of the mathematics needed to understand string theory had to be developed hand in hand with the theory.

First of all, string theory is built out of the ideas of quantum field theory and general relativity; therefore, it takes advantage of all the mathematics that appears in those frameworks, like complex analysis (which is calculus on steroids) and differential geometry. Because supersymmetry is very important in string theory, run-of-the-mill differential geometry isn't enough, and you need to use more refined ideas that marry supersymmetry to geometry. Moreover, we see in Chapter 12 that strings can split and merge, and attach to branes. Because of this, string theorists often need to understand whether a string can wind around some geometry, or what happens when branes wrap around spheres or more complicated geometrical objects.

Mathematicians are still developing whole branches of math, like algebraic geometry and K-theory, which help us understand how to deform geometries and characterize them.

Though string theory has already spawned innovative mathematics explorations, the fact that scientists don't have any complete version of M-theory implies to some skeptics that a key mathematical insight is missing — or that the theory simply doesn't exist.

Bootstrapping Our Way into String Theory

Some string theorists are trying to reverse the logic of using mathematical arguments to disprove string theory. Instead, they're trying to use abstract mathematical tools to show that string theory is *inevitable*.

The approach these physicists are taking is called "bootstrapping," as in the expression "pulling yourself up by your own bootstraps." In the context of theoretical physics, it means determining the features of a theory from general (symmetry) arguments.

The idea of bootstrapping is rather general, and it has been used in the context of quantum field theory since the 1960s. (Some of its ideas were even put forward by Werner Heisenberg in the 1940s!) But it has seen a huge revival over the last few years.

In the context of quantum gravity, the idea goes as follows. Suppose that you have *a reasonable theory of quantum gravity*, which therefore features gravitons. You take two gravitons and make them collide. Is the output of this collision compatible only with string theory or with other theories too?

In these terms, this seems like a tall order: How can we say anything about "a reasonable theory of quantum gravity" if we don't know what this theory is? But that's the point of bootstrapping here: We don't want to venture any guesses about the theory; we only want to use the few noncontroversial properties of quantum gravity on which (almost) everyone agrees.

For instance, regardless of what our theory is, we'll assume it's compatible with relativity — which, honestly, is as uncontroversial as you can get. At large enough distances, we know how gravity should behave. Additionally, certain theorems relate other reasonable (but admittedly more technical) properties of quantum field theory to the mathematical features of the function describing the scattering of the gravitons.

Putting together these relatively mild assumptions, it's possible in principle to exclude a large part of the potential outcomes of our hypothetical scattering experiment. We could do even better by considering more general scattering experiments. By carefully working out all these constraints, we can understand how the most general theory of quantum gravity may look.

Note that all this isn't meant to disprove string theory — we know that string theory is self-consistent enough to pass these tests. The question is, how much else is possible?

Bootstrapping is a very new way to try to prove string theory, and the jury is still very much out. However, the first indications seem to be that the theories which meet the bootstrap criteria are remarkably similar to string theory — at least as far as the thought experiments go.

Looking for Evidence in the Cosmic Laboratory: Exploring the Universe

The problem with conducting actual experiments in string theory is that it requires massive amounts of energy to reach the level where the Standard Model and general relativity break down. Although we address man-made attempts to explore this realm in the next section, here we look at the different route the field of string cosmology takes: attempting to look into nature's own laboratory, the universe as a whole, to find the evidence string theorists need to test their theories.

Using outer space rays to amplify small events

Among the various phenomena in the universe, two types produce large amounts of energy and may provide some insight into string theory: *gamma ray bursts* (GRBs) and *cosmic rays*.

Some physical events are hard to see because they

- Are very rare (like, possibly, proton decay)
- Are very small (like Planck-scale events or possible deviations in gravity's effects)
- Happen only at very high energies (like high-energy particle collisions)

Or some combination of those three things makes the event a challenge to witness. Scientists are unlikely to see these improbable events in laboratories on Earth, at least without a lot of work, so sometimes they look where they're more likely to find them. Because both GRBs and cosmic rays contain very high energies and take so long to reach us, scientists hope they can observe these hard-to-see events by studying the cosmic happenings.

For years, physicists have used this method to explore potential breakdowns in special relativity, but Italian physicist Giovanni Amelino-Camelia of the University of Rome realized in the mid-1990s that gamma ray bursts and cosmic rays could be used to explore the Planck length (and energy) scale.

Gamma ray bursts

Exactly what causes a GRB is disputed, but it seems to happen when massive objects, such as a pair of neutron stars or a neutron star and a black hole (the most probable theories), collide with each other. These objects orbit around each other for billions of years, but finally collapse together, releasing energy in the most powerful events observed in the universe (see Figure 14-1).

The name *gamma ray bursts* clearly implies that most of this energy leaves the event in the form of gamma rays (which are very high-energy photons), but not all of it does. These objects release bursts of light across a range of different energies (or frequencies — energy and frequency of photons are related).

According to Einstein, all the photons from a single burst should arrive at the same time because light (regardless of frequency or energy) travels at the same speed. By studying GRBs, we may be able to determine if this is true.

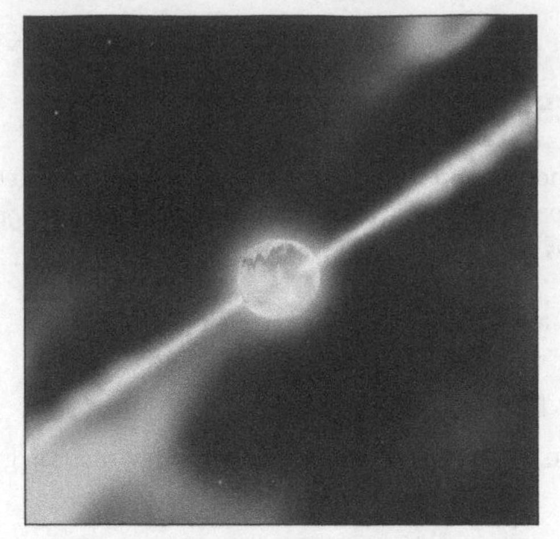

FIGURE 14-1: When some stars die, they release massive bursts of energy.

Courtesy of NASA/Swift/Sonoma State University/A. Simonnet

Calculations based on Amelino-Camelia's work have shown that photons of different energy that have traveled for billions of years could, due to (estimated and possibly overoptimistic) quantum gravity effects at the Planck scale, have differences of about 1 one-thousandth of a second (0.001s).

The Fermi Gamma-ray Space Telescope (formerly the Gamma-ray Large Area Space Telescope, or GLAST) was launched in June 2008 as a joint venture between NASA, the U.S. Department of Energy, and French, German, Italian, Japanese, and Swedish government agencies. Fermi is a low-Earth-orbit observatory with the precision required to detect differences this small.

So far, there's no evidence that Fermi has identified Planck-scale breakdown of general relativity. While it has made many intriguing discoveries in its observations of pulsars, GRBs, and other astrophysical phenomena, none of the discoveries indicated a breakdown of fundamental physics related to relativity.

If Fermi (or some other means) does detect a Planck-scale breakdown of relativity, that will only increase the need for a successful theory of quantum gravity, because it will be the first experimental evidence that the theory does break down at these scales. String theorists would then be able to incorporate this knowledge into their theories and models, perhaps narrowing the string theory landscape to regions that are more feasible to work with.

Cosmic rays

Cosmic rays are produced when particles are sent out by astrophysical events to wander the universe alone, some traveling at close to the speed of light. Some stay bound within the galactic magnetic field, while others break free and move between galaxies, traveling billions of years before colliding with another particle. These cosmic rays can be more powerful than our most advanced particle accelerators.

First of all, cosmic rays aren't really rays. They're stray particles in mostly three forms: 90 percent free protons, 9 percent alpha particles (two protons and two neutrons bound together — the nucleus of a helium atom), and 1 percent free electrons (beta-minus particles, in physics-speak).

Astrophysical events — everything from solar flares to binary star collisions to supernovae — regularly spit particles out into the vacuum of space, so our planet (and, in turn, our bodies) are constantly bombarded with them. The particles may travel throughout the galaxy, bound by the magnetic field of the galaxy as a whole, until they collide with another particle. (Higher energy particles, of course, may even escape the galaxy, though this takes a very long time.)

Fortunately, the atmosphere and magnetic field of Earth protect us from the most energetic of these particles so we aren't continuously dosed with intense (and lethal) radiation. The energetic particles are deflected or lose energy, sometimes colliding in the upper atmosphere to split apart into smaller, less energetic particles. By the time they get to us, we're struck with the less intense version of these rays and their offspring.

Cosmic rays have a long history as experimental surrogates. When Paul Dirac predicted the existence of antimatter in the 1930s, no particle accelerators could reach that energy level, so the experimental evidence of its existence came from cosmic rays.

As the cosmic ray particles move through space, they interact with the CMBR. This microwave energy that permeates the universe is pretty weak, but for the cosmic ray particles, moving at nearly the speed of light, the CMBR appears to be highly energetic. (This is an effect of relativity because energy is related to motion.)

In 1966, Soviet physicists Georgiy Zatsepin and Vadim Kuzmin, as well as Kenneth Greisen working independently at Cornell University, revealed that these collisions would have enough energy to create particles called *mesons* (specifically, pi-mesons, or *pions*). The *GZK cutoff energy* used to create the pions had to come from somewhere (because of conservation of energy), so the cosmic rays would lose energy. This placed an upper bound on how fast the cosmic rays could, in principle, travel.

In fact, the GZK cutoff energy needed to create the pions would be about 10^{19} electronvolts (eV), which is about one-billionth of the Planck energy of 10^{19} GeV.

The problem is, while most cosmic ray particles fall well below this threshold, some rare events have had *more* energy than this threshold — around 10^{20} eV. The most famous of these observations occurred in 1991 at the University of Utah's Fly's Eye cosmic ray observatory on the U.S. Army's Dugway Proving Ground.

Research since then indicates that the GZK cutoff does indeed exist. The rare occurrence of particles above the cutoff is a reflection of the fact that, very occasionally, these particles reach Earth before they come in contact with enough CMBR photons to slow them down to the cutoff point.

These observations later came into apparent conflict with Japan's Akeno Giant Air Shower Array (AGASA) project, which identified nearly ten times as many of these events. The AGASA results implied a potential failure of the cutoff, which could have had implications for a breakdown in relativity. However, by 2017 it was realized that the apparent contradiction is resolved if you notice that many of these high-energy particles aren't protons (thus violating one of the hypotheses of the GZK calculation).

In any case, the existence of such energetic particles provides one means of exploring these energy ranges, well above what current particle accelerators can reach, so string theory may have a chance of an experimental test using high-energy cosmic rays, even if they're incredibly rare.

Analyzing dark matter and dark energy

One other astronomical possibility to get results to support string theory comes from the two major mysteries of the universe: dark matter and dark energy. These concepts are discussed at length in Chapters 9 and 16.

The most obvious way that dark matter could help string theory is if it's found that the dark matter is actually supersymmetric particles, such as the *photino* (the superpartner of the photon) and other possible particles.

Another dark matter possibility is a theoretical particle called an *axion*, originally developed outside string theory as a means of conserving certain symmetry relationships in quantum chromodynamics. Many string theories contain the axion, so it could be a way to prove the theory, although the properties suggested don't really match what cosmologists are looking for.

Some of the most significant work in cosmology and astrophysics today involves attempts to detect dark matter, and there seems to be a lot of it in the universe. So there's some hope that physicists will make headway on its composition within the foreseeable future.

Detecting cosmic superstrings

Cosmic strings were originally proposed in 1976 by Tom Kibble of Imperial College London. At that time, they had nothing to do with the fundamental superstrings of string theory. Rather, Kibble reasoned that in the aftermath of the big bang, as the universe went through a rapid cooling phase, defects may have remained behind. These defects in quantum fields are similar to what happens when you rapidly freeze water into ice, creating a white substance that is full of defects.

For a while in the 1980s, some scientists thought cosmic strings might be the original seed material for galaxies, but the CMBR data doesn't indicate this to be true. Years later, string theory would resurrect the notion of cosmic strings in a new form.

According to some string theory models, superstrings created in the big bang may have expanded along with the universe itself, creating *cosmic superstrings*. An alternate explanation identifies these cosmic superstrings as remnants from the collision of two branes.

REMEMBER

Cosmic superstrings would be incredibly dense objects. Narrower than a proton, a single meter of a cosmic superstring could weigh about the same as North America. As they vibrate in space, they could generate massive gravity waves rippling out through space-time.

One way of seeing cosmic superstrings would be through *gravitational lensing*, where the string's gravity bends the light of a star (see Figure 14-2). This might mean that we see one star in two different locations, each equally bright.

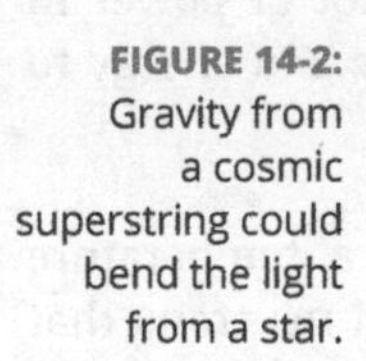

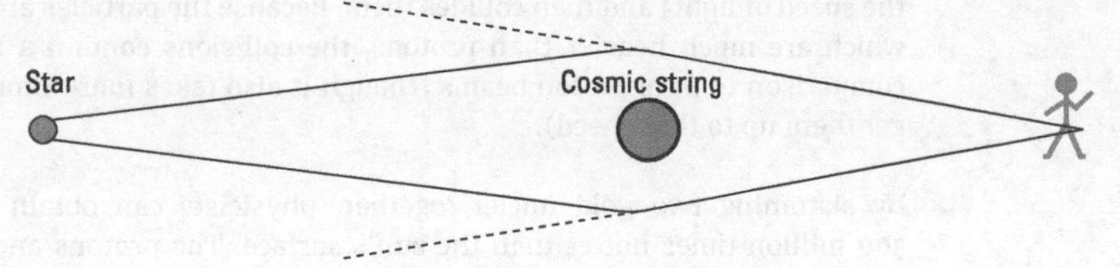

FIGURE 14-2: Gravity from a cosmic superstring could bend the light from a star.

According to Joe Polchinski, the best way to look for cosmic superstrings is to observe pulsars (like the ones that Fermi is detecting, mentioned earlier in this chapter). *Pulsars* are like astronomical lighthouses, spinning as they fire regular beams of electromagnetic radiation into the universe, which follow a predictable pattern. The gravity from a cosmic superstring could cause ripples in space-time that alter this pattern in a way that should be detectable here on Earth.

Looking for Evidence Closer to Home: Using Particle Accelerators

Although it would be nice if nature gave us the experimental results we need, scientists are never content to wait for a lucky break, which is why they proceed with experiments in apparatus that they control. For high-energy particle physics, this means particle accelerators.

A *particle accelerator* is a device that uses powerful magnetic fields to accelerate a beam of charged particles to incredibly fast speeds and then collides it with a beam of particles going the other way (this is why an *accelerator* is also called a *collider*). Scientists can then analyze the results of the collision.

The most powerful particle accelerator at the moment is the Large Hadron Collider (LHC) in Switzerland, which we discuss in some detail at the end of this chapter. Although the LHC is definitely the most famous of the particle accelerators, it's not the only one where important research is taking place.

Accelerating heavy ions at the RHIC

The Relativistic Heavy Ion Collider (RHIC) is a particle accelerator at Brookhaven National Laboratory in New York. It went online in 2000, after a decade of planning and construction.

The RHIC's name comes from the fact that it accelerates heavy ions — that is, atomic nuclei stripped of their electrons — at relativistic speeds (99.995 percent the speed of light) and then collides them. Because the particles are atomic nuclei, which are much heavier than protons, the collisions contain a lot of power in comparison to pure proton beams (though it also takes more time and energy to get them up to that speed).

By slamming two gold nuclei together, physicists can obtain a temperature 300 million times hotter than the sun's surface. The protons and neutrons that normally make up the nuclei of gold break down at this temperature into a plasma of quarks and gluons.

This *quark-gluon plasma* is predicted by quantum chromodynamics (QCD), but the problem is, the plasma is supposed to behave like a gas. Instead, it behaves like a liquid. As we note in Chapter 13, string theory is able to explain some of this behavior using insights from the holographic correspondence. In this way, the quark-gluon plasma may be described by an equivalent theory in the higher-dimensional universe: a black hole, in this case!

These results are far from conclusive, but theorists are looking at the behavior of these collisions to find ways to apply string theory to make more sense of the existing physical models (QCD, in this case), which is a powerful tool to help gain support for string theory.

At the time of this writing, the RHIC is still running experiments, but it's entering its final years of operation. These facilities are so massive that physicists don't want them to go to waste, so the plan is for many of the structures and components in the RHIC to be repurposed into the development of the upcoming Electron-Ion Collider (EIC).

This new design would still involve a beam of ions — atoms that have been stripped of their electrons so only the nuclei remain — spinning through the accelerator, but instead of colliding with other ions, they would collide with a beam of high-energy electrons that's heading in the opposite direction.

The development of the EIC was announced in January 2020 by the U.S. Department of Energy's Office of Science, one of the primary funding sources for Brookhaven National Laboratory. Timelines on these projects are difficult to predict, so it's hard for anyone to know for sure when the EIC will begin running.

Colliders of the future

Particle accelerators are so massive that there are no set designs for them; each particle accelerator is its own prototype. Because they require massive investments, particle accelerators are really international collaborations that span the globe. The design and implementation of new accelerators falls to the International Committee for Future Accelerators (ICFA), which identifies the strongest research needs of the scientific community.

One proposed accelerator is the International Linear Collider (ILC), which is an electron-positron collider. One benefit of this type of collider is that electrons and positrons are a lot less messy when they collide because they're fundamental particles, not composite particles like protons.

Electrons and positrons were also used by the Large Electron-Positron Collider, the collider that preceded the LHC at the European Organization for Nuclear Research (known as CERN) in Geneva. The problem with these particles is that they're quite light, which makes it harder to get to high energies unless you make the collider very big. And indeed, the plan is to make the ILC very long, around ten times longer than the biggest existing collider, the linear accelerator at Stanford University.

The ILC hasn't been approved, and many proposals for it have come and gone over the years. Despite proposals to build it at the Fermi National Accelerator Laboratory in the United States or through CERN in Europe, designers have focused on a location in Japan for the last decade or so.

Another apparatus proposed at CERN is the Compact Linear Collider (CLIC). Like the ILC, this accelerator would collide one electron beam with a positron beam. The CLIC would use a new two-beam accelerator, where one beam accelerates a second beam. The energy from a low-energy (but high-current) beam into a high-energy (but low-current) beam could allow for accelerations up to 3 teraelectronvolts (TeV) in a much shorter distance than traditional accelerators.

The plans for both the ILC and the CLIC are unified in an organization known as the Linear Collider Collaboration, designated by the ICFA to oversee the research and development efforts on making the two colliders a reality.

LHC finds a boson, but no superpartners yet

The LHC is a massive apparatus, built underground at the CERN particle physics facility on the border of Switzerland and France. (CERN is the European particle physics center that was, in 1968, the birthplace of string theory.) The accelerator itself is about 27 kilometers (17 miles) in circumference (see Figure 14-3). The 9,300 magnets in the collider can accelerate protons into collisions up to 13 TeV, well beyond our previous experimental limitations for collisions of similar particles.

As the LHC was being designed and built, physicists speculated about the discoveries it could make once it was up and running. A few of the optimistic options included:

- Microscopic black holes, which would support predictions of extra dimensions
- Supersymmetric particle (sparticle) creation
- Experimental confirmation of the Higgs boson, the final Standard Model particle to remain unobserved
- Evidence of curled-up extra dimensions

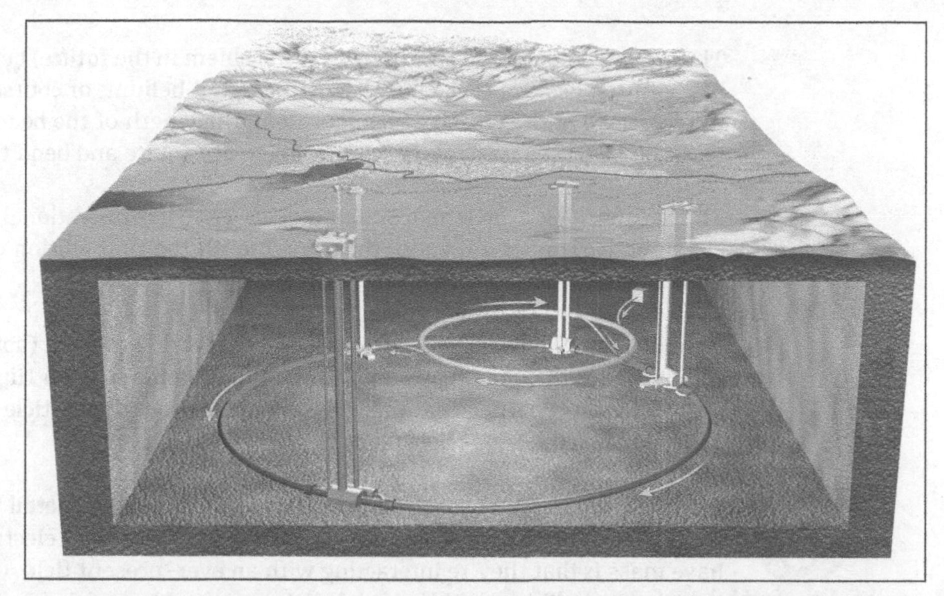

FIGURE 14-3: The Large Hadron Collider is built in a circular tunnel with a 17-mile circumference.

Courtesy CERN Press Office

As we note throughout this book, several of these discoveries would have profound implications for string theory. Understandably, theorists working in the field were enthusiastic about the possibility of getting experimental results that would shine some light on these phenomena. If there was new physics to be discovered, many felt that we would begin to see glimpses of it in the LHC.

So, at this point, what has the LHC found — or failed to find?

Discovering the Higgs boson

The first test run of the LHC was certainly dramatic, but from a scientific standpoint, it was anticlimactic. On September 10, 2008, the LHC officially came online by running a beam the full length of the tunnel. On September 19, a faulty electrical connection caused a rupture in the vacuum seal, resulting in a leak of 6 tons of liquid helium. And by "leak," we mean an explosion.

Fortunately, the LHC beam tunnel is 27 kilometers long and buried underground, so a malfunction in one area of the beam — even one that results in a major explosion — won't physically harm anyone.

The repairs (and upgrades to prevent the problem in the future) began with redoubled efforts. Besides replacing all the lost liquid helium, of course, scientists had to replace 50 of the massive magnets along the length of the beam pipe. For context, the LHC uses 1,230 of these magnets to accelerate and bend the proton beam.

The repairs took a little more than a year. The first operational run of the LHC began on November 20, 2009, the same month the first edition of this book was published, in fact!

On July 4, 2012, physicists from all around the world gathered (both in person and remotely) for the opening of the International Conference on High Energy Physics, where they learned that the LHC had indeed found a particle that was widely believed to be the Higgs boson.

REMEMBER

We cover the Higgs boson, and the related Higgs field, in detail in Chapter 8, so we'll keep it brief here. Basically, the explanation for why electrons and quarks have mass is that they're interacting with an ever-present field, commonly called the *Higgs field.* This was theorized decades ago, and one of the consequences of the proposal was that if you have collisions of sufficient energy, the Higgs field would manifest in the form of a previously unobserved type of particle, called the *Higgs boson.* The energies of the collisions at the LHC allowed for the discovery of a new particle in the right energy range to be the Higgs boson.

The prospect of discovering the Higgs boson, which was the one missing element of the Standard Model of particle physics, was certainly a major reason for building the LHC. And with that discovery out of the way, physicists began looking deeper into the mysteries of the universe, to either find evidence to support existing theories (like string theory) or to find unanticipated results that point toward entirely new physics.

Looking for superpartners

When the LHC was being built, string theorists' biggest hope was that it would help with the experimental verification of the existence of superpartners. *Superpartners* are the new fundamental particles hypothesized under the theory of supersymmetry, which is one of the principles at the core of all current variations of string theory.

Unfortunately, as of this writing, no announcement similar to the 2012 discovery of the Higgs boson has been made about a groundbreaking discovery of superpartners at the LHC. And it certainly hasn't been for a lack of searching.

NEW, EXCITING MATTER FROM THE LHC

The LHC has made all manner of significant discoveries in the realm of particle physics during its operation, of course, but they aren't immediately helpful to string theorists looking either for new evidence supporting existing theories or for something so revolutionary that it points them in the direction of brand-new theoretical areas to be explored. (After all, the only thing more enticing to most physicists than confirming a theory is a completely inexplicable new mystery of physics.)

The majority of the findings out of the LHC have been various new and exotic forms of high-energy hadrons. This makes sense, because the whole point of the LHC is that it's colliding hadrons together at high energy. Recall that *hadrons* are composite particles made up of quarks bound together by the strong nuclear force. When they collide at the LHC, it's with high enough energy to break the bonds of the strong nuclear force.

As the hadrons break apart into their constituent quarks, these now-liberated quarks find each other and come together in novel ways, including in a formation known as the *tetraquark,* a particle made up of four quarks. (The prefix *tetra-* means "four," which is why every piece in the video game Tetris is made up of four blocks.)

Exploring these new particles and their properties is exciting for particle physicists. For instance, theoretical physicists are scratching their heads trying to understand how these four quarks fit together. Is the tetraquark more like a molecule or an atom? (Have a look at Figure 14-4.) Unfortunately, this discovery doesn't seem to help with understanding string theory — at least for now.

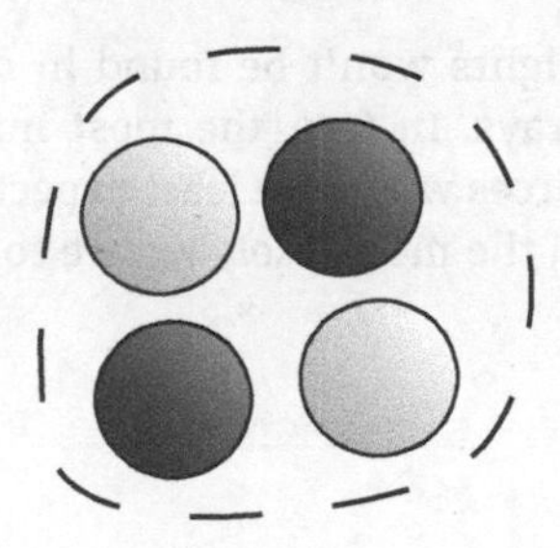

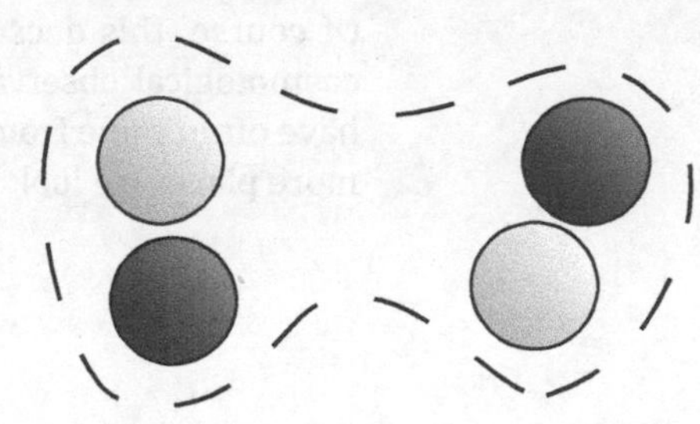

FIGURE 14-4: Two possible descriptions of the tetraquark. Left: Four quarks sit together. Right: Two pairs of quarks form a "meson molecule."

Source XXXX

It isn't just that the LHC hasn't found any superpartners yet — some of the things scientists at CERN have discovered cast doubts on supersymmetry itself. In particular, a 2013 finding showed measured very precisely the decays of B mesons, which is expected to be very sensitive to the existence of supersymmetry. They found that in every billion collisions, only about three B mesons decay into muons, which is precisely what they expected from the standard model *without supersymmetry*.

Instead, most supersymmetric models would predict many more such decays. As always, the key word here is *most*: It is still possible to come up with supersymmetric models that are compatible with such a low rate of decay, so that we cannot sound the death knell for supersymmetry. (This is a common theme in this story: There seems to always be some way to put supersymmetry back in the game.)

In other words, even though we think the Standard Model has to be an incomplete description of nature at its most fundamental level — because, after all, its quantum parts and its relativity parts conflict in various ways — the evidence from more than a decade of collisions at the LHC has acted largely to confirm the validity of the Standard Model.

Again, as we've seen time after time, the great virtue and the great vice of these theoretical approaches is their versatility. Even with the B meson decay rates being unexpectedly higher than what is predicted by supersymmetry, theorists are able to go back and rework the math to propose new approaches for supersymmetry that are consistent with the observed rates. Why aren't we seeing any superpartners at the current energy? Well, there's no shortage of variants where it would take higher energies to find them.

The problem is, it's not like there's another LHC on deck waiting to be built after the current LHC shuts down. To some realistic degree, we are reaching the limit of the energy levels that we may study with particle accelerators. This is because, to reach higher and higher energies, we need bigger and bigger facilities — and the LHC is already enourmous! If superpartners do exist at substantially higher energies than we can reach at the LHC, we might not be in a position where we're able to build an accelerator to find them.

Of course, this doesn't mean new insights won't be found in other places, like cosmological observations or cosmic rays. In fact, the most important insights have often come from the places or sources where we least expected them. But the more places we look for these insights, the more likely we are to find them.

4 The Unseen Cosmos: String Theory on the Boundaries of Knowledge

IN THIS PART . . .

Investigate extra dimensions.

Unravel the nature of our universe.

Explore the possibility of time travel.

IN THIS CHAPTER

- Understanding the meaning of dimensions
- Adding a time dimension to the space dimensions
- Bending dimensions as a mathematical pastime
- Offering alternatives to extra dimensions

Chapter 15
Making Space for Extra Dimensions

One of the most fascinating aspects of string theory is the requirement of extra dimensions to make the theory work. String theory requires nine space dimensions, while M-theory seems to require ten space dimensions. Under some theories, some of these extra dimensions may actually be long enough to interact with our own universe in a way that can be observed.

In this chapter, you get a chance to explore and understand the meaning of string theory's extra dimensions. First, we introduce the concept of dimensions in a very general way, talking about different approaches mathematicians have used to study 2- and 3-dimensional space. Then we tackle the idea of time as the fourth dimension. We analyze the ways in which the extra dimensions may manifest in string theory and whether the extra dimensions are really necessary.

What Are Dimensions?

Any point in a mathematical space can be defined by a set of coordinates, and the number of coordinates required to define that point is the number of dimensions the space possesses. In a 3-dimensional space like you're used to, for example, every point can be uniquely defined by precisely three coordinates, or three pieces of information: length, width, and height. Each dimension represents a degree of freedom within the space.

Though throughout this book we talk about dimensions in terms of space (and time), the concept of dimensions extends far beyond that. For example, the matchmaking website `eHarmony.com` provides a personality profile that claims to assess its users on 29 dimensions of personality. In other words, it uses 29 pieces of information as parameters for its dating matches.

TIP

We don't know the details of eHarmony's system, but one of us has some experience with using dimensions on other dating sites. Say you want to find a potential romantic partner. You're trying to target a specific type of person by entering different pieces of information: gender, age range, location, annual income, education level, number of kids, and so on. Each of these pieces of information narrows down the "space" that you're searching on the dating site. If you have a complete space consisting of every single person who has a profile on the dating site, when your search is over, you've narrowed your results to only the people who are within the ranges you specified.

Say Jennifer is a 30-year-old woman in Dallas with a college degree and one child. Those coordinates "define" Jennifer (at least to the dating site), and searches that sample those coordinates will include Jennifer as one of the "points" (if you think of each person on the site as a point) in that section of the space.

The problem with this analogy is that you end up with a large number of points within the dating site space that have the same coordinates. Another woman, Andrea, may enter essentially identical information as Jennifer. Any search of the sample space that brings up Jennifer also brings up Andrea. By contrast, in the physical space that we live in, each point is unique.

REMEMBER

Each dimension — in both mathematics and in the dating site example — represents a *degree of freedom* within the space. By changing one of the coordinates, you move through the space along one of the dimensions. For example, you can exercise a degree of freedom to search for someone with a different educational background or a different age range or both.

TIP

When scientists talk about the number of dimensions in string theory, they mean the degrees of freedom required for these theories to work without going haywire. In Chapter 10, we explain that the bosonic string theory required 25 space dimensions to be consistent, plus one for time, for a total of 26 space-time dimensions. Later, superstring theory required 10 space-time dimensions. M-theory seems to require 11 space-time dimensions.

2-Dimensional Space: Exploring the Geometry of Flatland

Many people think of geometry (the study of objects in space) as a flat, 2-dimensional space that contains two degrees of freedom: up or down and right or left. Throughout most of modern history, this field of mathematics has focused on the study of Euclidean geometry or Cartesian geometry.

Euclidean geometry: Think back to high school geometry

Probably the most famous mathematician of the ancient world was Euclid, who has been called the father of geometry. Euclid's 13-volume book, *Elements*, is the earliest known book to have included all the existing knowledge of geometry at the time of its writing (around 300 BCE). For nearly 2,000 years, virtually all geometry could be understood just by reading *Elements*, which is one reason it was dubbed the most successful math book ever written.

REMEMBER

In *Elements*, Euclid started off by presenting the principles of plane geometry — that is, the geometry of shapes on a flat surface, as in Figure 15-1. An important consequence of Euclidean plane geometry is that if you take the measure of all three angles inside a triangle, they add up to 180 degrees.

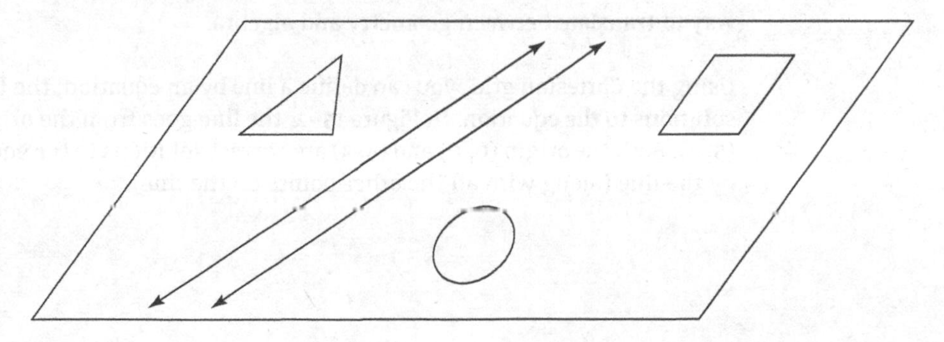

FIGURE 15-1: In Euclidean geometry, all figures are flat, as if drawn on a sheet of paper.

Later in the volumes, Euclid extended into 3-dimensional geometry of solid objects, such as cubes, cylinders, and cones. The geometry of Euclid is the geometry typically taught in school to this day.

Cartesian geometry: Merging algebra and Euclidean geometry

Modern analytic geometry was founded when French mathematician and philosopher René Descartes placed algebraic figures on a physical grid. Figure 15-2 shows this Cartesian grid. By applying concepts from Euclidean geometry to the equations depicted on the grids, mathematicians can obtain insights into geometry and algebra.

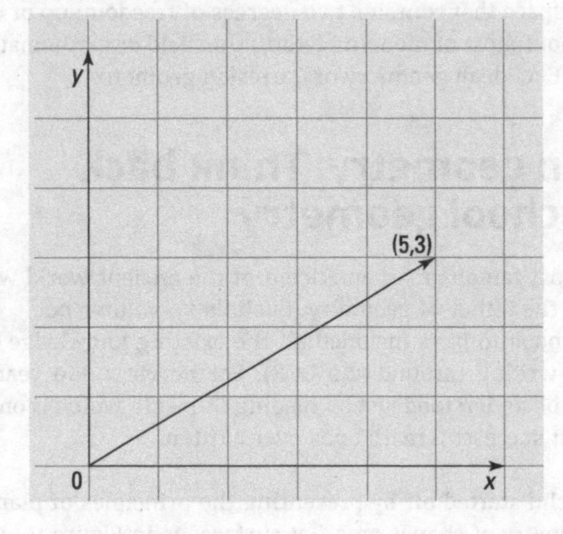

FIGURE 15-2: In Cartesian geometry, lines are drawn and analyzed on a grid of coordinates.

Around the same time that Galileo was revolutionizing astronomy, Descartes was revolutionizing mathematics. Until his work, the fields of algebra and geometry were separate. His idea was to display algebraic equations graphically, providing a way to translate between geometry and algebra.

Using the Cartesian grid, you can define a line by an equation; the line is the set of solutions to the equation. In Figure 15-2, the line goes from the origin to the point (5, 3). Both the origin (0, 0) and (5, 3) are correct solutions to the equation depicted by the line (along with all the other points on the line).

BOOKS OF MANY DIMENSIONS

The book *Flatland: A Romance of Many Dimensions,* written by Edwin A. Abbott in 1884, is a classic in the mathematics community for explaining the concept of multiple dimensions. In the book, A. Square lives in a flat world and gains perspective when he encounters a sphere passing through his world who pulls him out of it so he can briefly experience three dimensions.

Flatland appears to have been part of a growing popular culture interest in extra dimensions during the late 1800s. Lewis Carroll had written a story in 1865 titled "Dynamics of a Particle," which included 1-dimensional beings on a flat surface, and the idea of space going crazy is clearly a theme in Carroll's *Alice's Adventures in Wonderland* (1865) and *Through the Looking Glass* (1872). Later, H. G. Wells used the concepts of extra dimensions in several stories, most notably in *The Time Machine* (1895), where time is explicitly described as the fourth dimension a full decade before Einstein presented the first inkling of relativity.

Various independent sequels to *Flatland* have been written through the years to expand on the concept. These include Dionys Burger's *Sphereland* (1965), Ian Stewart's *Flatterland* (2001), and Rudy Rucker's *Spaceland* (2002). A related book is the 1984 science-fiction novel *The Planiverse,* where scientists in our world establish communication with a Flatland-like world.

Because the grid is 2-dimensional, the space the grid represents contains two degrees of freedom. In algebra, the degrees of freedom are represented by variables, meaning an equation that can be shown on a 2-dimensional surface has two variable quantities, often x and y.

Three Dimensions of Space

When looking at our world, we see three dimensions: up and down, left and right, back and forth. If you're given a longitude, a latitude, and an altitude, you can determine any location on Earth, for example.

A straight line in space: Vectors

Expanding on the idea of Cartesian geometry, you find that it's possible to create a Cartesian grid in three dimensions as well as two, as Figure 15-3 shows. In such a grid, you can define an object called a *vector,* which has both a direction and a length. In 3-dimensional space, every vector is defined by three quantities.

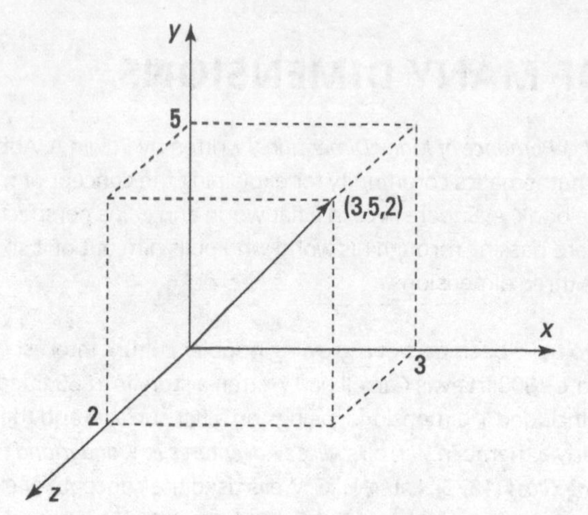

FIGURE 15-3: It takes three numbers to define a vector (or location) in three dimensions.

TIP

Vectors can also exist in one, two, or more than three dimensions. (Technically, you can even have a zero-dimensional vector, although it will always have zero length and no direction. Mathematicians call such a case "trivial.")

Treating space as containing a series of straight lines is probably one of the most basic operations that can take place within a space. One early field of mathematics that focuses on the study of vectors, *linear algebra*, allows you to analyze vectors and things called *vector spaces* of any dimensionality. (More advanced mathematics can cover vectors in more detail and extend into nonlinear situations.)

TECHNICAL STUFF

One of the major steps of working with vector spaces is to find the *basis* for the vector space, a way of defining how many vectors you need to define *any* point in the entire vector space. For example, a 5-dimensional space has a basis of five vectors. One way to look at superstring theory is to realize that the directions a string can move can only be described with a basis of ten distinct vectors, so the theory describes a 10-dimensional vector space.

Twisting 2-dimensional space in three dimensions: The Mobius strip

In the classic book *Flatland*, the main character is a square (literally — he has four sides of equal length) who gains the ability to experience three dimensions. Having access to three dimensions, you can perform actions on a 2-dimensional surface in ways that seem very counterintuitive. A 2-dimensional surface can actually be twisted in such a way that it has no beginning and no end!

Figure 15-4 shows the best known example of this, the *Mobius strip*. The Mobius strip was created in 1858 by German mathematicians August Ferdinand Mobius and Johann Benedict Listing.

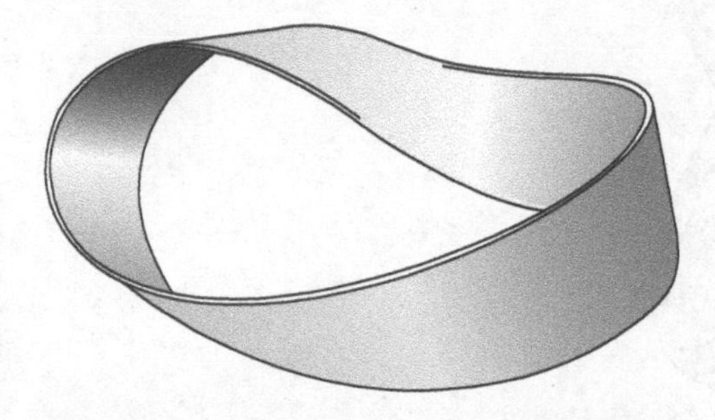

FIGURE 15-4: A Mobius strip is twisted so it has only one continuous surface.

TIP

You can create your own Mobius strip by taking a strip of paper — kind of like a long bookmark — and giving it a half-twist. Then take the two ends of the strip of paper and tape them together. Place a pencil in the middle of the surface and draw a line along the length of the strip without taking your pencil off the paper.

A curious thing happens as you continue along. Without taking your pencil from the paper, you've drawn the line on every part of the surface, and it eventually meets up with itself. There's no "back" of the Mobius strip that somehow avoids the pencil line. You've drawn a line along the entire shape without lifting your pencil.

In mathematical terms (and real ones, given the result of the pencil experiment), the Mobius strip has only one surface (it is called *unorientable*). There's no "inside" and "outside" of the Mobius strip, the way there is on a bracelet. Even though the two shapes may look alike, they are mathematically very different entities.

The Mobius strip does, of course, have an end (or boundary) in terms of its width. In 1882, German mathematician Felix Klein expanded on the Mobius strip idea to create the *Klein bottle:* a shape that has no inside or outside surface, but also has no boundaries in any direction. Take a look at Figure 15-5 to understand the Klein bottle. If you traveled along the "front" of the path (with the x's), you'd eventually reach the "back" of that path (with the o's).

TIP

If you were an ant living on a Mobius strip, you could walk its length and eventually get back to where you started. Walking its width, you'd eventually run into the "edge of the world." An ant living on a Klein bottle, however, could go in any direction and, if it walked long enough, eventually find itself back where it started. (Traveling along the o path eventually leads back to the x's.) The difference between walking on a Klein bottle and walking on a sphere is that the ant wouldn't just walk along the outside of the Klein bottle, like it would on a sphere, but it would cover both surfaces, just like on the Mobius strip.

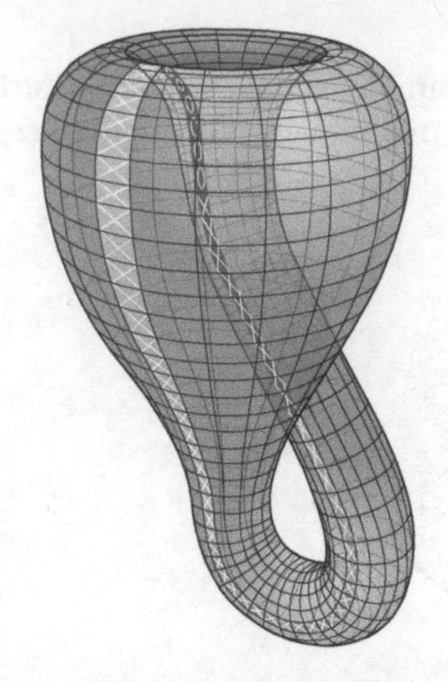

FIGURE 15-5: A Klein bottle has no boundary (edge).

More twists in three dimensions: Non-Euclidean geometry

The fascination with strange warping of space in the 1800s was perhaps nowhere as clear as in the creation of *non-Euclidean geometry*, in which mathematicians began to explore new types of geometry that weren't based on the rules laid out 2,000 years earlier by Euclid. One version of non-Euclidean geometry is Riemannian geometry, but there are others, such as projective geometry.

The reason for the creation of non-Euclidean geometry is based in Euclid's *Elements* itself, in his "fifth postulate," which was much more complex than the first four postulates. The fifth postulate is sometimes called the *parallel postulate*, and though it's worded fairly technically, one consequence is important for string theory's purposes: A pair of parallel lines never intersects.

Well, that's all well and good on a flat surface, but on a sphere, for example, two parallel lines can and do intersect. Lines of longitude — which look parallel to each other under Euclid's definition when they're on a flat map — intersect at both the north and south poles of a globe. Lines of latitude, which also look parallel, don't intersect at all. Mathematicians weren't sure what a "straight line" on a sphere even meant!

One of the greatest mathematicians of the 1800s, Carl Friedrich Gauss, turned his attention to ideas about non-Euclidean geometry. (Some earlier thoughts on the matter had been kicked around over the years by other mathematicians, including Nikolai Lobachevsky and János Bolyai.) Gauss passed the majority of the work off to his former student, Bernhard Riemann.

Riemann worked out how to perform geometry on a curved surface — a field of mathematics called *Riemannian geometry.* Figure 15-6 depicts one consequence of his work — that the angles of a triangle sometimes do *not* add up to 180 degrees. (We encounter this concept in our discussion of the hyperbolic plane in Chapter 13.)

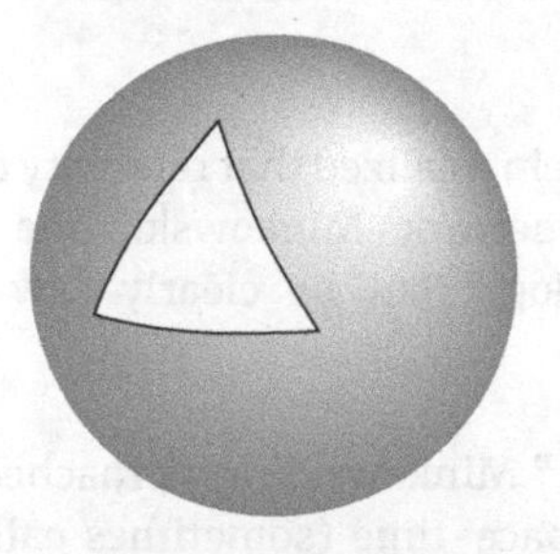

FIGURE 15-6: Sometimes the angles of a triangle don't measure up to 180 degrees.

When Albert Einstein developed general relativity as a theory about the geometry of space-time, it turned out that Riemannian geometry was exactly what he needed.

THE MATHEMATICS OF ARTWORK

Understanding and manipulating space is a key feature of artwork, which often attempts to reflect a 3-dimensional reality on a 2-dimensional surface. This is probably most notable in the work of Pablo Picasso and M. C. Escher, who manipulated space in such a way that the manipulation itself is part of the artistic message.

Most artists try to manipulate space so it's not noticed. One of the most common examples of this is perspective, a technique developed during the Renaissance that involves creating an image that matches the way the eye perceives space and distance. Parallel railroad tracks appear to meet at the horizon, even though they never meet in reality. On a 2-dimensional surface, the basis for the railroad tracks is a triangle that does, in fact, have a corner at the horizon line.

This is the basis of the mathematical field of non-Euclidean geometry called *projective geometry,* where you take one 2-dimensional space and project it in a precise mathematical way onto a second surface. There is an exact 1-to-1 correspondence between the two spaces, even though they look completely different. The two images represent different mathematical ways of looking at the same physical space — one of them an infinite space and one a finite space.

Four Dimensions of Space-Time

In Einstein's general theory of relativity, the three space dimensions connect to a fourth dimension: time. The total package of four dimensions is called *space-time*, and in this framework, gravity is seen as a manifestation of space-time geometry. We tell the story of relativity in Chapter 6, but some dimension-related points are worth revisiting.

Hermann Minkowski, not Albert Einstein, realized that relativity could be expressed in a 4-dimensional space-time framework. Minkowski, one of Einstein's old teachers, had called him a "lazy dog," but he clearly saw the brilliance of relativity.

In a 1908 talk titled "Space and Time," Minkowski first broached the topic of creating a dimensional framework of space-time (sometimes called a "Minkowski space"). The Minkowski diagrams, introduced in Chapter 6, are an attempt to graphically represent this 4-dimensional space on a 2-dimensional Cartesian grid. Each point on the grid is a "space-time event," and understanding the ways these events relate to each other is the goal of analyzing relativity in this way.

TIP

Even though time is a dimension, it's fundamentally different from the space dimensions. Mathematically, you can generally exchange "left" for "up" and end up with results that are fairly consistent. If you exchange "left one meter" for "one hour from now," however, it doesn't work out so well. Minkowski divided the dimensions into *spacelike dimensions* and *timelike dimensions*. One spacelike dimension can be exchanged for another, but it can't be exchanged with a timelike dimension. (In Chapter 17, you find out about some ideas involving extra timelike dimensions in our universe.)

TECHNICAL STUFF

The reason for this distinction is that Einstein's equations are written in such a way that they result in a term defined by the space dimensions squared minus a term defined by the time dimension squared. (Because the terms are squared, each term has to be positive, no matter what the value of the dimension.) The space dimensional values can be exchanged without any mathematical problem, but the minus sign means that the time dimension can't be exchanged with the space dimensions.

Adding More Dimensions to Make a Theory Work

For most interpretations, superstring theory requires a large number of extra space dimensions to be mathematically consistent; M-theory requires ten space dimensions. With the introduction of branes as multidimensional objects in string theory, it becomes possible to construct and imagine wildly creative geometries for space — geometries that correspond to different possible particles and forces. It's unclear, at present, whether those extra dimensions exist in a real sense or are just mathematical artifacts.

REMEMBER

The reason string theory requires extra dimensions is that trying to eliminate them results in much more complicated mathematical equations. It's not impossible (as you see later in this chapter), but most physicists haven't pursued these concepts in a great deal of depth, leaving science (perhaps by default) with a theory that requires many extra dimensions.

As we mention earlier, from the time of Descartes, mathematicians have been able to translate between geometric and physical representations. Mathematicians can tackle their equations in virtually any number of dimensions they choose, even if they can't visually picture what they're talking about.

TIP

One of the tools mathematicians use in exploring higher dimensions is analogy. If you start with a zero-dimensional point and extend it through space, you get a 1-dimensional line. If you take that line and extend it into a second dimension, you end up with a square. If you extend a square through a third dimension, you end up with a cube. If you then were to take a cube and extend it into a fourth dimension, you'd get a shape called a *hypercube*.

A line has two "corners," but extending it to a square gives it four corners, while a cube has eight corners. By continuing to extend this algebraic relationship, you get a hypercube, a 4-dimensional object with 16 corners; a similar relationship can be used to create analogous objects in additional dimensions. Such objects are obviously well outside what our minds can picture.

TIP

Humans aren't psychologically wired to be able to picture more than three space dimensions. A handful of mathematicians (and possibly some physicists) have devoted their lives to the study of extra dimensions so fully that they may be able to actually picture a 4-dimensional object, such as a hypercube. Most mathematicians can't (so don't feel bad if you can't).

Whole fields of mathematics — linear algebra, abstract algebra, topology, knot theory, complex analysis, and others — exist with the sole purpose of trying to take abstract concepts, frequently with large numbers of possible variables, degrees of freedom, or dimensions, and make sense of them.

These sorts of mathematical tools are at the heart of string theory. Regardless of the ultimate success or failure of string theory as a physical model of reality, it has motivated mathematicians to grow and explore new questions in new ways, and for that alone, it has proved useful.

Sending Space and Time on a Bender

Space-time is viewed as a smooth "fabric," but that smooth fabric can be bent and manipulated in various ways. In relativity, gravity bends our four space-time dimensions, but in string theory, more dimensions are bound up in other ways. In relativity and modern cosmology, the universe has an inherent curvature.

The typical approach to string theory's extra dimensions has been to wind them up in a tiny, Planck length–sized shape. This process is called *compactification*. In the 1980s, it was shown that the extra six space dimensions of superstring theory could be compactified into so-called Calabi-Yau spaces.

TECHNICAL STUFF

Since then, other methods of compactification have been offered, most notably G2 compactification, spin-bundle compactification, and flux compactification. For the purposes of this book, the details of the compactification don't matter.

TIP

To picture compactification, think of a garden hose. If you were an ant living on the hose, you'd live on an enormous (but finite) universe. You can walk very far in either of the length directions, but if you go around the curved dimension, you can only go so far. However, to someone very far away, your dimension — which is perfectly expansive at your scale — seems like a very narrow line with no space to move except along the length.

This is the principle of compactification — we can't see the extra universes because they're so small that nothing we can do can ever distinguish them as a complex structure. If we got close enough to the garden hose, we'd realize that something was there, but scientists can't get close to the Planck length to explore extra compactified dimensions.

Of course, some recent theories have proposed that the extra dimensions may be larger than the Planck length and theoretically in the range of experiment.

THE WRAPAROUND UNIVERSE

Some cosmologists have considered some extreme cases of space warping in our own universe, theorizing that the universe may be smaller than we think. A new field of cosmology called *cosmic topology* attempts to use mathematical tools to study the overall shape of the universe.

In his 2008 book, *The Wraparound Universe,* cosmologist Jean-Pierre Luminet proposes the idea that our universe wraps around so it has no particular boundary, sort of like the Klein bottle in Figure 15-5. Any direction you look, you may be seeing an illusion, as if you were standing in a funhouse full of mirrors that appeared to go on forever. Distant stars may actually be closer than expected, but their light travels a larger path along the wraparound universe to reach us.

In this scenario, the horizon problem from Chapter 9 ceases to be an issue because the universe is small enough to have become uniform within the time frame of its existence. Inflation is consistent with the wraparound universe hypothesis, but many of the problems it fixes are solved in other ways.

Still other theories state that our region of the universe manifests only four dimensions, even though the universe as a whole contains more. Other regions of the universe may exhibit additional dimensions. Some radical theories even suppose that the universe as a whole is curved in strange ways.

Are Extra Dimensions Really Necessary?

Though string theory implies extra dimensions, that doesn't mean the extra dimensions need to exist as dimensions of space. Some work has been done to formulate a 4-dimensional string theory where the extra degrees of freedom aren't physical space dimensions; but the results are incredibly complex, and it doesn't seem to have caught on.

Several groups have performed this sort of work because some physicists are uncomfortable with the extra space dimensions that seem to be required by string theory. In the late 1980s, a group worked on an approach called free fermions. Other approaches that avoid introducing additional dimensions include the covariant lattice technique, asymmetric orbifolds, the 4-D *N*-2 string (what's in a name?), and non-geometric compactifications. These are variations on the theme of string theory all run into their own complications and perhaps get less publicity than the original compactification idea. Probably, even among string theorists, the geometric approach of compactifying extra dimensions remains the dominant approach.

One early technically complex (and largely ignored) approach to 4-dimensional string theory was proposed by S. James Gates Jr. of the University of Maryland at College Park (with assistance from Warren Siegel of Stony Brook University's C. N. Yang Institute for Theoretical Physics). This work is by no means the dominant approach to 4-dimensional string theory, but its benefit is that it can be explained and understood (in highly simplified terms) without a doctorate in theoretical physics.

Offering an alternative to multiple dimensions

In his approach, Gates essentially trades dimensions for charges. This creates a sort of dual approach that's mathematically similar to the approach in extra space dimensions, but doesn't actually require the extra space dimensions, nor does it require guessing at compactification techniques to eliminate the extra dimensions.

This idea dates back to a 1938 proposal by British physicist Nicolas Kemmer. Kemmer proposed that the quantum mechanical properties of charge and spin were different manifestations of the same thing. Specifically, he said that the neutron and proton were identical, except that they rotated differently in some extra dimension, which resulted in a charge on the proton and no charge on the neutron. The resulting mathematics, which analyzes the physical properties of these particles, is called an *isotopic charge space* (originally developed by Werner Heisenberg and Wolfgang Pauli, and then used by Kemmer). Though this is an "imaginary space" (meaning the coordinates are unobservable in the usual sense), the resulting mathematics describes properties of protons and neutrons, and is at the foundation of the current Standard Model.

Gates's approach was to take Kemmer's idea in the opposite direction: If you wanted to get rid of extra dimensions, perhaps you could view them as imaginary and think of them as charges. (The word "charge" in this sense doesn't really mean electrical charge, but a new property to be tracked, like "color charge" in quantum chromodynamics.) The result is to view vibrational dimensions of the heterotic string as "left charge" and "right charge."

When Gates applied this concept to the heterotic string, the trade didn't come out even — to give up six space dimensions, he ended up gaining more than 496 right charges!

In fact, together with Siegel, Gates was able to find a version of heterotic string theory that matched these 496 right charges. Furthermore, their solution showed that the left charges would correspond to the family number. (There are three known generations, or families, of leptons, as Figure 8-1 in Chapter 8 shows: the electron, muon, and tau families. The family number indicates which generation the particle belongs to.)

This may explain why there are multiple families of particles in the Standard Model of particle physics. Based on these results, a string theory in four dimensions could require extra particle families! In fact, it would require many more particle families than the three that physicists have seen. These extra families (if they exist) could include particles that make up the unseen dark matter in our universe.

Weighing fewer dimensions against simpler equations

The usefulness of this 4-dimensional approach is hindered by the sheer complexity of the resulting equations that describe the role of all those extra charges. These 4-dimensional constructions have, to date, shown meager predictive power. In a sense, they move all the difficulties of compactifying the six extra dimensions in the complexity of correctly assigning the values of a large number of charges. However, in practice, the techniques to handle compactifications seem more powerful, or are better established, than those to handle the extra charges. Hence, most string theorists stick to the approach of compactification of extra dimensions.

This goes back to the principle of Occam's razor, which says that a scientist shouldn't make a theory unnecessarily complex. The simplest explanation that fits the facts is the one physicists tend to gravitate toward.

In this case, Occam's razor cuts both ways. The simpler mathematical equation of 10-dimensional string theory requires stipulating a large number of space dimensions that no one has ever observed, which would certainly seem to go against Occam's razor. But the type of isotopic charge coordinates used in Gates's approach is exactly the same as the ones that provide the mathematical foundations of the Standard Model — where the isotopic dimensions aren't observed.

In the end, the 4-dimensional approaches to string theory are an intriguing way of understanding how complex string theory can be. One of the most basic aspects of string theory has been the idea that it requires extra space dimensions, but this work shows that string theory doesn't necessarily require even that, provided that an additional structure (the extra charges) is introduced.

IN THIS CHAPTER

» Looking back beyond the big bang theory

» Tying black holes to string theory

» Knowing where the universe has been and where it may be going

» Tackling the question of how the universe supports life

Chapter 16

Our Universe — String Theory, Cosmology, and Astrophysics

Though string theory started as a theory of particle physics, much of the significant theoretical work today is in applying the startling predictions of string theory and M-theory to the field of cosmology. Chapter 9 covers some of the amazing facts scientists have discovered about our universe, especially in the last century.

In this chapter, we return to these same ideas from the background of string theory. We explain how string theory relates to our understanding of the big bang, the theory of the universe's origin. We then discuss what string theory has to say about another mystery of the universe: black holes. From there, we cover what string theory reveals about how the universe changes over time and how it may change in the future. Finally, we return to the question of why the universe seems perfectly tuned to allow for life and what, if anything, string theory (along with the anthropic principle) may have to say about it.

The Start of the Universe with String Theory

According to the traditional big bang theory, if you extrapolate the expanding universe backward in time, the entire known universe would have been compacted down into a singular point of incredibly immense density. In fact, that extrapolation backward in time is the reason the theory developed in the first place. It reveals nothing, however, about whether anything existed a moment before that point. Under the big bang theory — formulated in a universe of quantum physics and relativity — the laws of physics result in meaningless infinities at that moment. String theory may offer some insights into what may have come before and what caused the big bang.

What was before the bang?

String theory offers the possibility that we are "stuck" on a brane with three space dimensions. These brane world scenarios, such as the Randall-Sundrum models, offer the possibility that before the big bang, something was already here: namely, collections of strings and branes. Because these entities occupy dimensions beyond those of our observable universe, our own universe would then be a portion of space and time that came into being through the interaction of these structures in a higher-dimensional space.

The search for an eternal universe

Scientists were originally very upset by the big bang theory because they believed in an eternal universe, meaning the universe had no starting point (and, on average, didn't change over time). This had been the dominant view of the universe since Aristotle, really, even as we moved from a view of static heavens where stars were mounted on celestial spheres, to one where the universe was governed by gravitation.

Einstein initially believed in the idea of a constant and eternal universe, though he did abandon it when evidence suggested otherwise. British astronomer Fred Hoyle devoted most of his career to trying to prove the universe was eternal. Today, some physicists continue to look for ways to explain what, if anything, existed before the big bang.

Some cosmologists say that the question of what happened at or before the big bang is inherently unscientific because science currently has no way of extending its physical theories past the singularity at the dawn of our universe's timeline. It's not even clear that science would *ever* be able to really answer this question, so that's another concern cosmologists have about such matters being within the realm of science. Others point out that if we never ask the questions, we'll never discover a way to answer them.

Though string theory isn't yet ready to answer such questions, that hasn't stopped cosmologists from beginning to ask them and offer possible scenarios. In these scenarios, which are admittedly vague, the pre–big bang universe (which likely isn't confined to only three space dimensions) is a conglomerate of *p*-branes, strings, anti-strings, and anti-*p*-branes. In many cases, these objects are still "out there" somewhere beyond our own 3-brane, perhaps even impacting our own universe (as in the case of the Randall-Sundrum models).

One of these models was a pre–big bang model presented by Gabriele Veneziano — the same physicist who came up with the 1968 dual resonance model that sparked string theory. In this model, our universe is a black hole in a more massive universe of strings and empty space. Prior to the current expansion phase, there was a period of contraction.

Though probably not completely true according to today's major models, this work by Veneziano (and similar ideas by others) has an impact on most of the superstring cosmology work today, because it pictures our known universe as just a subset of the universe, with a vast "out there" beyond our knowledge.

The old-fashioned cyclic universe model

One idea that was popular in the 1930s was that of a *cyclic universe*, in which the matter density was high enough for gravity to overcome the expansion of the universe. The benefit of this model was that it allowed the big bang to be correct, but the universe could still be eternal.

In this cyclic model, the universe would expand until gravity began to pull it back, resulting in a "big crunch" where all matter returned to the primordial "superatom" — and then the cycle of expansion would start all over again.

The problem is, the second law of thermodynamics dictates that the entropy, or disorder, in the universe would grow with each cycle. If the universe went through an infinite number of cycles, the amount of disorder in the universe would be infinite — every bit of the universe would be in thermal equilibrium with every other bit of the universe. In a universe where every region has exactly the same structure, no one region has more order than any other, so all regions have the maximum amount of disorder allowed. (If the universe had gone through a finite number of cycles, scientists still run into the problem of how the whole thing started; they just pushed it back a few cycles. This kind of defeated the whole purpose of the model, so the model assumed an infinite number of cycles.)

String theory, however, might just have a way of bringing back the cyclic model in a new form.

What banged?

The big bang theory doesn't offer any explanation for what started the original expansion of the universe. This is a major theoretical question for cosmologists, and many are applying the concepts of string theory in their attempts to answer it. One controversial conjecture is a cyclic universe model called the *ekpyrotic universe theory*, which suggests that our universe is the result of branes colliding with each other.

The banging of strings

Well before the introduction of M-theory or brane world scenarios, there was a string theory conjecture of why the universe had the number of dimensions we see: A compact space of nine symmetrical space dimensions began expanding in three of those dimensions. Under this analysis, a universe with three space dimensions (like ours) is the most likely space-time geometry. How convenient. Lucky us!

In this idea, initially posed in the 1980s by Robert Brandenberger and Cumrun Vafa, the universe began as a soup of tightly wound strings with all ten dimensions equally confined to the Planck length. Brandenberger and Vafa wanted to turn around the usual logic of string theory: Rather than starting from ten large dimensions and making six of them small, they investigated the possibility of having ten small dimensions and a mechanism that makes four of them big.

Brandenberger and Vafa argued that it was the dynamics of the strings that kept the dimensions tiny, and their interactions may allow some of them to blow up. This expansion would be the result of a collision of strings that could undo each other and pop out of existence. The collision annihilates the strings, which, in turn, unleash the dimensions they were confining. The dimensions thus begin expanding, as in the inflationary and big bang theories.

Instead of thinking about strings and anti-strings (an *anti-string* winds in the opposite direction of a string), picture a room that has a bunch of cables attached to random points on the walls. Imagine that the room wants to expand, with the walls and floor and ceiling trying to move away from each other — but they can't because of the cables. Now imagine that the cables can move, and every time they intersect, they can recombine.

Picture two taut cables stretching from the floor to the ceiling that intersect to form a tall, skinny X. They can recombine to become two loose cables: one attached to the floor and one attached to the ceiling. If these are the only two cables stretching from floor to ceiling, then after this interaction, the floor and ceiling are free to move apart from each other (see Figure 16-1).

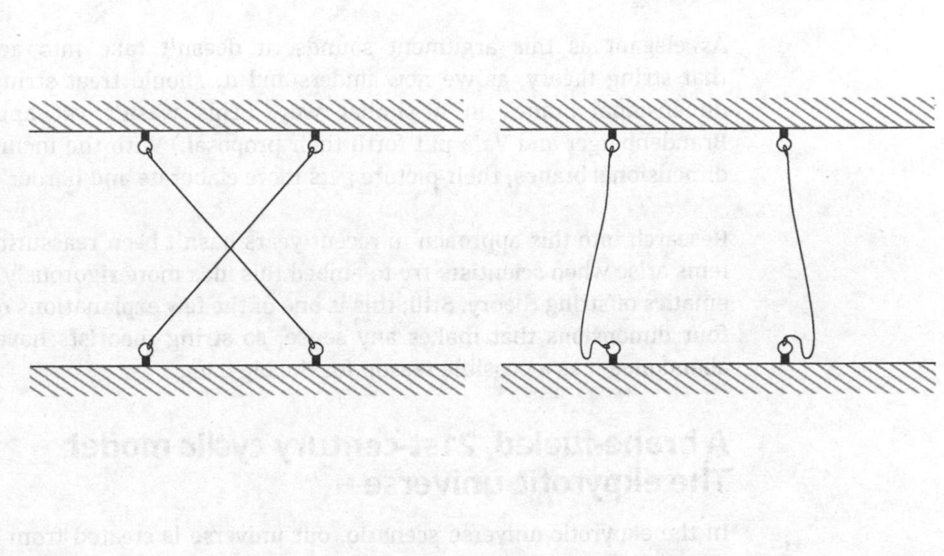

FIGURE 16-1: In this thought experiment, the floor and ceiling want to move away from each other, but they can't because the taut cables are holding them tightly in place in the image on the left. If the two cables can interact where they touch, recombining into two different cables, the new cables will have slack that allows the ceiling and floor to move away from each other until the new cables become taut, as the right image shows.

In the Brandenberger and Vafa scenario, this dimension (up/down), as well as two others, is free to grow large. The final step of the argument is explaining why this works for three space dimensions, but not for more. The answer is quite elegant: In four or more space dimensions, two moving strings will typically never meet.

To understand this, think about two points on a circle moving randomly, either clockwise or counterclockwise: They will probably meet after a while. But two points moving randomly in a square are far less likely to hit each other, and in more and more dimensions, the likelihood of a collision goes down substantially. Physicists and mathematicians can precisely estimate the probability of such a collision occurring (they've been studying this problem for points since the discovery of Brownian motion — see Chapter 8), and this reasoning can be repeated for strings. It turns out that while strings in up to three space dimensions are likely to collide with each other, in four or more dimensions, they are likely to miss each other.

In other words, the very geometry of string theory implies that this scenario would lead to us seeing fewer than four space dimensions: Dimensions of four or more are less likely to go through the string/anti-string collisions required to "liberate" them from the tightly bound configuration. The higher dimensions continue to be bound up by the strings at the Planck length and are therefore unseen.

As elegant as this argument sounds, it doesn't take into account the fact that string theory, as we now understand it, should treat strings and higher-dimensional branes in a similar way. (This wasn't yet appreciated when Brandenberger and Vafa put forth their proposal.) With the inclusion of higher-dimensional branes, their picture gets more elaborate and harder to interpret.

Research into this approach in recent years hasn't been reassuring. Many problems arise when scientists try to embed this idea more rigorously into the mathematics of string theory. Still, this is one of the few explanations of why there are four dimensions that makes any sense, so string theorists haven't completely abandoned it as a possible reason for the big bang.

A brane-fueled, 21st-century cyclic model: The ekpyrotic universe

In the ekpyrotic universe scenario, our universe is created from the collision of two branes. The matter and radiation of our universe comes from the kinetic energy created when the branes collided.

The ekpyrotic universe scenario was proposed in a 2001 paper by Paul Steinhardt of Princeton University, Burt Ovrut of the University of Pennsylvania, and Neil Turok, formerly of Cambridge University and currently at the Perimeter Institute for Theoretical Physics in Waterloo, Ontario, along with Steinhardt's student Justin Khoury.

The theory builds on the ideas in some M-theory brane world scenarios that show the extra dimensions of string theory may be extended, perhaps even infinite in size. They also probably aren't expanding (or at least string theorists have no reason to think they are) the way our own three space dimensions are. When you play the video of the universe backward in time, these dimensions don't contract.

REMEMBER

Imagine that within these dimensions you have two infinite 3-branes. Some mechanism (such as gravity) draws the branes together through the infinite extra dimensions, and they collide with each other. Energy is generated, creating the matter for our universe and pushing the two branes apart. Eventually, the energy from the collision dissipates, and the branes are drawn back together to collide yet again.

The ekpyrotic model is divided into various *epochs* (periods of time), based on which influences dominate:

- The big bang
- The radiation-dominated epoch
- The matter-dominated epoch
- The dark energy-dominated epoch
- The contraction epoch
- The big crunch

The story up until the contraction epoch is essentially identical to the predictions made by regular big bang cosmology. The radiation that's spawned by the brane collision (the big bang) means the radiation-dominated epoch is fairly uniform (save for quantum fluctuations), so inflation may be unnecessary. After about 75,000 years, the universe becomes a particle soup during the matter-dominated epoch. Today and for many years, we are in the dark energy–dominated epoch, until the dark energy decays and the universe begins contracting once again.

Because the theory involves two branes colliding, some have called this the "big splat" theory or the "brane smash" theory, which is certainly easier to pronounce than *ekpyrotic.* The word "ekpyrotic" comes from the Greek word "ekpyrosis," which was an ancient Greek belief that the world was born out of fire. (Burt Ovrut reportedly thought it sounded like a skin disease.)

Some feel the ekpyrotic universe model has a lot going for it — it solves the flatness and horizon problems like inflationary theory does, while also providing an explanation for why the universe started in the first place — but the creators are still far from proving it. One benefit is that this model avoids the problem of previous cyclic models because each universe in the cycle is larger than the one before it. Because the volume of the universe increases, the total entropy of the universe in each cycle can increase without ever reaching a state of maximum entropy.

While this is an appealing theory, it has proven difficult so far to find firm evidence for it. (You may notice that a lack of firm evidence is a common thread for many of these theories).

TIP

There's obviously much more detail to the ekpyrotic model than we've included here. If you're interested in this fascinating theory, we highly recommend Paul J. Steinhardt and Neil Turok's popular book, *Endless Universe: Beyond the Big Bang*. In addition to the lucid and nontechnical discussion of complex scientific concepts, their descriptions offer a glimpse inside the realm of theoretical cosmology, which is well worth the read.

Explaining Black Holes with String Theory

One major mystery of theoretical physics that requires explanation is the behavior of black holes, especially how they evaporate and whether they lose information. We introduce these topics in Chapter 9, but with the concepts of string theory in hand, you may be able to further your understanding of them.

Black holes are defined by general relativity as smooth entities, but at very small scales (such as when they evaporate down to the Planck length in size), quantum effects need to be taken into account. Resolving this inconsistency is the sort of thing that string theory should be good at, if it's true.

String theory and the thermodynamics of a black hole

When Stephen Hawking described the Hawking radiation emitted by a black hole, he had to use his physical and mathematical intuition, because quantum physics and general relativity aren't reconciled. He treated the black hole as approximately classical (in terms of physics) and the radiation it emitted as quantum. One of the major successes of string theory is in offering a complete quantum description of (some) black holes.

REMEMBER

Hawking radiation takes place when radiation is emitted from a black hole, causing it to lose mass. Eventually, the black hole evaporates into nothing (or, according to some, almost nothing).

Stephen Hawking's incomplete argument

Hawking's paper on the way a black hole radiates heat (also called *thermodynamics*) begins a line of reasoning that doesn't quite work all the way through to the end. In the middle of the proof, there's a disconnect because there's no theory of quantum gravity that would allow the first half of his reasoning (based on general relativity) to connect with the second half of his reasoning (based on quantum mechanics).

The reason for the disconnect is that performing a detailed thermodynamics analysis of a black hole involves examining all the possible quantum states of the black hole. But black holes are described with general relativity, which treats them as smooth — not quantum — objects. While this might make sense for a fairly big black hole (and even then, there may be pitfalls), it's certainly troubling in the final stages of the evaporation, when the black hole becomes small. Without a theory of quantum gravity, there seems to be no way to analyze the specific thermodynamic nature of a black hole.

In Hawking's paper, this connection was made by means of his intuition, but not in the sense that most of us probably think of intuition. The intuitive leap he took was in proposing precise mathematical formulas, called *greybody factors,* even though he couldn't absolutely prove where they came from.

Most physicists agree that Hawking's interpretation makes sense, but a theory of quantum gravity would show whether a more precise process could take the place of his intuitive leap.

String theory may complete the argument

Work by Andrew Strominger and Cumrun Vafa on the thermodynamics of black holes is seen by many string theorists as the most powerful evidence in support of string theory. By studying a problem that is mathematically equivalent to black holes — a dual problem — they precisely calculated the black hole's thermodynamic properties in a way that matched Hawking's analysis.

TIP

Sometimes, instead of simplifying a problem directly, you can create a *dual problem,* which is essentially identical to the one you're trying to solve but is much simpler to handle. Strominger and Vafa used this tactic in 1996 to calculate the entropy in a black hole.

In their case, they found that the dual problem of a black hole described a collection of 1-branes and 5-branes. These "brane constructions" are objects that can be defined in terms of quantum mechanics. They found that the results matched precisely with the result Hawking anticipated 20 years earlier.

Now, before you get too excited, the Strominger and Vafa results work only for very specific types of black holes, called *extremal black holes*. These extremal black holes have the maximum amount of electric or magnetic charge that's allowed without making the black hole unstable. An extremal black hole has the odd property of possessing entropy but no heat or temperature. (*Entropy* is a measure of disorder, often related to heat energy, within a physical system.)

At the same time Strominger and Vafa were performing their calculations, Princeton student Juan Maldacena was tackling the same problem (along with thesis advisor Curtis Callan). Within a few weeks of Strominger and Vafa, they had confirmed the results and extended the analysis to black holes that are *almost* extremal. Again, the relationship holds up quite well between these brane constructions and black holes, and analyzing the brane constructions yields the results Hawking anticipated for black holes. Further work has expanded this analysis to even more generalized cases of black holes.

REMEMBER

To get this analysis to work, gravity has to be turned down to zero, which certainly seems strange in the case of an object that's, quite literally, defined by gravity. Turning off the gravity is needed to simplify the equations and obtain the relationship. String theorists conjecture that by ramping up the gravity again, you'd end up with a black hole, but string theory skeptics point out that without gravity, you really don't have a black hole.

Still, even a skeptic can't help but think that there must be some sort of relationship between the brane constructions and black holes because they both follow the Hawking thermodynamics analysis created 20 years earlier. What's even more amazing is that string theory wasn't designed to solve this specific problem, yet it did. The fact that the result falls out of the analysis is impressive, to say the least.

String theory and the black hole information paradox

One of the important aspects of the thermodynamics of black holes relates to the *black hole information paradox*. This paradox may well have a solution in string theory, either in the string theory analyses described in the previous section or in the holographic principle.

Hawking had said that if an object falls into a black hole, the only information that's retained are the quantum mechanical properties of mass, spin, and charge. All other information is stripped away.

The problem with this is, quantum mechanics is built on the idea that information can't be lost. If information can be lost, then quantum mechanics isn't a secure theoretical structure. Hawking, as a relativist, was more concerned about maintaining the theoretical structure of general relativity, so he was okay with the information being lost if it had to be.

REMEMBER

The fact that information is conserved at a fundamental level is one of the main assumptions of quantum mechanics. All the successes derived from quantum theory, from early atomic spectroscopy to the latest particle collider experiments, are built on this bedrock. It would be extremely hard to violate this fundamental principle and still come up with a consistent quantum theory that explains all those phenomena.

In 2004, after a debate that lasted more than 20 years, Hawking announced that he no longer believed that any black hole information was forever lost to the universe. In admitting this, he lost a 1997 bet with physicist John Preskill. The payoff was a baseball encyclopedia, from which information could be retrieved easily. Who said physicists don't have a sense of humor?

One reason for Hawking's change of mind was that he redid some of his earlier calculations and found it was possible that as an object fell into a black hole, it would disturb the black hole's radiation field. The information about the object could seep out, though probably in mangled form, through the fluctuations in this field.

Another way to approach the problem of black hole information loss is through the holographic principle of Gerard 't Hooft and Leonard Susskind, or the related AdS/CFT correspondence developed by Juan Maldacena. (Both principles are discussed in Chapter 13.) If these principles hold for black holes, it may be possible that all the information within a black hole is also encoded in some form on the surface area of the black hole.

The controversy over the black hole information paradox is described in detail in Susskind's 2008 book, *The Black Hole War: My Battle with Stephen Hawking to Make the World Safe for Quantum Mechanics*.

The Evolution of the Universe

Other questions that scientists hope string theory can answer involve the way the universe changes over time. The brane world scenarios described earlier in this book offer some possibilities, as do the various concepts of a multiverse. Specifically, string theorists hope to understand the reason for the increased expansion of our universe as defined by dark matter and energy.

The swelling continues: Eternal inflation

Some cosmologists have worked hard on a theory called *eternal inflation*, which helps contribute to the idea of a vast multiverse of possible universes, each with different laws (or different solutions to the same law, to be precise). In eternal inflation, island universes spring up and disappear throughout the universe, spawned by the very quantum fluctuations of the vacuum energy itself. This is seen by many as further evidence for the string theory landscape and the application of the anthropic principle.

The *inflation theory* says that our universe began on a hill (or ledge) of potential vacuum energies. Here, the vacuum energy is the fuel for the expansion of the universe. The universe began to roll down that hill, burning the vacuum energy and expanding exponentially fast, until we settled into a valley of vacuum energy and expansion slowed down considerably. The question that eternal inflation tries to answer is, *why did we start on that hill?*

Seemingly, the universe began with a random starting point on the spectrum of possible energies, so it's only through sheer luck that we were on the hill, and in turn, luck caused us to go through the right amount of inflation to distribute mass and energy the way it's distributed.

Or, alternately, there are a vast number of possibilities, many of which spring into existence, and we can possibly exist only in the ones that have this specific starting condition. (This is, in essence, the *anthropic principle*.)

In either case, the particles and forces of our universe are determined by the initial location on that hill and the laws of physics that govern how the universe will change over time.

In 1977, Sidney Coleman and Frank De Luccia described how quantum fluctuations in an inflating universe create tiny bubbles in the fabric of space-time. These bubbles can be treated as small universes in their own right. For now, the key is that they do form. Some of these bubbles will undergo inflation, and some will not; moreover, those that undergo inflation will themselves create new bubbles, which sets off a never-ending process.

Cosmologist Andrei Linde has been the one to most strenuously argue that this finding, in combination with Alan Guth's inflationary universe theory, demands eternal inflation — the creation of a vast population of universes, each with slightly different physical properties. He has been joined by Guth himself and Alexander Vilenkin, who helped hammer out the key aspects of the theory.

REMEMBER

The eternal inflation model says that these *bubble universes* (Guth prefers "pocket universes," while Susskind calls them "island universes") spring up, somehow getting physical laws among the possible solutions dictated by the string theory landscape (through some as-yet-unknown means). The bubble universe then undergoes inflation. Meanwhile, the space around it continues to expand — and it expands so quickly that information about the inflating bubble universe can never reach another universe. Our own universe is one of these bubble universes, but one that finished its inflationary period long ago.

The hidden matter and energy

Two mysteries of our universe are dark matter and dark energy (Chapter 9 contains the basics about these concepts). *Dark matter* is unseen matter that holds stars together in galaxies, while *dark energy* is unseen vacuum energy that pushes different galaxies farther apart from each other. String theory holds several possibilities for both.

A stringy look at dark matter

String theory provides a natural candidate for dark matter in supersymmetric particles, which are needed to make the theory work but which scientists have never observed. Alternatively, it's possible that dark matter somehow results from the gravitational influence of nearby branes.

Probably the simplest explanation of dark matter would be a vast sea of supersymmetric particles residing inside galaxies, but we can't see them (presumably because of some unknown properties of these new particles). Supersymmetry implies that every particle science knows about has a superpartner (see Chapter 10 if you need a refresher on supersymmetry). Fermions have bosonic superpartners, and bosons have fermionic superpartners. In fact, one popular candidate for the missing dark matter is the photino, the superpartner of the photon.

A computer simulation reported in the journal *Nature* in November 2008 offers one possible means of testing this idea. The simulation, performed by the international Virgo Consortium research group, suggests that dark matter in the halo of the Milky Way galaxy should produce detectable levels of gamma rays. This simulation indicates a direction to start looking for such telltale signs, at least.

Another possible dark matter candidate comes from the various brane world scenarios. Though the details still have to be worked out, it's possible that there are branes that overlap with our own 3-brane. Perhaps where we have galaxies, there are gravitational objects that extend into other branes. Because gravity is the one force that can interact across the branes, it's possible that these hyperdimensional objects create added gravity within our own 3-brane.

Finally, the 4-dimensional string theories discussed in Chapter 15 present yet another possibility because they require not only supersymmetry but a vast number of families of particles beyond the electron, muon, and tau families in our current Standard Model. Bringing string theory down to four dimensions seems to greatly expand the number of particles that physicists would expect to find in the universe, and (if they exist) these could account for dark matter.

A stringy look at dark energy

Even more intriguing than dark matter is dark energy, which is a positive energy that seems to permeate the entire universe and to be much more abundant than either ordinary matter or dark matter — but also much less abundant than physicists think it *should* be. Recent discoveries in string theory have allowed for this dark energy to exist within the theory.

Although string theory offers some (perhaps too many) possibilities for dark matter, it offers little explanation for dark energy. Theoretically, dark energy should be explained by the value of the vacuum energy in particle physics, where particles are continually created and destroyed. These quantum fluctuations grow immensely, leading to infinite values. We explain in Chapter 8 that to avoid these infinite values in quantum field theory, the process of renormalization is used. What is a little less obvious is that, after renormalization, some quantities are still expected to be large, even if finite. The energy of the vacuum is one such quantity.

REMEMBER

When physicists try to use their standard methods, based on the framework of quantum field theory and renormalization, to compute the value of the vacuum energy, they get a value that's off from the experimental value of dark energy by 10^{120}!

The real value is incredibly small, but not quite zero. Though the amount of dark energy in the universe is vast (according to recent data, it makes up about 73 percent of the universe), the intensity of dark energy is very small — so small that until 1998, scientists assumed the value was exactly zero. (The fact that a tiny density of dark energy results in a large total energy is a consequence of the fact that the universe is quite big indeed!)

The existence of dark energy (or a positive cosmological constant, depending on how you want to look at it) doesn't remove the many solutions of string theory relating to different possible physical laws. The number of solutions that include dark energy may be on the order of 10^{500}, or potentially much larger. This dark energy reflects a positive energy built into the very fabric of the universe, likely related to the energy of the vacuum itself.

To some, the ekpyrotic universe has an advantage over the inflationary model because it offers a reason for why we might observe such a value for dark energy in our universe: That's the part of the cyclic phase we're in. At times in the past, the dark energy may have been stronger, and at times in the future, it may be weaker. To many others, this reason isn't any more intellectually satisfying than the lack of a reason in other cosmological models. It still amounts to an accidental coincidence (or an application of the anthropic principle, discussed later in this chapter).

Outside of the ekpyrotic universe, there's little explanation for what's going on. The problem of offsetting the expected vacuum energy by such a large amount — enough to *almost*, but not quite, cancel it out — is seen by many physicists as too much chance to contemplate.

Many would rather turn to the anthropic principle to explain it. Others see that as waving a white flag of surrender, admitting that dark energy is just too tough a challenge to figure out.

The Undiscovered Country: The Future of the Cosmos

In cosmology, the past and the future are linked together, and the explanation for one is tied to the explanation of the other. With the big bang model in place, there are essentially three possible futures for our universe. Determining the solutions to string theory that apply to our universe might allow us to determine which future is most likely.

A universe of ice: The big freeze

In this model of the universe's future, the universe continues to expand forever. Energy slowly dissipates across a wider and wider volume of space, and eventually, the result is a vast, cold expanse of space as the stars die. This *big freeze* has always had some degree of popularity, dating back to the rise of thermodynamics in the 1800s.

The laws of thermodynamics tell us that the entropy, or disorder, in a system will always increase. This means that the heat will spread out. In the context of cosmology, it means that the stars will die and their energy will radiate outward. In this "heat death," the universe becomes a static soup of near-absolute zero energy. The universe as a whole reaches a state of thermal equilibrium, meaning that nothing interesting can really happen.

A slightly different version of the big freeze model is based on the more recent discovery of dark energy. In this case, the repulsive gravity of dark energy will cause clusters of a galaxy to move apart from each other, while on the smaller scale, those clusters will gather closer together, eventually forming one large galaxy.

Over time, the universe will be populated by large galaxies that are extremely far apart from each other. The galaxies will become inhospitable to life, and the other galaxies will be too far away to even see. This variant, sometimes called a "cold death," is another way the universe could end as a frozen wasteland. (This time-scale is incredibly vast, and humans likely won't still exist. So, no need to panic.)

From point to point: The big crunch

One model for the future of the universe is that the mass density of the universe is high enough that the attractive gravity will eventually overpower the repulsive gravity of dark energy. In this *big crunch* model, the universe contracts back into a microscopic point of mass.

This idea of a big crunch was once a very popular notion, but with the discovery of the repulsive dark energy, it seems to have gone out of favor. Because physicists are observing the expansion rate increase, it's unlikely that there's enough matter to overcome that and pull it all back together.

A new beginning: The big bounce

The ekpyrotic model (see the earlier section "A brane-fueled, 21st-century cyclic model: The ekpyrotic universe") brings the big crunch back, but with a twist. When the crunch occurs, the universe once again goes through a big bang period. This isn't the only model that allows for such a *big bounce* cyclic model.

In the ekpyrotic model, the universe goes through a series of big bangs, followed by expansion and then a contracting big crunch. The cycle repeats over and over, presumably without any beginning or end. Cyclic models of the universe are not original, going back not only to 1930s physics but also to some religions, including certain interpretations of Hinduism.

It turns out that string theory's major competitor — loop quantum gravity (explained in Chapter 19) — may also present a big bounce picture. The method of loop quantum gravity is to *quantize* (break up into discrete units) space-time itself. This avoids a singularity at the formation of the universe, which means it's possible that time extends back before the big bang moment. In such a picture, a big bounce scenario is possible.

Exploring a Finely Tuned Universe

One major issue in cosmology for years has been the apparent fine-tuning seen in our universe. The universe seems specially crafted to allow life. One of the major explanations for this is the anthropic principle, which many string theorists have recently begun adopting. Many physicists still feel that the anthropic principle is a poor substitute for an explanation of why these physical properties must have the values they do.

To a physicist, the universe looks as if it were made for the creation of life. British Astronomer Royal Martin Rees clearly illuminated this situation in his 1999 book, *Just Six Numbers: The Deep Forces That Shape the Universe*. In this book, Rees points out that there are many values — the intensity of dark energy, gravity, electromagnetic forces, atomic binding energies, to name just a few — that would, if different by even an extremely small amount, result in a universe that is inhospitable to life as we know it. (In some cases, the universe would have collapsed only moments after its creation, resulting in a universe inhospitable for *any* form of life.)

The goal of science has always been to explain why nature has to have these values. This idea was once posed by Einstein's famous question: *Did God have a choice in creating the universe?*

TIP

Einstein's religious views were complex, but what he meant by this question wasn't actually so much religious as scientific. In other words, he was wondering if there was a fundamental reason — buried in the laws of nature — why the universe turned out the way it did.

For years, scientists sought to explain the way the universe worked in terms of fundamental principles that dictate the way the universe was formed. However, with string theory (and eternal inflation), that very process has resulted in answers that imply the existence of a vast number of universes and a vast number of scientific laws that could be applied in those universes.

The major success of the anthropic principle is that it provided one of the only predictions for a small, but positive, cosmological constant prior to the discovery of dark energy. This was put forward in the 1986 book *The Anthropic Cosmological Principle*, by John D. Barrow and Frank J. Tipler, and cosmologists in the 1980s appeared to be at least open-minded about the possibility of using anthropic reasoning.

Nobel laureate Steven Weinberg made the big case for anthropic reasoning in 1987. Analyzing details of how the universe formed, he realized two things:

- If the cosmological constant were negative, the universe would quickly collapse.
- If the cosmological constant were slightly larger than the experimentally possible value, matter would have been pushed apart too quickly for galaxies to form.

In other words, Weinberg realized that if scientists based their analysis on what was required to make life possible, then the cosmological constant couldn't be negative and had to be very small. There was no reason, in his analysis, for it to be exactly zero. A little over a decade later, astronomers discovered dark energy, which fit the cosmological constant in precisely the range specified by Weinberg. Martin Rees appealed to this type of discovery in his explanation of how the laws in our universe end up with such finely tuned values, including the cosmological constant.

You may wonder if there's anything particularly anthropic about Weinberg's reasoning, however. You only have to look around to realize that the universe didn't collapse and galaxies were able to form. It seems like this argument can be made just by observation.

The problem is that physicists are looking not only to determine the properties of our universe, but to explain them from a set of rules as small as possible. To use this reasoning to explain the special status of our universe (that is, it contains us) requires something very important — a large number of *other* universes, most of which have properties that make them significantly different from us.

TIP

For an analogy, consider a scenario where you're driving along and get a flat tire. If you were the only person who had ever gotten a flat tire, you might be tempted to explain the reason why you, out of everyone on the planet, were the one to get the flat tire. Knowing that many people get flat tires every day, no further explanation is needed — you just happen to have been in one of many cars that happened to get a flat tire.

If there is only one universe, then having the fine-tuned numbers that Rees and others note is a miraculously fortunate turn of events. If there are billions of universes, each with random laws from hundreds of billions (or more) possible laws from the string theory landscape, then every once in a while a universe like ours will be created. No further explanation is necessary.

REMEMBER

The problem with the anthropic principle is that it tends to be a last resort for physicists. Scientists turn to the anthropic principle only when more conventional methods of argument have failed them, and the second they can come up with a different explanation, they abandon it.

This isn't to imply that the scientists applying the anthropic principle are anything but sincere. Those who adopt it seem to believe that the vast string theory landscape — realized in a multiverse of possible universes— can be used to explain the properties of our universe.

IN THIS CHAPTER

» Scientists are still trying to figure out why we travel through time

» Tricking time with relativity

» Need more time? Considering the possibility of a second time dimension

Chapter 17

Have Time, Will Travel

One of the most fascinating concepts in science fiction is the idea of traveling forward or backward in time, as in H. G. Wells's classic story *The Time Machine*. Scientists haven't been able to build a time machine yet, but some physicists believe it may someday be possible — and some (probably most) believe it will *never* be possible.

Time travel exists in physics because of possible solutions to Einstein's general theory of relativity, mostly resulting in singularities. These singularities would be eliminated by string theory, so in a universe where string theory dictates the physical laws, time travel will probably not be allowed — a result that many physicists find quite favorable to the alternative (though far less interesting).

In this chapter, we explore the notion of time and our travel through it — both in the normal, day-to-day method and in more unusual, speculative methods. We discuss the scientific meaning of time, in both classical terms and from the standpoint of special relativity. One possible method of time travel involves using cosmic strings. There's a possibility, which we explore, that more than one time dimension may exist. Finally, we explain one scenario for creating a physically plausible (though probably practically impossible) time machine using wormholes.

Temporal Mechanics 101: How Time Flies

We move through time every single day, and most of us don't even think about how fascinating that is. Scientists who have thought about it have constantly run into trouble in figuring out exactly what time means because time is such an abstract concept. It's something we're intimately familiar with, but so familiar that we almost never have to analyze it in a meaningful way.

Over the years, our view of time — both individually and from a scientific standpoint — has changed dramatically, from an intuition about the passage of events to a fundamental component of the mathematical geometry that describes the universe.

The arrow of time: A one-way ticket

Physicists refer to the one-way motion through time (into the future and never back to the past) using the phrase "arrow of time," introduced by Arthur Eddington (the guy who helped confirm general relativity) in his 1928 book, *The Nature of the Physical World.* The first note he makes on this concept is that "time's arrow" points in one direction, as opposed to directions in space, where you can reorient as needed. He then points out three key ideas about the arrow of time:

- Human consciousness inherently recognizes the direction of time.
- Even without memory, the world makes sense only if the arrow of time points into the future.
- In physics, the only place the direction of time shows up is in the behavior of a large number of particles, in the form of the second law of thermodynamics. (See the nearby sidebar, "Time asymmetries," for a clarification of the exceptions to this.)

The conscious recognition of time is the first (and most significant) evidence any of us has about the direction we travel in time. Our minds (along with the rest of us) "move" sequentially in one direction through time, and most definitely not in the other. The neural pathways form in our brains, which retain this record of events. In our minds, the past and future are distinctly different. The past is static and unchanging, but the future is fully undetermined (at least so far as our brains know).

As Eddington pointed out, even if you didn't retain any sort of memory, logic would dictate that the past must have happened before the future. This is probably true, although whether we could conceptualize a universe in which time flowed from the future to the past is a question that's open for debate.

Finally, though, we reach the physics of the situation: the second law of thermodynamics. According to this law, as time progresses, no *closed system* (that is, a system that isn't gaining energy from outside the system) can lose *entropy* (disorder) as time progresses. In other words, as time goes on, it's not possible for a closed system to become more orderly all by itself.

TIP

Intuitively, this is certainly the case. Consider a house that's been abandoned: It will grow disordered over time. For it to become more orderly, there has to be an introduction of work from outside the system. Someone has to mow the yard, clean the gutters, paint the walls, and so on. (This analogy isn't perfect because the abandoned house isn't a totally closed system. It gets energy and influence from outside — sunlight, animals, rainfall, and so on — but you get the idea.)

REMEMBER

In physics, the arrow of time is the direction in which entropy (disorder) increases. It's the direction of decay.

Oddly, these same ideas (the same in spirit, though not scientific) date all the way back to St. Augustine of Alexandria's *Confessions*, written in 400 BCE, where he said:

> What then is time? If no one asks me, I know: if I wish to explain it to one that asketh, I know not: yet I say boldly that I know, that if nothing passed away, time were not; and if nothing were, time present were not.

What Augustine is pointing out here is the inherent problem in explaining the slippery nature of time. We know exactly what time is — in fact, we are unable *not* to understand how it flows in our own lives — but when we try to define it in precise terms, it eludes us. His words "if nothing passed away, time were not" could, in a sense, describe how the second law of thermodynamics defines time's arrow. We know time passes because things change in a certain way as time passes.

Relativity, worldlines, and worldsheets: Moving through space-time

Understanding how time travel works within string theory would require a complete understanding of how the fabric of space-time behaves within the theory. So far, string theory hasn't exactly figured that out.

In general relativity, the motion of objects through space-time is described by a *worldline*. In string theory, scientists talk about strings (and branes) creating entire *worldsheets* as they move through space-time.

TECHNICAL STUFF

TIME ASYMMETRIES

Arthur Eddington's third observation about the arrow of time indicates that physical laws actually ignore the direction of time, except for the second law of thermodynamics. What this means is that if you take the time *t* in any physics equation and replace it with a time *–t*, and then perform the calculations to describe what takes place, you'll end up with a result that makes sense.

For gravity, electromagnetism, and the strong nuclear force, changing the sign on the time variable (called *T-symmetry*) allows the laws of physics to work perfectly well. In some special cases related to the weak nuclear force, this actually turns out not to be the case.

There is actually a larger type of symmetry, called *CPT symmetry,* which is always preserved. The C stands for *charge-conjugation symmetry,* which means that positive and negative charges switch. The P stands for *parity symmetry,* which involves basically replacing a particle for a complete mirror image — a particle that has been flipped across all three space dimensions. (CPT symmetry is a property of quantum theory in our 4-dimensional space-time, so at present we are ignoring the other six dimensions proposed by string theory.)

The total CPT symmetry, it turns out, appears to be preserved in nature. This is both an experimental fact as well as a mathematical property baked into the framework of quantum field theory. Notably, CPT symmetry is also one of the few cases of unbroken symmetry in our universe. In other words, an *exact* mirror image of our universe — one with all matter swapped for antimatter, reflected in all spatial directions, and traveling backward in time — would obey physical laws that are identical to those of our own universe in every conceivable way.

If CP symmetry is violated, then there must be a corresponding break in T-symmetry, so the total CPT symmetry is preserved. In fact, the handful of processes that violate T-symmetry are called *CP violations* (because the CP violation is easier to test than a violation in the time-reversal symmetry).

Worldlines were originally constructed by Hermann Minkowski when he created his Minkowski diagrams (refer to Figure 6-3 in Chapter 6). Similar diagrams return in the form of Feynman diagrams (refer to Figure 8-3 in Chapter 8), which demonstrate the worldlines of particles as they interact with each other through the exchange of gauge bosons.

In string theory, instead of the straight worldlines of point particles, it is the movement of strings through space-time that interests scientists, as the right side of Figure 17-1 shows.

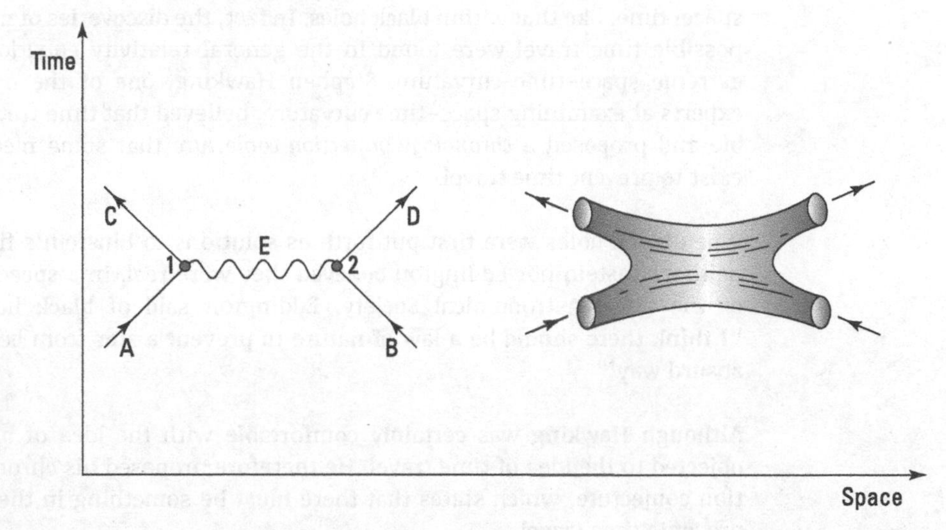

FIGURE 17-1: Instead of a worldline (left), a string creates a worldsheet (right) when it moves through space.

TIP

Notice that in the original Feynman diagram, shown on the left of Figure 17-1, there are sharp points where the worldlines intersect (representing the point where the particles interact). In the worldsheet, the virtual string exchanged between the two original strings creates a smooth curve that has no sharp points. This equates to the fact that string theory contains no infinities in the description of this interaction, as opposed to pure quantum field theory. (Removing the infinities in quantum field theory requires renormalization, which is problematic for the gravitational interaction.)

REMEMBER

One problem with both quantum field theory and string theory is that they are constructed in a way that gets placed inside the space-time coordinate system. General relativity, on the other hand, depicts a universe in which the space-time is dynamic. String theorists hope string theory will solve this conflict between the background-dependent quantum field theory and the background-independent general relativity so that, eventually, dynamic space-time will be derived out of string theory. One criticism (discussed in Chapter 18) is that string theory is, at present, still by and large background-dependent.

The competing theory, loop quantum gravity, incorporates space into the theory but is still mounted on a background of time coordinates. Loop quantum gravity is covered in more detail in Chapter 19.

Hawking's chronology protection conjecture: You're not going anywhere

The concept of time travel is often closely tied to infinities in the curvature of space-time, like that within black holes. In fact, the discoveries of mathematically possible time travel were found in the general relativity equations containing extreme space-time curvature. Stephen Hawking, one of the most renowned experts at examining space-time curvature, believed that time travel is impossible and proposed a *chronology protection conjecture* that some mechanism must exist to prevent time travel.

When black holes were first put forth as solutions to Einstein's field equations, neither Einstein nor Eddington believed they were real. In a speech to members of the Royal Astronomical Society, Eddington said of black hole formation, "I think there should be a law of nature to prevent a star from behaving in this absurd way!"

Although Hawking was certainly comfortable with the idea of black holes, he objected to the idea of time travel. He therefore proposed his chronology protection conjecture, which states that there must be something in the universe that prevents time travel.

Hawking's occasional collaborator, Oxford physicist Roger Penrose, made the much more guarded claim that all singularities would be protected by an event horizon, which would shield them from direct interaction with our normal space-time. His proposal, known as the *cosmic censorship conjecture*, would also potentially prevent many forms of time travel from being accessible to the universe at large.

One major reason time travel causes so much trouble for physics (and must therefore be prohibited, according to Einstein and Hawking) is that you could create a way of generating an infinite amount of energy. Say you have a portal into the past and shine a laser into it. You set up mirrors so the light coming out of the portal is deflected back around to go into the portal again, in tandem with the original beam you set up.

Now the total intensity of light coming out of the portal (in the past) would be (or would've been) twice the original laser light going in. This laser light is sent back through the portal, yielding an output of four times as much light as originally transmitted. This process could be continued, resulting in literally an infinite amount of energy being created instantaneously.

Obviously, such a situation is just one of many examples why physicists tend to doubt the possibility of time travel (with a few notable exceptions, which we cover throughout this chapter). If time travel is possible, then the predictive power of physics is lost because the initial conditions are no longer trustworthy! The predictions based on those conditions would therefore be completely meaningless.

Slowing Time to a Standstill with Relativity

In physics, time travel is closely linked to Einstein's theory of relativity, which allows motion in space to actually alter the flow of time. This effect, known as *time dilation*, was one of the earliest predictions of relativity. This sort of time travel is completely allowed by the known laws of physics, and it has been measured in many experiments, but it allows travel only into the future, not into the past.

In this section we explore the special cases in relativity that imply that time travel — or at least altered motion through time — may in fact be possible. Skip ahead to "General Relativity and Wormholes: Doorways in Space and Time" for more information about how the general theory of relativity relates to potential time travel.

TIP

Time dilation and black hole event horizons, both of which we explain in the following sections, present intriguing ways of extending human life. In science fiction, they've long provided the means for allowing humans to live long enough to travel from star to star (see the sidebar "The science fiction of time" at the end of this chapter.)

Time dilation: Sometimes even the best watches run slow

The most evident case of time acting oddly in relativity, and one that has been experimentally verified, is the concept of time dilation under special relativity. *Time dilation* is the idea that as you move through space, time itself is measured differently for the moving object than for the unmoving object. For motion that is near the speed of light with respect to some other unmoving reference, this effect is noticeable and allows a way to travel into the future faster than we normally do.

One experiment that confirms this strange behavior is based on unstable particles, pions and muons. Physicists know how quickly the particles would decay if they were sitting still, but when they bombard Earth in the form of cosmic rays, they're moving very quickly. Their decay rates don't match the predictions, but if you apply special relativity and consider time from the particle's point of view, the decay rate comes out as expected.

In fact, time dilation is confirmed by a number of experiments. In the Hafele-Keating experiments of 1971, atomic clocks (which are *very* precise) were flown on airplanes traveling in opposite directions. The time differences shown on the clocks, as a result of their relative motion, precisely matched the predictions from relativity. Also, global positioning system (GPS) satellites have to compensate for time dilation to function properly (and keep your cell phone's GPS on track). So time dilation is on very solid scientific ground.

TIP

Time dilation leads to one familiar notion of time travel. If you were to get into a spaceship that traveled very quickly away from Earth, time inside the ship would slow down in comparison to that on Earth. You could do a flyby of a nearby star and return to Earth at nearly the speed of light, and a few years would pass on Earth while possibly only a few weeks or months would pass for you, depending on how fast you were going and how far away the star was.

The biggest problem with this is how to accelerate a spaceship up to those speeds. Scientists and science fiction authors have made various proposals for such devices, but all are well outside the range of what we could feasibly build today or in the foreseeable future.

REMEMBER

It is hard to accelerate anything but the tiniest objects to close to the speed of light c. The speed of light is very large, so it takes a lot of energy to even get too close to it. Moreover, as you accelerate an object to high speeds, its *relativistic mass* also increases, which means it takes more and more energy to keep accelerating it: It takes much more energy to get from 10% to 20% of c than it does to go from 0% to 10%. This formula of relativistic mass increase is similar to the formula that describes time dilation, so that it takes an enormous energy to get close to c, where time-dilation becomes significant.

The question is how much time dilation you really need, though, especially for trips within only a few light-years of Earth.

Black hole event horizons: An extra-slow version of slow motion

One other case where time slows down, this time in general relativity, involves black holes. Recall that a black hole bends space-time itself, to the point where even light can't escape. This bending of space-time means that as you approach a black hole, time will slow down for you relative to the outside world.

If you were approaching a black hole and we were watching from far away (and could somehow watch "instantly," without worrying about the time lag from light speed), we would see you approach the black hole, slow down, and eventually come to rest to hover outside it. Through the window of our spaceship, we would see you sitting absolutely still.

You, on the other hand, wouldn't notice anything in particular — at least until the intense gravity of the black hole killed you. But until then, it certainly wouldn't "feel" like time was moving differently. You'd have no idea that as you glided past the black hole's event horizon (which you possibly wouldn't even notice), thousands of years were passing outside the black hole. These ideas are neatly illustrated in the movie *Interstellar* (see "The science fiction of time" sidebar at the end of this chapter).

As you find out in the next section, some believe that black holes may actually provide a means to more impressive forms of time travel as well.

General Relativity and Wormholes: Doorways in Space and Time

In general relativity, the fabric of space-time can occasionally allow for worldlines that create a *closed timelike curve*, which is relativity-speak for time travel. Einstein himself explored these concepts when he was developing general relativity but never made much progress on them. In the following years, solutions allowing for time travel were discovered.

The first application of general relativity to time travel was by Scottish physicist W. J. van Stockum in 1937. Van Stockum imagined (in mathematical form, because that's how physicists imagine things) an infinitely long, extremely dense rotating cylinder, like an endless barber's pole. It turns out that in this case, the dense cylinder actually drags space-time with it, creating a space-time whirlpool.

REMEMBER

This space-time whirlpool is an example of a phenomenon called *frame dragging*, which takes place when an object "drags" space (and time) along with it. This frame dragging is in addition to the normal bending of space-time due to gravity and is due to the movement of incredibly dense objects in space, such as neutron stars. It's similar to how an electric mixer causes the cake batter surrounding its beaters to swirl. This effect is frequently exploited to come up with time travel solutions.

In van Stockum's situation, you could fly up to the cylinder in a spaceship and set a course around the cylinder and arrive back at a point in time before you arrived at the cylinder. In other words, you can travel into the past along a closed timelike curve. (If you can't picture this path, don't feel bad. The path is in four dimensions, after all, and results in going backward in time, so it's clearly something our brains didn't evolve to picture.)

Another theory about time travel was proposed in 1949 by mathematician Kurt Gödel, Einstein's colleague and friend at Princeton University's Institute for Advanced Study. Gödel considered the situation where all of space — the entire universe itself — was actually rotating. You might ask, if *everything* was rotating, how would we ever know it? Well, it turns out that if the universe was rotating, then according to general relativity, we'd see laser beams curve slightly as they move through space (beyond the normal gravitational lensing, where gravity bends light beams).

The solution that Gödel arrived at was disturbing because it allowed time travel. It was possible to create a path in a rotating universe that ended before it began. In Gödel's rotating universe, the universe itself could function as a time machine.

TIP

So far, physicists haven't found any conclusive evidence that our universe is rotating. In fact, the evidence points overwhelmingly toward the idea that it's not. But even if the universe as a whole doesn't rotate, objects in it certainly do.

Taking a shortcut through space and time with a wormhole

In a solution called an *Einstein-Rosen bridge* (more commonly called a *wormhole* — see Figure 17-2), two points in space-time can be connected by a shortened path. In some special cases, a wormhole may actually allow for time travel. Instead of connecting different regions of space, the wormhole could connect different regions of time!

Wormholes were studied by Albert Einstein and his pupil Nathan Rosen in 1935. (Ludwig Flamm had first proposed them in 1916.) In this model, the singularity at the center of a black hole is connected to another singularity, which results in a theoretical object called a *white hole*.

TIP

While a black hole draws matter into it, a white hole spits matter out. Mathematically, a white hole is a time-reversed black hole. Because no one has ever observed a white hole, it's probable that they don't exist, but they are allowed by the equations of general relativity and haven't been completely ruled out yet.

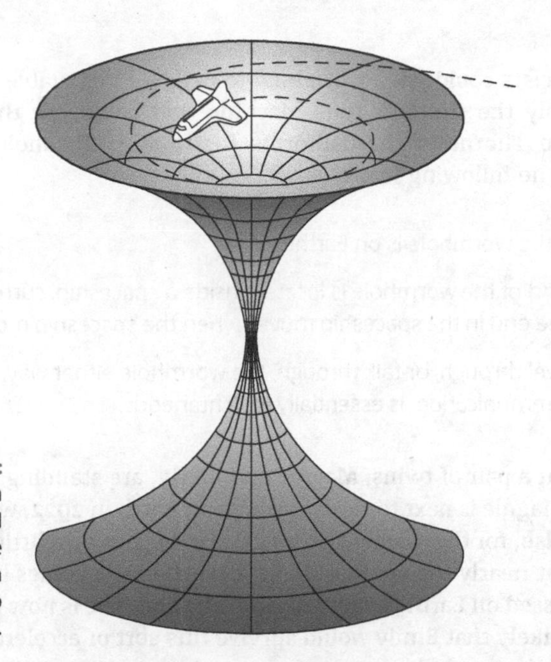

FIGURE 17-2: Traveling into a wormhole could get you from one location in space-time to another.

An object falling into a black hole could travel through the wormhole and come out the white hole on the other side in another region of space. Einstein argued that there were two flaws with using a wormhole for time travel:

- A wormhole is so unstable that it would collapse in upon itself almost instantaneously.
- Any object going into a black hole would be ripped apart by the intense gravitational force inside the black hole and would never make it out the other side.

Then, in 1963, New Zealand mathematician Roy Kerr calculated an exact solution for Einstein's field equations, representing a *Kerr black hole.* The special feature of a Kerr black hole is that it rotates. So far as scientists know, *all* objects in the universe rotate, including stars, so when a star collapses into a black hole, it's likely that it too will rotate.

In Kerr's solution, it's actually possible to travel through the rotating black hole and miss the singularity at the center, so you could come out the other side. The problem is, again, that the black hole would likely collapse as you're going through it. (We address this problem in the next section.)

Assuming physicists could get a wormhole to be large and stable enough to pass through, probably the simplest time machine that could use this method was theorized by Kip Thorne of the California Institute of Technology. Consider a wormhole with the following features:

- One end of the wormhole is on Earth.
- The other end of the wormhole is located inside a spaceship, currently at rest on Earth. The end in the spaceship moves when the spaceship moves.
- You can travel through, or talk through, the wormhole either way, and such travel, or communication, is essentially instantaneous.

Now assume that a pair of twins, Maggie and Emily, are standing at either end of the wormhole. Maggie is next to the wormhole on Earth in 2022, while Emily is on the spaceship (also, for the moment, in 2022). Emily goes on a little jaunt for a few days, traveling at nearly the speed of light, but when she comes back, thousands of years have passed on Earth because of time dilation (she is now in 5022). (Realistically, it's unlikely that Emily would survive this sort of acceleration, but we're imagining a stable wormhole that can be moved around, so we'll just assume for this scenario that the engineers of this amazing spaceship have come up with some way to keep Emily from being crushed by the acceleration and deceleration process.)

On Maggie's side of the wormhole (still 2022), only a few days have passed. In fact, the twins have regularly been sending messages about the strange sights Emily has witnessed over the few days of her journey. Emily (in 5022) is able to go through the wormhole to Maggie's location (in 2022), and voilà, she has traveled back in time thousands of years!

In fact, now that Emily has gone to the trouble of setting up the portal, Maggie (or anyone else) could just as easily travel from 2022 to 5022 (or vice versa) simply by stepping through it.

REMEMBER

Since Thorne's model, physicists have developed several wormhole-based time-travel scenarios. In fact, some physicists have shown that if a wormhole exists, it *has* to allow travel in time as well as space.

Overcoming a wormhole's instability with negative energy

The problem with using wormholes to travel in space or time is that they are inherently unstable. When a particle enters a wormhole, it creates fluctuations that cause the structure to collapse in upon itself. There are theories that a wormhole could be held open by some form of *negative energy*, which represents a case where the energy density (energy per volume) of space is actually negative.

Under these theories, if a sufficient quantity of negative energy could be employed, it might continue to hold the wormhole open while objects pass through it. This would be an absolute necessity for any of the previously discussed theories that allow a wormhole to become a time portal, but it isn't obvious how to achieve this.

Recently, physicist Juan Maldacena and his collaborators at Princeton have considered physical setups that would allow for stable wormholes. They have proposed that the negative energy needed to make a wormhole stable could be provided by quantum effects due to matter (in particular, fermions). However, while this type of wormhole would allow an alternate route of travel through space-time, it wouldn't allow time travel. In fact, it wouldn't be a shortcut at all! In the cases studied by Maldacena and his collaborators, going through the wormhole would take *longer* than traveling through ordinary space-time.

In some other models, it may be possible to relate dark energy and negative energy (both exhibit a form of repulsive gravity, even though dark energy is a positive energy), but these models are highly contrived. The good news (if you see possible time travel as good news) is that our universe appears to have dark energy in abundance, although the problem is that it looks like it's evenly distributed throughout the universe.

String theory and string-inspired models can provide potential sources of negative energy, but even in these cases, there's no guarantee that stable wormholes can occur. For instance, in certain models of the universe inspired by strings and branes, like the one proposed by Lisa Randall and Raman Sundrum (see Chapter 12), it seems to be possible to construct wormholes large and stable enough to allow a person to go through. However, this would be possible only if there was little energy in the universe overall (because energy makes wormholes unstable). The amount of radiation currently traveling around our universe would be more than sufficient to collapse any such wormhole.

All in all, constructing a stable wormhole that allows time travel seems exceedingly difficult, and perhaps impossible. This may seem disappointing, but most physicists would breathe a sigh of relief knowing that the paradoxes of time travel are averted.

Crossing Cosmic Strings to Allow Time Travel

Cosmic strings are theoretical objects that predate string theory, but in recent years, there's been some speculation that they may actually be enlarged strings left over from the big bang, or possibly the result of branes colliding. There has also been speculation that they can be used to create a time machine.

Regardless of their origin, if cosmic strings exist, they should have an immense amount of gravitational pull, which means they can cause frame dragging. In 1991, J. Richard Gott (who, with William Hiscock, solved Einstein's field equations for cosmic strings in 1985) realized that two cosmic strings could actually allow time travel.

The way this works is that two cosmic strings cross paths with each other in a certain way, moving at very high speeds. A spaceship traveling along the curves could take a very precise path (several of which were worked out by Curt Cutler in the months after Gott's publication) and arrive at its starting position, in both space and time, allowing for travel in time. Like other time machines, the spaceship couldn't travel farther back than when the cosmic strings originally got in position to allow the travel — in essence, the time travel is limited to when the cosmic string time machine was activated.

Gott's was the second time machine (following Kip Thorne's) to be published in a major journal in the early 1990s, and it sparked a wave of work in the area of time travel. In May 1991, Gott was featured in *Time* magazine. In the summer of 1992, physicists held a conference on time travel at the Aspen Center for Physics (the same place where, nearly a decade earlier, John Schwarz and Michael Green had determined that string theory could be consistent).

When Gott proposed this model, cosmic strings were thought to have nothing to do with string theory. In recent years, physicists have come to believe that cosmic strings, if they exist, may actually be very closely related to string theory.

A Two-Timing Science: String Theory Makes More Time Dimensions Possible

Because relativity showed time as one dimension of space-time and string theory predicts extra space dimensions, a natural question would be whether string theory also predicts (or at least allows for) extra time dimensions. According to physicist Itzhak Bars, this may actually be the case, in a field he calls *two-time physics*. Though still a marginal approach to string theory, understanding this potential extra dimension of time could lead to radically new insights into the nature of time.

Adding a new time dimension

With one time dimension, you have the arrow of time, but with *two* time dimensions, things become less clear. Given two points along a single time dimension, there's only one path between them. With two time dimensions, two points can potentially be connected by a number of different paths, some of which could loop back on themselves, creating a route into the past.

Most physicists have never looked into this possibility, for the simple fact that (in addition to making no logical sense) it wreaks havoc with the mathematical equations. Time dimensions have a negative sign, and if you incorporate even more of them, you can end up with negative probabilities of something happening, which is physically meaningless.

However, Itzhak Bars of the University of Southern California in Los Angeles discovered in 1995 that M-theory allowed for the addition of an extra dimension — as long as that extra dimension was timelike.

To get this to make any sense, Bars had to apply another type of gauge symmetry, which placed a constraint on the way objects could move. As he explored the equations, he realized that this gauge symmetry worked only if there were two extra dimensions: one extra time dimension and one extra space dimension. *Two-time relativity* has four space dimensions and two time dimensions, for a total of six dimensions. *Two-time M-theory*, on the other hand, ends up with 13 total dimensions: 11 space dimensions and two time dimensions.

The gauge symmetry that Bars introduced provided exactly the constraint he needed to eliminate time travel and negative probabilities from his theory. With his gauge symmetry in place, a world with six (or 13) dimensions should behave exactly like a world with four (or 11) dimensions.

Reflecting two-time physics onto a one-time universe

In a 2006 paper, Bars showed that the Standard Model is a shadow of his 6-dimensional theory. Just like a 2-dimensional shadow of a 3-dimensional object can vary depending on where the light source is placed, the 4-dimensional physical properties ("shadows") can be caused by the behavior of the 6-dimensional objects. The objects in the extra dimensions of Bars's two-time physics theory can have multiple shadows in the 4-dimensional universe (like ours), each of which corresponds to different phenomena. Different physical phenomena in our universe can result from the same fundamental 6-dimensional objects, manifesting in different ways.

TIP

To see how this works, consider a particle moving through empty space in six dimensions, with absolutely no forces affecting it. According to Bars's calculations, such activity in six dimensions relates to at least two shadows (two physical representations of this 6-dimensional reality) in the 4-dimensional world:

- An electron orbiting an atom
- A particle in an expanding universe

Bars believes that two-time physics can explain a puzzle in the Standard Model. Some parameters describing quantum chromodynamics (QCD) have been measured to be quite small, meaning that certain types of interactions are favored over others, but nobody knows why this is. Physicists have come up with a possible fix, but it involves predicting a new theoretical particle called an *axion*, which has never been observed.

According to Bars's predictions, two-time physics presents a 4-dimensional world in which QCD interactions are not at all lopsided, so the axion isn't needed. Unfortunately, the lack of discovery of an axion isn't really enough to be counted as experimental proof of two-time physics.

For that, Bars has applied two-time physics to supersymmetry. In this case, the superpartners predicted have slightly different properties than the superpartners predicted by other theories. If superpartners with the properties Bars suggests are observed at the Large Hadron Collider, it would be considered intriguing experimental evidence in favor of his claims. However, this hasn't happened, or at least not yet, and much more work would be needed to sway most physicists toward a "two-time universe" scenario.

Does two-time physics have any real applications?

Most physicists believe that these extra-dimensional results from Bars are just mathematical artifacts. However, history has shown that "mathematical artifacts" can frequently have a real existence. The first glimpse that antimatter might exist was, after all, a negative charge showing up in math equations. Bars himself seems to believe that his extra dimensions have as much physical reality as the four dimensions that we know exist, although we'll never experience these extra dimensions as directly.

Though two-time physics doesn't directly imply any time travel, if it's true, it means that time is inherently more complex than physicists have previously believed. Unraveling the mystery of two-time physics could well introduce new ways that time travel might manifest in our universe.

Sending Messages through Time

The original string theory, bosonic string theory, contained a massless particle called the *tachyon*, which travels faster than the speed of light. In Chapter 10, we explain how these particles are usually a sign that a theory has an inherent flaw — but what if they actually exist? Would they allow a means of time travel?

The short answer is that no one knows. The presence of tachyons in a theory means that things begin to go haywire, which is why physicists consider them to be a sign of fundamental instabilities in the theory. (These instabilities in string theory were fixed by including supersymmetry, creating superstring theory — see Chapter 10.)

If tachyons existed, then in theory it would be possible to send messages that travel faster than the speed of light. These particles could actually travel backward in time and, in principle, be detected.

To avoid this problem (because, remember, time travel can destroy all of physics!), physicist Gerald Feinberg presented the *Feinberg reinterpretation principle* in 1967, which says that a tachyon traveling back in time can be reinterpreted, under quantum field theory, as a tachyon moving forward in time. In other words, detecting tachyons is the same as emitting tachyons. There's just no way to tell the difference, which would make sending and receiving messages fairly challenging.

THE SCIENCE FICTION OF TIME

Talking about time travel without mentioning science fiction would leave an elephant in the chapter, so to speak. Here are some key science fiction novels and films related to the time travel concepts discussed in this chapter, although the list is by no means complete. Spoiler alert: Some plot details are revealed in the descriptions below . . . but these are all classics that have been around for years, so you're probably safe

Novels:

- *The Time Machine,* **by H. G. Wells (1895):** The first story with a man-made device to travel in time, where the travel was under the control of the traveler (as opposed to stories that preceded it like "Rip Van Winkle," *A Connecticut Yankee in King Arthur's Court,* and *A Christmas Carol,* where the time traveler had no control).
- *Tau Zero,* **by Poul Anderson (1967):** A spaceship is trapped accelerating closer and closer to the speed of light, unable to decelerate. The novel explores the effects of time dilation and the possible end of the universe.
- *Gateway,* **by Frederick Pohl (1977):** The sole survivor of a space accident has to come to terms with intense survivor's guilt for the crew he left behind. The plot's powerful climax (which we may now be ruining by telling you) relates to the idea that as you fall into a black hole, time slows down.

Films:

- *Somewhere in Time* **(1980):** Richard Collier (Christopher Reeve) is a playwright who travels to 1912 from 1980. The film takes the stance that the past has already happened and Collier was already part of the events of the past (or he's hallucinating, in which case this film has nothing to do with time travel and is far less interesting). For example, "before" he ever time travels, he finds his own signature in a hotel guest book from 1912. Based on a novel by Richard Matheson.
- *Back to the Future* **(1985):** Marty McFly (Michael J. Fox) travels from 1985 to 1955 and interferes with his parents' first meeting. The film explores the concept of time paradoxes and potential multiple timelines. There were two sequels, but the original film was by far the best.
- *Frequency* **(2000):** New York detective John Sullivan (Jim Caviezel) begins communicating with his father (Dennis Quaid) 30 years in the past over a ham radio, which is bouncing signals off strange sunspot activity. In this film, no material objects travel in time — only information in the form of radio waves. String theorist and author Brian Greene served as physics consultant and had a cameo in the film.

Not only do science fiction authors learn from scientists in developing their time-travel systems, but inspiration can flow the other way. Dr. Ronald Mallet, who is trying to build a time machine, was motivated throughout his life by science fiction accounts of time travel.

Nobel Prize-winning theoretical physicist Kip Thorne has developed his theories of time travel out of helping friends work out the details of their science fiction novels. (His Nobel was for his work detecting gravity waves.) His first theory on time travel was based on work performed to help astronomer Carl Sagan develop a realistic wormhole for his science fiction novel *Contact* in the 1980s.

Thorne later gained insights from science fiction author Robert Forward, culminating in his work on the 2014 science fiction epic *Interstellar,* which was praised widely among scientists (though certainly not universally) for having a scientifically accurate depiction of time travel. High praise indeed!

5

What the Other Guys Say: Criticisms and Alternatives

IN THIS PART . . .

Understand the major criticisms of string theory.

Consider loop quantum gravity and other avenues of research.

Discover additional ways to explain the nature of the universe.

IN THIS CHAPTER

» Considering what string theory does and doesn't explain

» Realizing that string theory may never explain our universe

» Should string theorists control physics departments and research funding?

Chapter 18
Taking a Closer Look at the String Theory Controversy

Although many physicists believe that string theory holds promise as the most likely theory of quantum gravity, there's a growing skepticism among some that string theory hasn't achieved the goals it set out for. The major thrust of the criticism is that whatever useful benefits there are to studying string theory, it's not actually a fundamental theory of reality, but only a useful approximation at best.

String theorists acknowledge some of these criticisms as valid and dismiss others as premature or even completely contrived. Whether or not the critics are right, they've been a part of string theory since the very first days and are likely to be around as long as the theory persists. Over recent years, the criticism has risen to such furor that it's being called "the string wars" across many science blogs and magazines.

In this chapter, we discuss some of the major criticisms of string theory. We begin with a brief recap of the history of string theory, from the eye of the skeptic, who focuses on the failures instead of the successes. After that, we look into whether string theory has the ability to actually provide any solid predictions about the universe. Next, you see how string theory critics object to the extreme amount of control that string theorists would hold over academic institutions and research plans. We then consider whether string theory possibly describes our own reality. And finally, we explain some of the major string theory responses to these criticisms and try to give our own view of the debate.

The String Wars: Outlining the Arguments

As long as it's been around, string theory has contended with criticisms. Some of string theory's early critics are among the most respected members of the physics community, including Nobel laureates such as Sheldon Glashow and the late Richard Feynman, both of whom were skeptical as far back as the first superstring revolution in the mid-1980s. Still, string theory has steadily grown in popularity for decades.

Recently, the criticisms against string theory have spilled into the popular media, making their way into cover stories featured in science magazines and even long articles in mainstream publications. The debate has been raging across podcasts, the internet, scientific conferences, the blogosphere, and anywhere else debates are allowed to rage.

Though the debate sounds passionate, none of the critics are really advocating that physicists completely abandon string theory. Instead, they tend to view string theory as a toy model for quantum gravity rather than a truly *fundamental theory* that describes the most basic level of reality itself. They are especially critical of string theorists' attempts to continue to promote the theory as a fundamental theory of reality.

REMEMBER

Here are some of the most significant criticisms levied against string theory (or the string theorists who espouse it):

- String theory is unable to make any useful prediction about how the physical world behaves, so it can't be falsified or verified.
- String theory is so vaguely defined and lacking in basic physical principles that any idea can be incorporated into it.
- String theorists put too much weight on the opinions of leaders and authorities within their own ranks, as opposed to seeking experimental verification.
- String theorists present their work in ways that falsely demonstrate that they've achieved more success than they actually have. (This isn't necessarily

an accusation of lying, but it may be a fundamental flaw in how success is measured by string theorists and the scientific community at large.)

- String theory gets more funding and academic support than other theoretical approaches (in large part because of the aforementioned reported progress).
- String theory doesn't describe our universe, but contradicts known facts of physical reality in a number of ways, requiring elaborate hypothetical constructions that have never been successfully demonstrated.

Behind many of these criticisms is the assumption that string theory, which has been around for over 50 years, should be a bit more fully developed than it actually is. None of the critics are arguing that the study of string theory should be abandoned; they generally want alternative theories to be pursued with greater intensity because of their belief that string theory is falling short of the mark.

To explore the validity of these claims and determine whether string theory is in fact unraveling, it's necessary to frame the debate by looking at where string theory has been and where it is today.

50 years and counting: Framing the debate from the skeptic's point of view

Even now, with criticism on the rise, it doesn't appear that the study of string theory and string-inspired models has been in decline. To help you understand why some physicists continue to study string theory, and why other physicists believe it isn't delivering as promised, we briefly recount the general trends in the history of string theory, focusing this time on its shortcomings. (This material is presented in significantly greater detail in Chapters 10 through 13.)

String theory started in 1968 as a theory (called the dual resonance model) to predict the interactions of hadrons (protons and neutrons), but failed at that. Instead of this model, quantum chromodynamics, which said that hadrons were composed of quarks held together by gluons, proved to be the correct model.

Analysis of the early version of string theory showed that it could be viewed as very tiny strings vibrating. In fact, this bosonic string theory had several flaws: Fermions couldn't exist, and the theory contained 25 space dimensions, tachyons, and too many massless particles.

These problems were "fixed" with the addition of supersymmetry, which transformed bosonic string theory into superstring theory. Superstring theory still contained nine space dimensions, though, so most physicists still believed it had no physical reality.

This new version of string theory was shown to contain a massless spin-2 particle that could be the graviton. Now, instead of a theory of hadron interactions, string theory was a theory of quantum gravity. But most physicists were exploring other theories of quantum gravity, and string theory languished throughout the 1970s.

The first superstring revolution took place in the mid-1980s, when physicists showed ways to construct string theory that made all the anomalies go away. In other words, string theory was shown to be consistent. In addition, physicists found ways to compactify the extra six space dimensions by curling them up into complex shapes that were so tiny they would never be observed.

The increased work on string theory had great results. In fact, the results were too good: Physicists discovered five distinct variations of string theory, each of which predicted different phenomena in the universe and none of which precisely matched our own.

In 1995, Edward Witten proposed that the five versions of string theory were different low-energy approximations of a single theory, called M-theory. This new theory contained ten space dimensions and strange objects called branes, which had more dimensions than strings.

A major success of string theory was achieved when it was used to construct a description for black holes that calculated the entropy correctly, according to the Bekenstein-Hawking predictions for black hole thermodynamics. This description applied only to specific types of simplified black holes, although there was some indication that the work might extend to more general black holes.

Building on the ideas of Jacob Bekenstein and Stephen Hawking, Gerard 't Hooft and Leonard Susskind formulated the holographic principle, stating that a theory with gravity in a D-dimensional universe is equivalent to a theory without gravity in one dimension less. This mind-bending idea was put on firm ground by Juan Maldacena within the framework of string theory, leading to the AdS/CFT correspondence (see Chapter 13).

A problem for string theory arose in 1998, when astrophysicists showed that the universe was expanding. In other words, the cosmological constant of the universe is positive, but all work in string theory had assumed a negative cosmological constant. (The positive cosmological constant is commonly referred to as dark energy.)

In 2003, a method was found to construct string theory in a universe that has dark energy, but there was a major problem with it: A vast number of distinct string theories were possible. Some estimates have indicated there are as many as 10^{500}

distinct ways to formulate the theory, which is such an absurdly large number that it can be treated as if it were basically infinity.

In response to these findings, Susskind proposed the application of the anthropic principle as a means of explaining why our universe had the properties it did, given the incredibly large number of possible configurations, which Susskind called the landscape.

This brings us to the current status of string theory, in very broad strokes. You can probably see some chinks in the theory's armor, where the criticisms seem to resonate particularly strongly.

A rise of criticisms

After evidence of dark energy was discovered in 1998 and the 2003 work increased the number of known solutions, the criticisms of string theory seemed to swell. The attempts to make the theory fit physical reality were growing a bit more strained in the eyes of some, and a discontent that had always existed under the surface began to seep out of the back rooms at physics conferences and onto the covers of major science magazines.

While innovative new variants — such as the Randall-Sundrum models and the incorporation of a positive cosmological constant — were rightly recognized as brilliant, some people believed that physicists had to come up with contrived explanations to keep the theory viable.

The growth in criticism became glaringly obvious to the general public in 2006 with the publication of two books criticizing — or outright attacking — string theory: Lee Smolin's *The Trouble with Physics: The Rise of String Theory, the Fall of a Science, and What Comes Next* and Peter Woit's *Not Even Wrong: The Failure of String Theory and the Search for Unity in Physical Law*. These books, along with the media fervor that accompanies any potential clash of ideas, put string theory on the public relations defensive even while many string theorists dismiss the Smolin and Woit claims as failed attempts to discredit string theory for their own aggrandizement.

The truth is likely somewhere in between. The criticisms have a bit more merit than some string theorists would give them, but aren't quite as destructive as Woit, at least, would have readers believe. (Smolin is a bit more sympathetic toward string theory, despite his book's subtitle.) None of the critics propose abandoning string theory entirely; they merely would like to see more scientists pursuing other areas of inquiry, such as those described in Chapters 19 and 20.

Is String Theory Scientific?

The first two criticisms cut to the core of whether string theory is successful as a scientific theory. Not every idea, not even one that's expressed in mathematical terms, is scientific. In the past, to be scientific, a theory had to describe something that is happening in our own universe. A theory that strays too far from this boundary enters the realm of speculation. Criticisms of string theory as unscientific tend to fall into two (seemingly contradictory) categories:

- String theory explains nothing.
- String theory explains too much.

Argument No. 1: String theory explains nothing

The first attack on string theory is that after about 40 years of investigation, it still makes no clear predictions. (Physicists would say it has no *predictive power.*) The theory makes no unique prediction that, if true, supports the theory and, if false, refutes the theory.

This is particularly striking in contrast to the historical evolution of particle physics, when theoreticians were able to confidently state something like, "A new particle with certain properties and a mass within this range must exist; otherwise, this model is false." This meant that experimentalists could devise an experiment to try to rule out the model. The experiment might be complicated, taking years to complete and possibly even requiring some new insight or technological development, but it was clear what type of evidence they would look for to disprove the prediction.

According to philosopher Sir Karl Popper, the trait of "falsifiability" is the defining trait of science. If a theory isn't falsifiable — if there is no way to make a prediction that gets a false result — then the theory isn't scientific.

If you subscribe to Popper's view (and many scientists don't), then string theory is certainly not scientific — at least not yet. The question is whether string theory is fundamentally unable to make a clear, falsifiable prediction or whether it merely hasn't done so yet, but will at some point in the future.

It's possible that string theorists will make a distinct prediction at some point. Part of the criticism, though, is that they really don't seem concerned about making a prediction. Some string theorists don't even seem to consider the lack of a currently testable prediction to be a shortcoming, so long as string theory remains consistent with the known evidence and continues producing interesting mathematical structures and insights.

This is what motivates the major critics of string theory, from Feynman in the 1980s to Smolin and Woit more recently, to complain that string theory has no contact with experiment and is fundamentally warping what it means to investigate something scientifically.

Argument No. 2: String theory explains too much

The second attack is based on the same problem — that string theory makes no unique prediction — but the emphasis this time is on the word "unique." There are so many implementations of string theory that even if it could be formulated in a way that it would make a prediction, it seems as if each version of string theory would make a slightly different prediction.

This is, in a way, almost worse than making no prediction at all. With no prediction, you can make the argument that more work and refinement needs to be done, new mathematical tools developed, and so on. With a nearly infinite number of predictions, you're stuck with a theory that's completely useless. Again, it has no predictive power, for the simple reason that you can never sort out the sheer volume of results.

Part of this argument relates back to the principle of Occam's razor. According to this principle, there is an *economy* in nature, which means that nature (as described by science) doesn't include things that aren't necessary. String theory includes extra dimensions, new types of particles, and possibly whole extra universes that have never been observed (and possibly *can never* be observed). String theorists argue that these things are necessary because the theory doesn't work without them . . . but concerns about the lack of other evidence for these things are a reasonable criticism.

New rules for the game: The anthropic principle revisited

One solution for so many predictions, proposed by physicist Leonard Susskind, is to apply the anthropic principle to focus on the regions of the string theory landscape that allow life to exist. According to Susskind, Earth clearly exists in a universe (or a region of the universe, at least) that allows life to exist, so selecting only theories that allow life to exist seems to be a reasonable strategy.

IS STRING THEORY SO DIFFERENT FROM QUANTUM FIELD THEORY?

The criticism that string theory allows for too many implementations, only one of which is correct, may seem a little bit inappropriate. After all, the same is true about quantum field theory. There are an infinite number of models that we can cook up within quantum field theory, with different types of particles and different rules for their interactions. And even when the particles and interactions are fixed, as is the case in the Standard Model, we still need to make an experiment to fix the parameters of the theory, like the mass of the particles. In this sense, quantum field theory is a framework that allows us to construct many models. Why shouldn't we think of string theory in the exact same way?

A big difference is that string theory initially appeared as a very rigid framework, which raised hopes of nailing down a single theory of everything. Another is that after it became clear that a huge landscape of solutions to string theory may exist, it seemed very hard to find a principle to organize them and fish out the correct one. This is in contrast to quantum field theory, where by now physicists have developed an intuition on how to construct models that provide an effective description of many phenomena.

Taking a theory that doesn't allow life to exist and considering it on equal footing with theories that do allow life to exist, when we know that life *does* exist, defies both scientific reasoning and common sense.

From this stance, the anthropic principle is a way of removing selection bias when looking at different possible string theories. Instead of looking only at the mathematical viability of a theory as if that were the only criteria, physicists can also make a selection based on the fact that we live here.

However, there's a bit of clever maneuvering within this discussion that shouldn't go unmentioned. It's not just that Susskind has said that we can use the anthropic principle to select which theories are viable in our universe, but he's gone further, indicating that the very fact that all these versions of string theory exist is a *good* thing. It provides a richness to the theory, making it more robust.

For nearly two decades, many physicists were trying to find a single version of string theory that included basic physical principles that dictated the nature of the universe. The properties of the fundamental particles of the Standard Model had to be measured in experiments and placed into the theory by hand. Part of the goal of string theory was to find a theory that, based on pure physical principles and mathematical elegance, would yield a single theory describing all of reality, including all those properties.

Instead, string theorists have found a virtually infinite number of different models (or, to be more precise, different string theory solutions) and have apparently discovered that no fundamental law describes the universe based on pure physical principles. Selection of the correct parameters for the theory is, once again, left to experiment.

But instead of interpreting this as a failure and indicating that we have no choice but to apply the anthropic principle to provide limitations on which options are available to us, Susskind takes lemons and turns them into lemonade by reframing the entire context of success. Success is no longer finding a single theory, but exploring as much of the landscape as possible.

In their book *Aristotle and an Aardvark go to Washington: Understanding Political Doublespeak Through Philosophy and Jokes,* authors Thomas Cathcart and Daniel Klein refer to this technique as the "Texas sharpshooter fallacy." Imagine the Texas sharpshooter who pulls out a pistol and fires at the wall and then walks up and draws the bull's-eye around the location where the shots landed.

In a (very critical) sense, this is what Susskind has done by changing the actual definition of success in string theory. He has (according to some) redefined the goal of the enterprise in such a way that the current work is exactly in line with the new goal. If this new approach is valid, yielding a way to correctly describe nature, it's brilliant. If it's not valid, then it's not brilliant. (For a more favorable interpretation of the anthropic principle, see Chapters 12 and 16.)

A similar moving target can be seen in the discussion of proton decay. Originally, experiments to prove grand unified theories (GUTs) anticipated that these experiments would detect the decay of a few protons every year. No proton decays have been found, however, which has caused theorists to revise their calculations to arrive at a lower decay rate. Except most physicists believe that these attempts are not valid and that these GUT approaches have been disproved. This after-the-fact change in what they're looking for isn't a valid approach to science — unless the decays are discovered at the new rate, of course (at which point the theoretical modification becomes a brilliant insight).

None of this is to imply that Susskind is being dishonest or manipulative in presenting the anthropic principle as an option that he believes in. He has very genuinely been led to this belief because of the growing number of mathematically viable string theory solutions, which leave him with no choice (except for abandoning string theory, which we get to in a bit).

After you accept that string theory dictates a large number of possible solutions, and you realize that modern theories of eternal inflation dictate that many of these solutions may well be borne out in some reality, there's very little choice, in Susskind's view, other than to accept the anthropic principle. And there's every indication that he went through some serious soul-searching before deciding to preach the anthropic message (so to speak).

Interpreting the string theory landscape

No longer are string theorists looking for a single theory, but they're now trying to pare down the vast options in the landscape to find the one, or the handful, that may be consistent with our universe. The anthropic principle can be used as one of the major selection criteria to distinguish theories that clearly don't apply to our universe.

The question that remains is whether string theorists (or any physicist) should be happy about this situation.

Certainly, some are not, but it seems like support for this approach has grown, particularly over the last decade. While string theory pioneers David Gross and Edward Witten had long been holdouts against using anthropic reasoning, their public comments in recent years have suggested that this opposition has largely been eroded.

Or, to quote Witten on his evolution, "Twenty years ago, I used to find the anthropic interpretation of the universe upsetting, in part because of the difficulty it might present in understanding physics. Over the years I have mellowed. I suppose I reluctantly came to accept that the universe was not created for our convenience in understanding it."

The anthropic principle seems unavoidable if there exists a vast multiverse, where many different regions of the string theory landscape are realized in the form of parallel universes. Some universes will exist where life is allowed, and we're one of them — get used to it.

Some string theorists who haven't accepted the anthropic arguments are hopeful that the theory's mathematical and physical features can rule out large portions of the landscape. String theorists are still divided over exactly which conclusions the theory allows and whether there might be some way to sort them out without applying the anthropic principle. More work must be done before anyone knows for sure.

Turning a Critical Eye on String Theorists

One of the major criticisms of string theory has to do not with the theory so much as with the theorists. The argument is that they are forming something of a "cult" of string theorists who have bonded together to promote string theory above all alternatives.

This criticism, which is at the heart of Smolin's *The Trouble with Physics*, is not so much a condemnation of string theory as a fundamental criticism of the way academic resources are allocated. One critique of Smolin's book has been that he is in part demanding more funding for the research projects he and his friends are working on, which he feels are undersupported. (Many of these alternative fields are covered in Chapters 19 and 20.)

Hundreds of physicists just can't be wrong

String theory is the most popular approach to a theory of quantum gravity, but that very phrase — most popular — is exactly the problem in the eyes of some. In physics, who cares (or who *should* care) how popular a theory is?

In fact, some critics believe that string theory is little more than a cult of personality. The practitioners of this arcane art, they say, have long ago forgone the regular practice of science, and now bask in the glory of seer-like authority figures like Edward Witten, Leonard Susskind, or Cumrun Vafa, whose words can no more be wrong than the sun can stop shining.

This is, of course, an exaggeration of the criticism, but not by much in some cases. String theorists spent more than two decades building a community of physicists who firmly believe that they are performing some of the deepest and most fundamental physics on the planet, even while achieving no conclusive evidence to definitively support their version of science as the right one, and the folks at the top of that community carry a lot of weight. (For a look at this behavior in non-physics contexts, see the nearby sidebar "Appeal to authority.")

John Moffat has joined Lee Smolin and Peter Woit in lamenting the "lost generation" of brilliant physicists who have spent their time on string theory, to no avail. He points out that the sheer number of physicists publishing papers on string theory, and in turn citing other string theorists, skews the indexes on which papers and scientists are truly the most important.

For example, Edward Witten is one of the most widely cited theoretical physicists, if not the most cited, with a good fraction of his papers being related to string theory. (The combined number of citations that one author receives is often used as a rough measure of the impact of their ideas.) If you look at it from Moffat's point of view, this isn't necessarily a result of Witten being the most important physicist of his generation, but rather a result of Witten writing papers that are fundamental to string theory and are in turn cited by the vast majority of people writing papers on string theory, which is a lot of papers.

Now the problem with this argument, at least when it comes to Witten specifically, is that his research has objectively had an enormous impact outside string theory — most notably in quantum field theory as well as in certain parts of mathematics. Certainly, his Fields Medal attests to his position as one of the most mathematically gifted theoretical physicists of his time. But the main point of the argument is that if Witten is an important physicist who has helped lead a generation of physicists down a road that ends in string theory as a failed theory of quantum gravity, then that would indeed make for a "lost generation" and a tragic waste of Witten's brilliance.

APPEAL TO AUTHORITY

Although it may seem odd to many people that scientists can be swayed by figures of authority, it's a fundamental part of human nature. The "appeal to authority" was cited by Aristotle, the father of rhetoric (the science of debate). It has been given the Latin name *argumentum ad verecundiam,* and evidence from psychology has borne out that it works. People are inclined to believe an authority figure, sometimes even over common sense.

Marketers know that one of the most persuasive ways to sell something is to get a testimonial. This is why speakers are introduced by someone else, for example. If another person gets up and lists the speaker's accomplishments, it means a lot more to the listeners than if the speaker stands up, introduces themself, and lists off their own accomplishments. This is the case even when the introducer knows nothing about the person except what they read off a card or teleprompter.

When the person who is providing the testimonial is perceived as an authority figure, it's even more potent. This is why some books have quotes from authorities on their covers and why politicians seek celebrity endorsements.

In the case of string theory, of course, the authority figures aren't just popular. They are experts in physics, and string theory in particular, so listening to their opinion on string theory is a bit more reasonable than listening to a single popular actor, musician, athlete, or clergy member on whom to vote into office. Ultimately, in science (as in the rest of life) people should use their own logic to evaluate the arguments put forward by the experts. Fortunately, scientists are trained to use their logic more intently than most of society . . . at least within their scientific field of expertise.

Nowadays, the fraction of physicists who are directly working on string theory is smaller than it was ten or twenty years ago, perhaps vindicating some of the criticism of Moffatt, Smolin, and Woit. However, a very large community of physicists is still exploring string-inspired research. The yearly *Strings* conference, whose scientific committee comprises prominent string theorists, consists mostly of seminars that are not on string theory in the strict sense. Still it attracts hundreds of participants every year who, presumably, identify themselves as string theorists at least to some degree.

Holding the keys to the academic kingdom

The theoretical physics and particle physics communities in many of the major physics departments, especially in the United States, lean heavily toward string theory as the preferred approach to a quantum gravity theory. In fact, the growing need for diverse approaches (such as those from Chapters 19 and 20) is maintained even by some string theorists, who realize the importance of including conflicting viewpoints.

In a debate with Lee Smolin on National Public Radio, Brian Greene acknowledged the need to work on areas other than string theory, pointing out that some of his own graduate students are working on different approaches to solving problems of quantum gravity.

Lisa Randall — whose own work has often been influenced by string theory — describes how, during the first superstring revolution, Harvard physicists remained more closely tied to the particle physics tradition and to experimental results, while Princeton researchers devoted themselves largely to the purely theoretical enterprise of string theory. In the end, every particle theorist at Princeton worked on string theory, which she identifies as a mistake — and one that continued until recently, though now most research has moved to "string-inspired topics" rather than to string theory per se. These stances indicate that if a "string theory cult" does exist, then Greene and Randall have apparently not been inducted into it. Still, the fact is that theoretical physics departments at several major universities are now dominated by string theory supporters, and some feel that other approaches are inherently marginalized by that.

This criticism is probably one of the fairest because science, like any other field of endeavor, *needs* criticism. Psychologists have shown that the phenomenon of "groupthink" takes hold in situations where the only people who are allowed a seat at the table are those who think alike. If you want to have a robust intellectual exchange — something that's at the heart of physics and other sciences — it's important that you include people who will challenge your viewpoints and not just agree with them.

Some critics of Smolin's book have speculated that he wants some sort of handout for himself and his buddies who aren't able to cut it in the normal grant application process. (In the other direction, Smolin and Woit have implied that similar economic interests are at the heart of the support for string theory.)

But if the institutes that determine how funding is allocated are dominated by people who believe that string theory is the only viable theory, then these alternate approaches won't get funded. Add to that the citation issues described earlier in this chapter, which possibly make string theory look more successful than it actually is, and there's room for valid criticism of how funding is allocated in physics.

Still, hope for these alternatives isn't lost. As popular as string theory is, we believe it's likely that most theoretical physicists want to find answers more than they want to be proved right. Physicists will gravitate (so to speak) toward the theories that provide them the best opportunity to discover a fundamental truth about the universe.

So long as these non-string theorists continue doing solid work in other areas, they have a hope of drawing recruits from the younger generation. Eventually, if string theorists don't find some way to make string theory succeed, it will lose its dominant position.

Does String Theory Describe Our Universe?

Now comes the real science question related to string theory: Does it describe our universe? The short answer is no, or at least not yet. It can be written in such a way to describe some idealized worlds that bear similarities to our world, but it can't yet describe our world.

Unfortunately, you have to know a lot about string theory to realize that. String theorists are rarely up-front about how far their theory is from describing our reality (when talking to public audiences, at least). It tends to be a disclaimer, woven into the details of their presentations or thrown in just near the end. In fact, you can read many of the books on string theory out there, and after turning the last page, you won't have ever been told explicitly that it doesn't describe our universe.

Congratulations on not choosing one of those books.

Making sense of extra dimensions

The world described by string theory has at least six more space dimensions than the three we know, for a total of nine space dimensions. In M-theory, there are at least ten space dimensions, and in the two-time M-theory, there are 11 space dimensions (with a second time dimension tacked on).

The problem is that physicists don't know where these extra dimensions are. In fact, the main reason for believing they exist is that the equations of string theory demand them. These extra dimensions have been compactified (in some models) in ways that their particular geometry generates certain features of our universe.

There are two major ways of dealing with the extra dimensions:

- The extra dimensions are compactified, probably at about the Planck scale (although some models allow for them to be larger).
- Our universe is "stuck" on a brane with three space dimensions (brane world scenarios).

There is another alternative: The extra dimensions may not exist. (This would be the approach suggested by applying Occam's razor.) Various physicists have developed approaches to string theory without extra dimensions, as discussed in Chapter 15, so abandoning the idea of extra dimensions doesn't even require an abandonment of string theory!

Space-time should be fluid

One of the hallmarks of modern physics is general relativity. The clash between general relativity and quantum physics is part of the motivation for looking for a string theory, but some critics believe that string theory is designed in such a way that it doesn't faithfully maintain the principles of general relativity.

Which principles of general relativity aren't maintained in string theory? Specifically, the idea that space-time is a dynamic entity that responds to the presence of matter around it. In other words, space-time is flexible. In physics terminology, general relativity is a *background-independent theory*, because the background (space-time) is incorporated into the theory. A *background-dependent theory* is one where objects in the theory are sort of "plugged in" to a space-time framework.

Right now, string theory is a background-dependent framework. Space-time is rigid, instead of flexible. If you are given a certain configuration of space-time, you can discuss how a given version of string theory would behave in that system.

The question is whether string theory, which right now can only be formulated in fixed space-time environments, can really accommodate a fundamentally dynamic space-time framework. How can you turn the rigid space-time of string theory into the flexible space-time of general relativity? The pessimist replies "You can't" and works on loop quantum gravity or some other approach to quantum gravity (see Chapter 19).

The optimist, however, believes that string theory still has hope. Even with a rigid background of space-time, it's possible to get general relativity as a limiting case of string theory. This isn't quite as good as getting a flexible space-time, but it means that string theory certainly doesn't exclude general relativity. Instead of getting the full high-definition version of space-time, though, you're left with something more like a flipbook, which treats each image as static but provides the overall impression of smooth motion.

String theory is a work in progress, and it's still hoped that physical and mathematical principles that will allow for the expression of a fully dynamic background in string theory will be developed. String theorists are forced to talk about the theory in a rigid space-time (background-dependent) only because they haven't yet found the mathematical language that will let them talk about it in a flexible space-time (background-independent). Some believe that Maldacena's AdS/CFT correspondence may provide a means of incorporating this background-independent language. It's also possible that the principles that allow this new language will come from an unexpected direction, such as the work described in Chapters 19 and 20.

Or, of course, such principles may not exist at all, and the skeptic's inclination to criticize string theory may therefore be justified.

The ever-elusive superpartners

In Chapter 14, we tell you about the search for superpartners at the Large Hadron Collider (LHC) in Switzerland. If you skimmed through that part, here's a recap: That search has so far been unsuccessful. None of the superpartners scientists were hoping to find at the LHC have actually been discovered in the experiments conducted there.

When the first edition of this book was written, during the construction of the LHC, the big hope was that these predicted superpartner particles would soon be discovered experimentally, providing solid evidence in support of supersymmetry at least, if not direct evidence of string theory.

The lack of that discovery has to be considered a failure of string theory, and a victory on the side of the critics, though obviously not one that completely negates the viability of the theory. As we've mentioned before, there are so many different string theory models, it stands to reason that only some of them would have superpartners at the energy levels being explored by the LHC. Not finding the superpartners just puts additional constraints on the string theory models that may apply to our universe. It's possible that string theory still applies to our universe, but we live in a universe where the superpartners have higher energies than what the LHC collisions can detect.

But, at best, we have to concede that this defense of string theory in light of the ever-elusive superpartners is far less satisfying than if scientists could just detect the superpartners in an experiment, at the LHC or elsewhere.

How finite is string theory?

One criticism that has arisen largely since Smolin's *The Trouble with Physics* is the notion that string theory isn't necessarily a finite theory. Remember that this is one of the key features in support of string theory: It removes the infinities that arise when you try to apply quantum physics directly to problems.

As Smolin describes things, the belief in string theory finiteness is largely based on a 1992 proof performed by Stanley Mandelstam, in which Mandelstam proved only that the first term of string theory (remember, string theory is an equation made up of an infinite series of mathematical terms) was finite. It has since been proved for the second term as well.

Still, even if every individual term is finite, string theory currently is written in a form (like quantum field theory) that has an infinite number of terms. Even if each term is finite, it's possible that the sum of all the terms will yield an infinite result. Because infinities are never witnessed in our universe, this would mean that string theory doesn't describe our universe.

REMEMBER

According to Smolin, the fact that string theory finiteness hasn't been proved isn't a flaw in string theory. The fact that most string theorists *thought* it had been proved finite when it wasn't is the flaw — not necessarily a flaw in string theory itself, but a flaw in the very way these scientists are practicing their science. The bigger issue at stake in this particular criticism is one of precision and intellectual honesty.

A String Theory Rebuttal

In light of all these criticisms, many of which have some measure of validity or logic to them, you may be wondering how anyone could continue working on string theory. How could some of the most brilliant physicists in the world devote their careers to exploring a field that is apparently a house of cards?

The short answer, stated in various forms by many string theorists over the years, is that they find it hard to believe that such a beautiful theory would not apply to the universe. String theory describes all the behavior of the universe from certain fundamental principles as the vibrations of 1-dimensional strings and compactification of extra dimensional geometries, and can be used in some simplified versions to solve problems that have meaning to physicists, such as black hole entropy.

Most string theorists are able to dismiss the idea that string theory should be further along than it is. String theory does, after all, explore energies and sizes beyond our current technology to test. And even in cases where experiment can guide theory, there are cases where 40 years wasn't enough time.

REMEMBER

The theory of light took much longer than 40 years to develop. In the late 1600s, Newton described light as tiny particles. In the 1800s, experiments revealed that it travels as waves. In 1905, Einstein proposed the quantum principles that led to wave-particle duality, which in turn resulted in the theory of quantum electrodynamics in the 1940s. In other words, the rigorous physical examination of light traces a path from Newton through to Feynman that covers about 250 years, filled with many false leads along the way. And the framework of quantum field theory, which was invented to make sense of electrodynamics, has had its own issues, not so different from some of those imputed to string theory (see "Is string theory so different from quantum field theory?" earlier in this chapter).

For that matter, it took more than 1,500 years for heliocentric models of Earth's motion to be accepted over geocentric models, even though anyone can look up at the sky! It's only because our modern world moves so fast that we feel we need quick and easy answers to something as complex as the fundamental nature of the universe.

So how do string theorists respond to the specific technical criticisms mentioned earlier?

What about the extra dimensions?

Extra dimensions are a logical possibility in any theory, but in most theories we're able to easily eliminate extra dimensions and get down to the number that it seems we need. In string theory, though, the problem is that dimensions show up in the mathematics of the theory, and we don't have a clear way to eliminate them without creating new issues.

This inability to remove the dimensions from the theory to be consistent with the observed universe is what leads to the landscape problem: the vast number of possible string theory models we face, without a clear way to distinguish which ones may apply to our universe.

String theorists, despite not having issues with extra dimensions per se, do recognize that the landscape problem is a major concern.

Space-time fluidity?

It's true that string theory is a background-dependent theory, where space-time is rigid instead of inherently flexible. It's a problem in principle that string theory is formulated starting from a given background (a desired solution) and building on that. Physicists call this a *perturbative* description.

However, that didn't stop string theorists from discovering truly non-perturbative properties of string theory, and ultimately quantum gravity, like branes, dualities, and holography.

Does string theory need to be finite?

Finally, we get to the issue of finiteness. The criticism here is that although the predictions in string theory are made up of many terms, each individually finite, there's no telling if their sum remains finite. String theorists and other experts in theoretical physics may find such criticism a bit disingenuous: After all, the same problem plagues quantum field theory. Despite some advances (which are known as *resurgence theory*), it's very much an issue for quantum field theory too. Hence, saying a lack of finiteness is a failure of string theory in particular isn't accurate.

If you wonder why this criticism doesn't come up very often in quantum field theory, it's because in most cases our approximate understanding, based on computing just a few terms of the sum, actually works just fine and matches well enough with experiments. Still, the more mathematically oriented physicists are working hard to better understand how to sum all the terms, starting with the relatively simple case of quantum theories — and even then, it's no easy task!

Trying to Make Sense of the Controversy

We have tried to summarize the main criticisms of string theory and the rebuttal by the string community. So, who's right and who's wrong? It is, of course, silly to try to give an absolute, objective answer to this question when many brilliant minds — more brilliant than your humble authors — disagree on so much. But it's also a bit unfair to you not to take a stance at all.

On the scientific side of things, it's quite clear that string theory is an unfinished project and has many shortcomings that need to be addressed, like the landscape problem and the difficulty of computing anything beyond the first few terms in a perturbative expansion (see Chapter 11 for a discussion of perturbation theory). However, it seems to us that the alternative theories on the market have many more issues than string theory and cannot yet boast any of its major accomplishments (like the description of at least some types of black holes and their entropy, or the detailed derivation of the holographic principle, at least in some special frameworks).

Of course, some may say that's only true because an enormous amount of manpower was invested in studying string theory over the last 40 or so years — no wonder string theory got so much further ahead than other theories! This is perhaps not entirely accurate for *all* the alternatives to string theory (loop quantum gravity, which you encounter in Chapter 19, has also been around for a while, though it has never received the same research attention as string theory), but it may be true for some genuinely new approaches that don't get fully developed.

In fact, maybe a revolutionary theory of quantum gravity, different from string theory, loop quantum gravity, or anything developed so far, is waiting to be discovered — only it's not being discovered because people keep working on the same old ideas. This is a fair point, and it's really a criticism of how the academic system works.

Historically, many of the most important discoveries in physics were made by young researchers: Dirac, Einstein, Fermi, Heisenberg, and Schrödinger were all in their 20s or 30s when they had their groundbreaking ideas. Nowadays, a physicist of that age would be a *postdoc*, an independent researcher under a two- or three-year contract at some university, after which they would be expected to relocate somewhere else, until eventually they find a tenured position.

On the one hand, this seems nice: A young researcher is exposed to the ideas of different groups before settling down. In practice, it means that a postdoc has one or two years to get results and become visible in the community before they need to start looking for a new job. This creates an incentive to work in fields that are "hot," spearheaded by senior researchers from top institutions (who are able to offer prestigious and well-paid jobs). Working on some fringe or long-shot idea may be tantamount to career suicide.

To be fair, this problem has little to do with string theory itself and more to do with how the modern academic world is structured, though perhaps it's more of an issue in a field where direct validation of an idea by experimentation may take years or decades, like in quantum gravity.

IN THIS CHAPTER

» Seeing how loop quantum gravity is more focused than string theory

» Knowing what predictions loop quantum gravity makes

» Evaluating the similarities between loop quantum gravity and string theory

Chapter 19
Loop Quantum Gravity: String Theory's Biggest Competitor

Though string theory is often promoted as the "only consistent theory of quantum gravity" (or something along those lines), some would disagree with this categorization. Foremost among them are the researchers in a field known as *loop quantum gravity* (sometimes abbreviated *LQG*). We discuss other approaches to quantum gravity in Chapter 20.

In this chapter, we introduce you to loop quantum gravity, an alternative theory of quantum gravity. As string theory's major competitor, loop quantum gravity hopes to answer many of the same questions by using a different approach. We start by describing the basic principles of loop quantum gravity and then present some of the major benefits of this approach over string theory. We lay out some of the preliminary predictions of loop quantum gravity, including possible ways to test it. Finally, we consider whether loop quantum gravity suffers from the same issues we discuss in Chapter 18.

Taking the Loop: Introducing Another Road to Quantum Gravity

Loop quantum gravity is string theory's biggest competitor. It gets less press than string theory, in part because it has a fundamentally more limited goal: a quantum theory of gravity. Loop quantum gravity performs this feat by trying to quantize space itself — in other words, treat space like it comes in small chunks.

By contrast, string theory starts with methods of particle physics and frequently hopes to not only provide a means of creating a quantum theory of gravity but also explain all of particle physics, unifying gravity with the other forces at the same time. Along the way it makes some striking predictions, like extra dimensions and the holographic principle, which generate much excitement among scientists and in the public.

It's no wonder that loop quantum gravity has more trouble getting press.

The great background debate

The key insight of quantum physics is that some quantities in nature come in multiples of discrete values, called *quanta*. This principle has successfully been applied to all of physics, except for gravity. This is the motivation for the search for quantum gravity.

Alternately, the key insight from general relativity is that space-time is a dynamic entity, not a fixed framework. String theory is a *background-dependent theory* (built on a fixed framework; see Chapter 18 for more on this), so it doesn't currently account for the dynamic nature of space-time at the heart of relativity.

According to LQG researchers, a theory of quantum gravity must be background-independent, a theory that explains space and time instead of being plugged into an already existing space-time stage. This follows the way in which general relativity, which is background-independent from the get-go, is formulated.

REMEMBER

Loop quantum gravity tries to achieve this goal by looking at the smooth fabric of space-time in general relativity and contemplating the question of whether, like regular fabric, it might be made up of smaller fibers woven together. The connections between these quanta of space-time may yield a background-independent way of looking at gravity in the quantum world.

What is looping anyway?

Loop quantum gravity's key insight is that you can describe space as a *field;* instead of a bunch of points, space is a bunch of lines. The *loop* in loop quantum gravity has to do with the fact that as you view these field lines (which don't have to be straight lines, of course), they can loop around and through each other, creating a *spin network.* By analyzing this network of space bundles, you can supposedly extract results that are equivalent to the known laws of physics.

The foundation of LQG was laid in 1986, when Abhay Ashtekar rewrote general relativity as a series of field lines instead of a grid of points. The result turns out not only to be simpler than the earlier approach but also similar to a gauge theory.

There's one problem, though: Gauge theories are background-dependent theories (they are inserted into a fixed space-time framework), but that won't work for LQG, because the field lines themselves represent the geometry of space. You can't plug the theory into space if space is already part of the theory!

In order to proceed, physicists working in this area had to look at quantum field theory in a whole new way so it could be approached in a background-independent setting. Much of this work was performed by Ashtekar, Lee Smolin, Ted Jacobson, and Carlo Rovelli, who can reasonably be considered among the progenitors of loop quantum gravity.

As LQG developed, it became clear that the theory represented a network of connected *quantum space bundles,* often called "atoms" of space. The failure of previous attempts to write a quantum theory of gravity was that space-time was treated as continuous, instead of being quantized itself. The evolution of these connections is what provides the dynamic framework of space —although it has yet to be proved that loop quantum gravity actually reduces to the same predictions as those given by relativity.

Each atom of space can be depicted with a point (called a *node*) on a certain type of grid. The grid of all these nodes, and the connections between them, is called a *spin network.* (Spin networks were originally developed by Oxford physicist Roger Penrose back in the 1970s.) The graph around each node can change locally over time, as shown in Figure 19-1 (which depicts the initial state [a] and the new state it changes into [b]). The idea is that the sum total of these changes will end up matching the smooth space-time predictions of relativity on larger scales. (That last bit is the major part that has yet to be proved.)

TIP

Now, when you look at these lines and picture them in three dimensions, the lines exist inside space — but that's the wrong way to think about it. In LQG, the spin network with all these nodes and grid lines — the entire spin network — is actually space itself. The specific configuration of the spin network is the geometry of space.

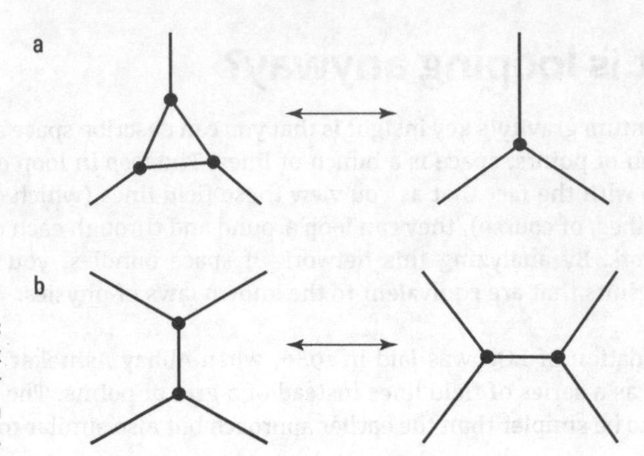

FIGURE 19-1: The spin network evolves over time through local changes.

The analysis of this network of quantum units of space may result in more than physicists bargained for, because recent studies have indicated that the Standard Model particles may be implicit in the theory. This work has largely been pioneered by Greek theoretical physicist Fotini Markopoulou and Australian theoretical particle physicist Sundance O. Bilson-Thompson. In Bilson-Thompson's model, the loops may braid together in ways that can create the particles, as indicated in Figure 19-2. (These results are entirely theoretical, and it remains to be seen how they work into the larger LQG framework as it develops, or whether they have any physical meaning at all.)

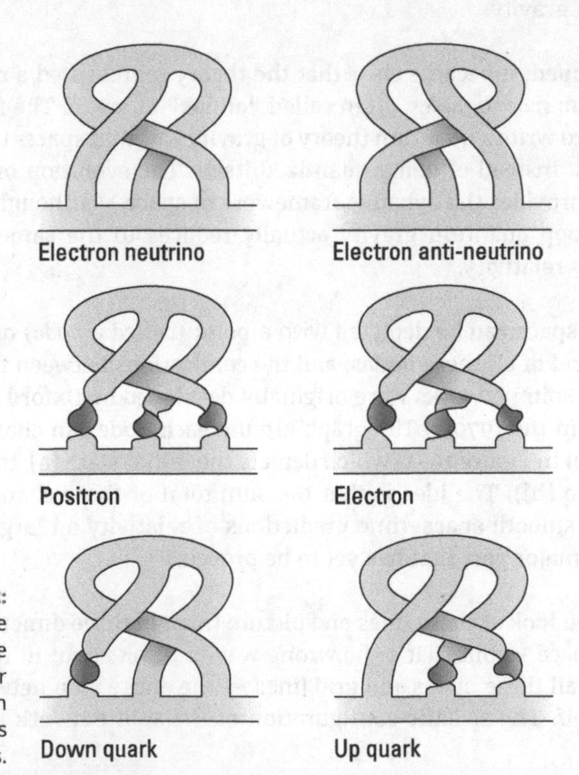

FIGURE 19-2: Braids in the fabric of space may account for the known particles of physics.

Making Predictions with Loop Quantum Gravity

Loop quantum gravity makes some definite predictions, which may mean that it can be tested well before string theory can be. As string theory's popularity is being brought into question, the amount of research into LQG may end up growing.

Gravity exists (Duh!)

Oddly enough, because LQG was born out of general relativity, one question has been whether science can get general relativity back out of the theory. In other words, can scientists use loop quantum gravity to actually match Einstein's classical theory of gravity on large scales? The answer is, yes, in some special cases (similar to string theory).

For example, work by Carlo Rovelli and his colleagues has shown that LQG contains gravitons, at least in the low-energy version of the theory, and also that two masses placed into the theory will attract each other in accord with Newton's law of gravity. Further theoretical work is needed to get solid correlations between LQG and general relativity.

Black holes contain only so much space

Loop quantum gravity's major success has been in matching Jacob Bekenstein's prediction of black hole entropy as well as Stephen Hawking's radiation predictions (both described in Chapter 9). As we mention in Chapters 11 and 16, string theory has been able to make some predictions about special types of black holes, which is also consistent with the Bekenstein-Hawking theories. So, at the very least, if scientists are able to create miniature black holes in the Large Hadron Collider and observe Hawking radiation, then it would certainly not rule out either of the theories.

However, the picture given by LQG is very different from that of classical black holes. Instead of an infinite singularity, the quantum rules say there's only so much space inside the black hole. Some LQG theorists hope they can predict tiny adjustments to Hawking's theory that, if experimentally proven true, would support LQG above string theory.

One prediction is that instead of a singularity, the matter falling into a black hole begins expanding into another region of space-time, consistent with some earlier predictions by Bryce DeWitt and John Archibald Wheeler. In fact, singularities at the big bang are also eliminated, providing another possible eternal universe model. (For more on eternal universe models, see Chapter 16.)

Gamma ray burst radiation travels at different speeds

Many of the experiments from Chapter 14 that may test whether the speed of light varies would also be consistent with loop quantum gravity. For example, it's possible that gamma ray burst radiation doesn't all travel at the same speed, as classical relativity predicts. As the radiation passes through the spin network of quantized space, the high-energy gamma rays would travel slightly slower than the low-energy gamma rays. Again, these effects would be magnified over the vast distances traveled to possibly be observed by powerful telescopes (though they have not yet been observed).

Finding Favor and Flaw with Loop Quantum Gravity

As with string theory, loop quantum gravity is passionately embraced by some physicists and dismissed by others. The physicists who study it believe that its predictions (described in the preceding section) are far better than those made by string theory. One major argument in support of LQG is that it's seen by its adherents as a finite theory, meaning the theory itself doesn't inherently admit infinities. These same researchers also tend to dismiss the flaws as being the product of insufficient work (and funding) devoted to the theory. String theorists, in turn, view them as much victims of "groupthink" as critics view string theorists.

The benefit of a finite theorem

One major benefit of loop quantum gravity is that the theory has been proved finite in a more definitive sense than string theory has. Lee Smolin, one of the key (and certainly most high-profile) researchers of LQG, describes in his book *The Trouble with Physics* three distinct ways that the theory is finite (with string theorists' objections in parentheses):

- The areas and volumes in loop quantum gravity are always in finite, discrete units. (String theorists would say this isn't a particularly meaningful form of finiteness.)

» In the Barrett-Crane model of loop quantum gravity, the probabilities for a quantum geometry to evolve into different histories are always finite. (This sounds just like unitarity, which is a property of string theory and all quantum field theories.)

» Including gravity in a loop quantum gravity theory that contains matter theory, like the Standard Model, involves no infinite expressions. If gravity is excluded, you have to do some tinkering to avoid them. (String theorists believe that this claim is premature and there are substantial problems with the proposed LQG models that yield this result.)

As we explain in Chapter 18, some questions exist (largely brought up by LQG theorists) about whether string theory is actually finite — or, more specifically, whether it has been rigorously proved finite. It must be noted that the same objections apply for most quantum field theories, including the Standard Model. Yet, from the theoretical side of things, the loop quantum gravity people view this uncertainty as a major victory over string theory. String theorists would argue that Smolin's statements about its finiteness still don't prove that LQG can't result in an infinite solution when experimental data is put into the theory.

Spending some time focusing on the flaws

Many of the flaws in loop quantum gravity are the same flaws in string theory. Their predictions generally extend into realms that aren't quite testable yet (although LQG is a bit closer to being able to be experimentally tested than string theory probably is). Also, it's not really clear that loop quantum gravity is any more falsifiable than string theory.

For example, the discovery of supersymmetry or extra dimensions won't disprove loop quantum gravity any more than the failure to detect them will disprove string theory. (The only discovery we think LQG would have a hard time overcoming would be if black holes are observed and Hawking radiation proves to be false, which would be a problem for *any* quantum gravity theory, including string theory.)

The biggest flaw in loop quantum gravity is that it has yet to successfully show that you can take a quantized space and extract a smooth space-time out of it. In fact, the entire method of adding time into the spin network seems somewhat contrived to some critics, although whether it's any more contrived than the entirely background-dependent formulation of string theory remains to be seen.

The quantum theory of space-time in loop quantum gravity is really just a quantum theory of space. The spin network described by the theory cannot yet incorporate time. Some, including Smolin, believe that time will prove to be a necessary and fundamental component of the theory, while Rovelli believes the theory will ultimately show that time doesn't really exist, but is just an emergent property without a real existence on its own.

Finally, while string theory remains far from being experimentally checked, it has produced some genuinely new insights into the nature of quantum gravity, especially in the context of dualities and holography. LQG is also hardly testable, and so far it seems to be struggling to bring new insights of the same magnitude — at best reproducing known features of gravity.

So Are These Two Theories the Same with Different Names?

One viewpoint is that both string theory and loop quantum gravity may actually represent the same theory approached from different directions. The parallels between the theories are numerous:

- » String theory began as a theory of particle interactions, but was shown to contain gravity. Loop quantum gravity began as a theory of gravity, but was shown to contain particles.
- » In string theory, space-time can be viewed as a mesh of interacting strings and branes, much like the threads of a fabric. In loop quantum gravity, threads of space are woven together, creating the apparently "smooth" fabric of space-time.
- » Some string theorists believe the compactified dimensions represent a fundamental quantum unit of space, while LQG starts with units of space as an initial requirement.
- » Both theories (provided certain assumptions are made) calculate the same entropy for black holes.

One way to view the differences is that string theory, which began by applying principles from particle physics, may point toward a universe in which space-time emerges from the behavior of these fundamental strings. LQG, on the other hand, began by applying general relativity principles and results in a world where space-time is fundamental, but matter and gravity may emerge from the behavior of these fundamental units.

At one time, Lee Smolin was one of the major supporters of the viewpoint that string theory, M-theory, and loop quantum gravity were different approximations of the same underlying fundamental theory. Over the last couple of decades, he has grown largely disillusioned with string theory (at least compared to his earlier conciliatory stance), becoming a prominent advocate of pursuing other avenues of inquiry. Some string theorists believe that the methods used by LQG will eventually be carried over to string theory, allowing for a background-independent version of string theory.

Despite the possible harmony between the two fields, at the moment they are competitors for research funding and attention. String theorists have their conferences, and loop quantum gravity people have their conferences, and rarely shall the two conferences meet. (Except for Smolin, who seems to have rather enjoyed flitting over to the string theory side of things over the years.) All too often, the groups seem unable to speak to each other in any meaningful way (see the nearby sidebar "The 'Big Bang' breakup"). Panel discussions between experts in the two fields, some of which can be found online, often are not very constructive or informative, as each side fails to find common ground with the other.

Part of the problem is one of sociology. Many string theorists, even in research papers, use phrases that make it clear they consider string theory not only to be their preferred theory but to be the only (or, in cases where they're being more generous, the "most promising") theory of quantum gravity. By doing this, they often dismiss LQG as even being an option. Some string theorists have indicated in interviews that they are completely unaware of any viable alternatives to string theory! (This is because string theorists aren't yet convinced that the alternatives are actually viable.)

Hopefully, these physicists will find a way to work together and use their results and techniques in ways that provide real insights into the nature of our universe. But so far, loop quantum gravity, like string theory, is still stuck on the drawing board.

THE "BIG BANG" BREAKUP

The conflict between loop quantum gravity and string theory enthusiasts made it into popular culture in an episode of the syndicated American television sitcom *The Big Bang Theory,* which focuses on two physicist roommates, Leonard and Sheldon. In the second episode of the second season, Leonard has begun a relationship with a physicist colleague, Leslie Winkle, a rival of Sheldon (or "nemesis," as he thinks of her). Leslie, you see, is a researcher in loop quantum gravity, while Sheldon is a string theorist.

(continued)

(continued)

In the climactic scene of the episode, Sheldon and Leslie get into a "string war" of their own, slinging theoretical physics barbs at each other. Their conflict is over which theory — loop quantum gravity or string theory — has the greatest probability of successfully achieving a quantum theory of gravity. The argument ends with Leonard being placed in the middle, forced to point out that they are two untested theories of quantum gravity, so he has no way to choose. Leslie is shocked and appalled by this response, immediately ending her relationship with Leonard.

Although the dispute is obviously played up for comedy purposes, among the physics community the funniest thing about it was how much truth there actually was to the scenario. When physicists get into passionate debates about loop quantum gravity versus string theory, all too often the first casualty seems to be reasonable discourse.

IN THIS CHAPTER

» Some physicists are working in areas other than string theory — honest!

» Working around the need for a theory of quantum gravity

» Seeking new mathematical approaches while solving string theory problems

Chapter 20
Considering Other Ways to Explain the Universe

In the event that string theory proves false, or that there's no "theory of everything" at all, some unexplained phenomena in the universe still require explanation. These issues mostly lie in the realm of cosmology, and include the flatness problem, dark matter, dark energy, and the details of the early universe.

Even though string theory is currently the dominant path being explored to answer most of these problems, some physicists have begun looking in other directions, beyond the theory of loop quantum gravity described in Chapter 19. These rebels (and, at times, outcasts) have refused, in many cases, to stick with the mainstream theoretical community in adopting the principles of string theory and have proposed new directions of inquiry that are sometimes extremely radical — though possibly no more radical, in their own ways, than string theory was in the 1970s.

In this chapter, we explain some of the alternative approaches that physicists are looking into in an effort to explain the problems they want to resolve. First, we explore some alternate quantum gravity theories, none of which are quite as fully developed as either string theory or loop quantum gravity. Next, we show how physicists have suggested modifying the existing law of general relativity to take into account the facts that don't fit with Einstein's original model. It's possible that some of the ideas from this chapter will ultimately be incorporated into string theory, or perhaps take its place entirely.

Taking Other Roads to Quantum Gravity

Though string theorists like to point out that theirs is the most developed theory to unite general relativity and quantum physics (at times they even seem clueless that alternatives exist), it sometimes seems like nearly every physicist has come up with some plan to combine the two — they just don't have the support that string theorists have.

TIP

Most of these alternate theories start with the same idea as loop quantum gravity: that space is made up of small, discrete units that somehow work together to provide the space-time we all know and love (relatively speaking, that is). Despite the fact that scientists don't know much about these units of space, some theorists can analyze how they might behave and use that information to generate useful models.

Here are some examples of these other quantum gravity approaches:

- **Causal dynamical triangulations (CDT):** CDT models space-time as being made up of tiny building blocks called *4-simplices,* which are identical and can reconfigure themselves into different curvature configurations.
- **Quantum Einstein gravity (or "asymptotic safety"):** Quantum Einstein gravity assumes that there's a point where "zooming in" on space-time stops increasing the force of gravity, which instead goes to some finite value.
- **Quantum graphity:** In the quantum graphity model, gravity didn't exist in the earliest moments of the universe because space itself doesn't exist on the small length and high energy scales involved in the early universe.
- **Tensor models:** This approach also tries to describe space-time as made up of microscopic building blocks, which are glued together by the rules of quantum field theory.

TIP

Of course, any of these approaches may advance either string theory or loop quantum gravity instead of leading off in a new direction. Some of the principles may prove fruitful, but only when applied in the framework of one of the other theories. Only time will tell what insights, if any, come out of these other theories and if they can be applied to give meaningful results.

CDT: If you've got the time, I've got the space

The causal dynamical triangulations approach consists of taking tiny building blocks of space, called *4-simplices* (sort of like multidimensional triangles), and using them to construct the space-time geometry. The result is a sequence of geometric patterns that are causally related in a sequence where one construction follows another (in other words, one pattern causes the next pattern). This system was developed by Renate Loll of Radboud University in Nijmegen, Netherlands, and also by her colleagues Jan Ambjørn and Jerzy Jurkiewicz.

One of the most important aspects of CDT is that time becomes an essential component of space-time because Loll includes the causal link as a crucial part of the theory. Relativity tells us that time is distinctly different from space (as we mention in Chapter 15, the time dimension has a negative in front of it in relativity), but Stephen Hawking and others have suggested that the difference between time and space can perhaps be ignored.

Loll then takes her causally linked configurations of 4-simplices and sums over all possible configurations of the shapes. (Richard Feynman used a similar approach in quantum mechanics, summing over all possible paths to obtain quantum physics results.) The result is classical space-time geometry! This is something that works very well in a small number of dimensions (like one dimension for space and one for time), but it's a bit trickier to get it to work in our three-plus-one dimensions.

TIP

If true, CDT shows that it's impossible to ignore the difference between space and time. The causal link of changes in space-time geometry — in other words, the "time" part of space-time — is absolutely necessary to get the classical space-time geometry that is governed by general relativity and matches what science knows of standard cosmological models.

REMEMBER

At the tiniest scales, though, CDT shows that space-time is only 2-dimensional. The model turns into a fractal pattern, where the structures repeat themselves at smaller and smaller scales, and there's no proof that real space-time behaves that way.

CDT's biggest flaw in comparison to string theory is that it doesn't tell us anything about where matter comes from, whereas matter arises naturally in string theory from the interactions of fundamental strings.

Quantum Einstein gravity: Too small to tug

Quantum Einstein gravity, developed by Martin Reuter of the University of Mainz in Germany, tries to apply the quantum physics processes that worked on other forces to gravity. Reuter believes that at small scales, gravity may have a cutoff point where its strength stops increasing. (This notion was proposed by Steven Weinberg in the 1970s under the more common name *asymptotic safety*.)

One reason to think that gravity stops increasing at small scales is that this is what quantum field theory tells us the other forces do. At very small scales, even the strong nuclear force drops to zero. This is called *asymptotic freedom*, and its discovery earned David Gross, David Politzer, and Frank Wilczek the 2004 Nobel Prize. The force of gravity wouldn't go to zero but rather to some finite strength (stronger than we usually see), an idea known as asymptotic safety.

Weinberg and others weren't able to pursue the idea at the time because the mathematical tools to calculate the cutoff point for gravity in general relativity didn't exist until Reuter developed them in the 1990s. Though the method is approximate, Reuter has a great deal of confidence.

Quantum Einstein gravity, like CDT, comes up with a fractal pattern to small-scale space-time, and the number of dimensions drops to two. Reuter himself has noted that this could mean his approach is fundamentally equivalent to CDT because they both have these rather distinctive predictions at small scales.

The idea of asymptotic safety is really a very conservative solution to the problem of quantum gravity. Unlike the other approaches that introduce some radically new physics that would take over from general relativity at high energies (or, equivalently, at short distances), it proposes a well-defined strongly interacting theory of gravity at high energies in which the usual general relativity is simply augmented by some extra interactions for the graviton.

The two major problems with this approach are technical, but deep. First, we need to make sense of this finite strength gravity theory, which isn't easy. Second, the techniques used by Reuter and his collaborators work very well when we don't distinguish between time and space. Treating time properly is a recurrent challenge in quantum gravity.

Quantum graphity: Disconnecting nodes

Quantum graphity was developed by Fotini Markopoulou, a founding member of the Perimeter Institute. In some ways, this is loop quantum gravity taken to its extreme — at extremely high energies, all that exists is the network of nodes.

This model is based on a suggestion by John Archibald Wheeler about a *pre-geometric phase* to the universe, which Markopoulou took literally. The nodes in the pre-geometric phase would all touch each other, but as the universe cooled, they would disconnect from each other and become separated, resulting in the space that we see today. (Physicists working on string theory have also found this sort of pre-geometric phase, so it's not unique to Markopoulou's approach.)

It's also possible that quantum graphity can explain the horizon problem (the problem that distant parts of the universe seem to be the same temperature). In the quantum graphity model, all points used to be in direct contact, so inflation proves to be unnecessary. (See Chapter 9 for more about the horizon problem and how inflation solves it.) Still, inflation is a much more well-defined and accepted theory, while activity on quantum graphity has slowed down since Markopoulou abandoned physics, leaving many unanswered questions on the viability of this approach.

Tensor models: gluing the space-time together

In the tensor models approach, the space-time emerges from randomly gluing together, according to specific rules, little pieces of space-time, like triangles in two dimensions or tetrahedra (triangular pyramids) in higher dimensions. The main selling point of this approach is that the gluing of the bits, and the probability of obtaining a given result, is performed in the formalism of quantum field theory. This is the current way in which physicists understand quantum physics, which is a big plus because it saves them from having to develop a new setup from scratch.

The idea of gluing bits of space-time to get the geometry of the universe in one space and one time dimension was initiated in the 1980s. The gluing of triangles was related to the multiplication of matrices, and the resulting quantum theories were called *matrix models*. The notion of promoting this model to higher dimensions came a little later through experimentation by Laurent Freidel, Daniele Oriti, and Razvan Gurau, who all worked on these ideas at the Perimeter Institute.

While it still seems tricky to obtain smooth geometries that resemble those of our universe from this approach, group field theory has generated great interest in theoretical physics, with many applications even in string-inspired topics, like the Gurau-Witten model, which is used to understand certain holographic setups.

THE PERIMETER INSTITUTE

If you follow theoretical physics, it isn't long until you hear about the Perimeter Institute for Theoretical Physics, located in Waterloo, Ontario, Canada. The Perimeter Institute was established in 1999 by Mike Lazaridis, who was founder and co-CEO of Research in Motion, the makers of the BlackBerry handheld device. Lazaridis decided to help foster research and innovation in Canada by starting the Perimeter Institute, which is devoted purely to theoretical physics research.

Many of the prominent critics of string theory who are working on other approaches — Lee Smolin, John Moffat, Fotini Markopoulou, and others — have called it home, so it's easy to believe that the Perimeter Institute seeks out anti-string theorists. The faculty at the Perimeter Institute includes physicists who work on string theory or closely related topics, like Davide Gaiotto and Freddy Cachazo, and its longtime director was Neil Turok, a cosmologist, and co-creator of the ekpyrotic model (based based on string theory principles). The Perimeter Institute, despite having been founded relatively recently, has quickly risen to prominence as one of the top institutes worldwide.

The Perimeter Institute's goal is to foster innovation, and its physicists work in a number of areas: cosmology, particle physics, quantum foundations, quantum gravity, quantum information theory, and superstring theory. It's one of the only places where string theorists and leaders in other quantum gravity approaches regularly work together under one roof. More information on the Perimeter Institute can be found at `www.perimeterinstitute.ca`.

Newton and Einstein Don't Make All the Rules: Modifying the Law of Gravity

Instead of trying to develop theories of quantum gravity, some physicists are looking at the existing law of gravity and trying to find specific modifications that will make it work to explain the current mysteries of cosmology. These efforts are largely motivated by attempts to find alternatives to the cosmological theories of inflation, dark matter, and dark energy.

These approaches don't necessarily resolve the conflicts between quantum physics and general relativity, but in many cases they make the conflict less important. The approaches tend to result in singularities and infinities falling out of the theories, so there just isn't as much need for a theory of quantum gravity.

DSR: Twice as many limits as ordinary relativity

One intriguing approach is *doubly special relativity* or *deformed special relativity* (abbreviated as *DSR* either way you slice it), originally developed by Giovanni Amelino-Camelia. In special relativity, the speed of light is constant for all observers. In DSR theories, all observers also agree on one other thing: the distance of the Planck length.

TIP

In Einstein's relativity, the constancy of the speed of light places an upper speed limit on everything in the universe. In DSR theories, the Planck length represents a lower limit on distance. Nothing can go faster than the speed of light, and nothing can be smaller than a Planck length. The principles of DSR may be applicable to various quantum gravity models, such as loop quantum gravity, though so far there's no proof for it.

MOND: Disregarding dark matter

Some physicists aren't comfortable with the idea of dark matter and have proposed alternative explanations to resolve the problems that make physicists believe dark matter exists. One of these explanations, which involves looking at gravity in a new way on large scales, is called *modified Newtonian dynamics (MOND)*.

The basic premise of MOND is that at low values, the force of gravity doesn't follow the rules laid out by Newton more than 300 years ago. The relationship between force and acceleration in these cases may turn out not to be exactly linear, and MOND predicts a relationship that will yield the results observed based on only the visible mass for galaxies.

In Newtonian mechanics (or, for that matter, in general relativity, which reduces to Newtonian mechanics at this scale), the gravitational relationships between objects are precisely defined based on their masses and the distance between them. When the amount of visible matter for galaxies is put into these equations, physicists get answers that show the visible matter just doesn't produce enough gravity to hold the galaxies together. In fact, according to Newtonian mechanics, the outer edges of the galaxies should be rotating much faster, causing the stars farther out to fly away from the galaxy.

Because scientists know the distances involved, the assumption is that somehow the amount of matter has been underestimated. A natural response to this (and the one that most physicists have adopted) is that there must be some other sort of matter that isn't visible to us: dark matter.

TIP

There is one other alternative: The distances and matter are correct, but the relationship between them is incorrect. MOND was proposed by Israeli physicist Mordehai Milgrom in 1981 as a means of explaining the galactic behavior without resorting to dark matter.

Most physicists have ruled out MOND because the dark matter theories seem to fit the facts more closely. Milgrom, however, hasn't given up, and in 2009 he made predictions about slight variations in the path of planets based on his MOND calculations. It remains to be seen if these variations will be observed.

VSL: Light used to travel even faster

In two separate efforts, physicists have developed a system where the speed of light actually wouldn't be constant, as a means of explaining the horizon problem without the need for inflation. The earliest model of the *variable speed of light (VSL)* was proposed by John Moffat (who later incorporated the idea into his modified gravity theory), and a later model was developed by João Magueijo and Andreas Albrecht.

PROVING DARK MATTER WRONG?

In August 2008, a group of astrophysicists published a paper titled "A Direct Empirical Proof of the Existence of Dark Matter." The "proof" they speak of came from an impact between two galaxy clusters. Using NASA's Chandra X-Ray Observatory, they were able to see *gravitational lensing* (the gravity of the collision caused light to bend, kind of how light bends when it passes through a lens), which let them determine the center of the collision. The center of the collision did *not* match the center of the visible matter. In other words, the center of gravity and the center of visible matter didn't line up. That's pretty conclusive evidence for there being nonvisible matter, right?

In the world of theoretical physics, nothing is quite that easy these days. By September, physicist John Moffat and others were beginning to cast doubt on whether dark matter was the only explanation. Using his own *modified gravity (MOG)* theory, Moffat performed a calculation on a simplified 1-dimensional version of the collision.

Most physicists accept the NASA findings, including more recent findings from the Wilkinson Microwave Anisotropy Probe (WMAP) and other observations, as conclusive evidence that dark matter exists. But there remain those who are unconvinced and search for other explanations.

The horizon problem is based on the idea that distant regions of the universe can't communicate their temperatures because they are so far apart, light hasn't had time to get from one to the other. The solution proposed by inflation theory is that the regions were once much closer together, so they could communicate (see Chapter 9 for more on this).

REMEMBER

In VSL theories, another alternative is proposed: The two regions could communicate because light traveled faster in the past than it does now.

Moffat proposed his VSL model in 1992, allowing for the speed of light in the early universe to be very fast — about 100,000 trillion trillion times the current values. This would allow for all regions of the observable universe to easily communicate with each other.

To get this to work out, Moffat had to make a conjecture that the *Lorentz invariance* — the basic symmetry of special relativity — was somehow spontaneously broken in the early universe. Moffat's prediction results in a period of rapid heat transfer throughout the universe that gives rise to the same effects as an inflationary model.

In 1998, physicist João Magueijo came up with a similar theory, in collaboration with Andreas Albrecht. Their approach, developed without any knowledge of Moffat's work, was very similar — which they acknowledged upon learning of his conjecture. Magueijo and Albrecht's work was published a bit more prominently than Moffat's (largely because they were more stubborn about pursuing publication in the prestigious *Physical Review D*, which had rejected Moffat's earlier paper). Their work inspired others, such as the late Cambridge physicist John Barrow, to investigate this idea.

One hoped-for piece of support for VSL approaches included some research that was believed to have indicated that the fine-structure constant may not have always been, well, constant. The *fine-structure constant* is a ratio made up from Planck's constant, the charge on the electron, and the speed of light. It's a value that shows up in some physical equations. If the fine-structure constant has changed over time, then at least one of these values (and possibly more than one) has also been changing.

The spectral lines emitted by atoms are defined by Planck's constant. Scientists know from observations that these spectral lines haven't changed, so it's unlikely that Planck's constant has changed. (Thanks to John Moffat for clearing that up.) Still, any change in the fine-structure constant could be explained by varying either the speed of light or the electron charge (or both).

Although this evidence would have been a great boon to VSL theories had it come out in their favor, the evidence turned against them: More recent detailed studies show that the fine-structure constant appears to be *constant* over time, not changing. This means that one of the major reasons for interest in VSL approaches has been eliminated.

Nevertheless, the VSL model, though likely mistaken, was a worthwhile and rigorous form of science. It was built around a solid theoretical idea that made clear experimental predictions. Of course, once those experimental predictions fail to be observed in the real world, it's a good reason to believe that the theoretical underpinnings are false.

While it might seem like a VSL theory would be completely different from a string theory approach, physicists Elias Kiritsis and Stephon Alexander independently developed VSL models that could be incorporated into string theory, and Alexander later worked with Magueijo on refining these concepts (even though Magueijo is critical of string theory's lack of contact with experiment).

TIP

These proposals are intriguing, but the physics community in general remains committed to the inflation model. Both VSL and inflation require some strange behavior in the early moments of the universe, but it's unclear that inflation is inherently more realistic than VSL. It's possible that further evidence of varying constants will ultimately lead to support for VSL over inflation, but the best evidence to date gives us little reason to believe that the speed of light has changed over time.

MOG: The bigger the distance, the greater the gravity

John Moffat's work in alternative gravity has resulted in his *modified gravity (MOG) theories*, in which the force of gravity increases over distance, as well as the introduction of a new repulsive force at even larger distances. Moffat's MOG actually consists of three different theories that he developed over the span of three decades, trying to make them simpler, more elegant, and more accessible for other physicists to work on.

The work began in 1979, when Moffat developed *nonsymmetric gravitational theory (NGT)*, which extended work that Einstein tried to apply to create a unified field theory in the context of a non-Riemannian geometry. Einstein's work failed to unify gravity and electromagnetics like he wanted, but Moffat believed it could be used to generalize relativity itself.

Over the years, NGT ultimately proved inconclusive. It's possible that its predictions (such as the idea that the sun deviated from a perfectly spherical shape) were incorrect or that the deviation was too small to be observed.

In 2003, Moffat developed an alternative with the unwieldy name *Metric-Skew-Tensor Gravity (MSTG)*. It was a symmetric theory (easier to deal with) that included a "skew" field for the nonsymmetric part. This new field was in fact a fundamentally new force — a fifth fundamental force in the universe.

Unfortunately, MSTG remained too mathematically complicated in the eyes of many, so in 2004 Moffat developed *Scalar-Tensor-Vector Gravity (STVG)*. In STVG, he again had a fifth force resulting from a vector field, called a *phion field*. The phion particle was the gauge boson that carried the fifth force in the theory.

REMEMBER

According to Moffat, all three theories give essentially the same results for weak gravity fields, like those we normally observe. The strong gravitational fields needed to distinguish the theories are the ones that always give scientists problems and have motivated the search for quantum gravity theories in the first place. They can be found at the moment of the big bang or during the stellar collapses that may cause black holes.

There are indications that STVG yields results very similar to Milgrom's MOND theory (refer to the earlier section "MOND: Disregarding dark matter" for a fuller explanation of MOND). Moffat has proposed that MOG may explain dark matter and dark energy, and that black holes may not actually exist in nature.

TIP

While these implications are amazing, the work is still in the very preliminary stages, and it will likely be years before it (or any of the other theories) are developed enough to have any hope of seriously competing with the entrenched viewpoints.

Massive gravity and bimetric theory: making the graviton heavy

The *graviton* is the particle responsible for the gravitational interaction. If we look at Einstein's gravity, it appears that the graviton must not have any mass and must move at the speed of light, much like the photon. Over the years there have been several attempts to allow for the graviton to have a mass, but they all lead to mathematical inconsistencies.

Therefore, when, in 2011, Claudia de Rham, Gregory Gabadadze, and Andrew Tolley showed a way to make sense of a theory with a massive graviton, many theoretical physicists were thrilled. This mass would be very tiny, but it may be enough to change in a very fundamental way the properties of gravity (and the problem of quantizing it).

One conceptual drawback of the *massive gravity* theories is that they require a reference geometry, called a *reference metric* (the metric is what defines the geometry in general relativity). This is not desirable in gravity because we would like all geometry to come out naturally from the equations of the theory, as is the case in general relativity. For this reason, physicists have tried to treat the "true" metric and the reference metric on the same footing. The result is a *bimetric theory* with both a massive (heavy) and a massless (light) graviton. These models are rather new, and the jury is still out on whether they are compatible with experiments.

Rewriting the Math Books and Physics Books at the Same Time

Revolutions in physics have frequently had an assist from earlier revolutions in mathematics. One of the problems with string theory is that it advanced so quickly, the mathematical tools didn't actually exist. Physicists have been forced (with the aid of some brilliant mathematicians) to develop the tools as they go.

Einstein got help in developing general relativity from Riemannian geometry, developed years earlier, and frequently corresponded with mathematicians like Gregorio Ricci-Curbastro. Quantum physics was built on a framework of new mathematical representations of physical symmetries, group representation theory, developed by the mathematician Hermann Weyl.

In addition to developing the physics needed to address problems of quantum gravity, some physicists and mathematicians have tried to focus on developing whole new mathematical techniques. The question remains, though, how (and if) these techniques can be applied to the theoretical frameworks to get meaningful results.

Compute this: Quantum information theory

One technique that is growing in popularity as a means of looking at the universe is *quantum information theory,* which deals with all elements in the universe as pieces of information. This approach was originally proposed by John Archibald Wheeler with the phrase "It from bit," indicating that all matter in the universe can be viewed as essentially pieces of information. (A *bit* is a unit of information stored in a computer.)

Overall, this approach basically treats the universe as a giant computer — in fact, a universe-sized quantum computer. The major benefit of this system is that, for a computer scientist, it's easy to see how random information sent through a series of computations results in complexity growing over time. The complexity within our universe could thus arise from the universe performing logical operations — calculations, if you will — upon the pieces of information (be they loops of space-time or strings) within the universe.

In this picture, however, the universe isn't an ordinary computer like your laptop, but a *quantum computer*. A quantum computer operates at scales so tiny that it can exploit the weird properties of quantum mechanics in the way it performs operations. While you won't find a quantum computer at your favorite retailer, they are becoming a technological reality in many research centers, though their computing power is quite limited for now.

Many physicists are interested in looking at the universe using the rules of quantum computations and trying to exploit this point of view to uncover the weird behavior of gravity — black holes, the big bang, you name it. This approach has recently been energized by a large research network funded by former mathematician and current hedge fund manager, billionaire, and philanthropist Jim Simons. The collaboration, called "It from Qubit," twists Wheeler's phrase to emphasize the quantumness of it all.

Looking at relationships: Twistor theory

For decades, the brilliant mathematical physicist Sir Roger Penrose has been exploring his own mathematical approach — *twistor theory*. Penrose developed the theory out of a strong general relativity approach (the theory requires only four dimensions). He maintains a belief that any theory of quantum gravity will need to include fundamental revisions to the way physicists think about quantum mechanics, something with which most particle physicists and string theorists disagree.

One of the key aspects of twistor theory is that the relation between events in space-time is crucial. Instead of focusing on the events and their resulting relationships, twistor theory focuses on the causal relationships, and the events become by-products of those relationships.

All of the light rays in space-time create *twistor space*, which is the mathematical universe in which twistor theory resides. In fact, there are some indications that objects in twistor space may result in objects and events in our universe.

TIP

The major flaw of twistor theory is that even after all these years (it was originally developed in the 1960s), it still exists only in a world absent of quantum physics. The space-time of twistor theory is perfectly smooth, so it allows no discrete structure of space-time. It's a sort of anti-quantum gravity, which means it doesn't provide much more help than general relativity in resolving the issues that string theorists (or other quantum gravity researchers) are trying to solve.

As with string theory, Penrose's twistor theory has provided some mathematical insights into the existing theories of physics, including some that lie at the heart of the Standard Model of particle physics.

Edward Witten and other string theorists have begun to investigate ways that twistor theory may relate to string theory. One approach has been to have the strings exist, not in physical space, but in twistor space. So far, it hasn't yielded the relationships that would provide fundamental breakthroughs in either string theory or twistor theory, but it has resulted in great improvements of calculational techniques in quantum chromodynamics.

Uniting mathematical systems: Noncommutative geometry

Another mathematical tool being developed is the *noncommutative geometry* of French mathematician Alain Connes, a winner of the prestigious Fields Medal. This system involves treating geometry in a fundamentally new way, using mathematical systems where the commutative principle doesn't hold.

In mathematics, two quantities *commute* if operations on those quantities work the same way no matter in what order you treat them. Addition and multiplication are both commutative because you get the same answer no matter in what order you add two numbers or multiply them.

However, mathematicians are a diverse bunch, and some mathematical systems define addition and multiplication differently, so the order does matter. As weird as it sounds, in these systems multiplying 5 by 3 could give a different result than multiplying 3 by 5. (We don't recommend using this excuse to argue with a teacher over your scores on a math test.) It's probably not surprising to discover that these noncommutative mathematical systems come up frequently in the bizarre world of quantum mechanics — in fact, this feature is the mathematical cause of the uncertainty principle described in Chapter 7.

The tools of noncommutative geometry have been used in many approaches, but Connes seeks a more fundamental unification of algebra and geometry that can be used to build a physical model where the conflicts are resolved by features inherent in the mathematical system.

Noncommutative geometry has had some success, because the Standard Model of particle physics seems to pop out of it in the simplest versions. The goal of the committed mathematicians working with Connes is eventually to be able to replicate all of physics (including possibly string theory), though that's likely still a long way off. (Are you beginning to see a pattern here?)

Mathematics All the Way Down: Are We Living in a Simulation?

One last radical idea that seems worth mentioning, quite different from both string theory and its main alternatives, is the idea that the world we're directly experiencing isn't physical reality at all, but rather some sort of virtual simulation of reality. You're familiar with the general idea if you've ever seen *The Matrix* (or, sadly, any of the sequels).

This isn't just an idea out of science fiction, although it isn't an idea that comes particularly out of science either. The philosopher Nick Bostrom has made a plausible argument, based on mathematical probabilities, that if it's physically possible to simulate a consciousness, then there's fairly good reason to suspect that you might be a simulated consciousness yourself.

Though Bostrom's argument itself is framed a little differently, consider it from this perspective:

- Assume it's possible to create a simulation of society where the people living in the simulated world aren't aware that they're simulations.
- Assume that people in the "real" society have an interest in running such simulated worlds.
- If that's the case, then the probability that you're a simulation is the "Number of simulations" divided by the "Number of simulations + 1."

Now it's possible that the assumptions made in the first two points are invalid. Maybe it's impossible to make a simulation of society, for example. Or maybe it's possible, but people just don't have any interest in doing it. (But, really, have you *met* humanity? C'mon, we'd be running simulated worlds all day long if we could.) If either of those are the case, and the "Number of simulations" is zero, then the probability you're living in a simulation is zero.

But if we just created one simulation, then there are two worlds to choose from — the real world and the simulated world — and the random probability of which world you're in is one-half. And the number just goes up from there.

So if it's possible to create simulations, and those simulations all contain billions of simulated minds, then very quickly we can see that the probability that any random mind is a simulated mind is very close to one. In other words, unless you think that the probability of our two assumptions is extremely low, you have to at least give some credibility to the idea that you might be a simulated mind.

And if that's the case, then the physical reality we're observing is a simulated physical reality. Maybe the difficulties addressed throughout this book occur because the world literally isn't built at the level of detail where we can resolve questions about our universe. The inconsistencies could exist because there just isn't information at that scale.

Recall that in discussing the holographic principle (see Chapter 13), we refer to "pixels" when talking about the size of information, and that seems relevant here in our speculations about a computer simulation. There's a certain level of information about a computerized world where you can't drill down any deeper because you've reached the lowest level of resolution. The simulation may do fine for things like calculating planetary orbits, or flight paths, or thermodynamic interactions, but then it just kind of goes haywire in the realm of quantum gravity because no one programmed the simulation to deal with it.

Now we're certainly not saying we're living in a simulation. That isn't what Bostrom's saying either. But it's interesting that we can't definitively rule it out. And, as Bostrom's reasoning suggests, we have at least a mathematical reason to give it some serious credibility.

And be honest: Is it really any crazier than some of the other things we've had to consider in thinking about string theory and its alternatives?

6

The Part of Tens

IN THIS PART . . .

Explore ten tests for a theory of quantum gravity.

Discover ten concepts that a theory of everything should explain.

IN THIS CHAPTER

» Constructing a new theory of quantum gravity that physicists find plausible

» Basic (and not-so-basic) hurdles that a theory needs to clear

» Using the theory to reproduce known physics

Chapter 21
Ten Tests for a Theory of Quantum Gravity

As we discuss throughout this book, right now physicists have two different theories — quantum mechanics and general relativity — that do an excellent job of describing the physical behavior of systems within certain well-defined limitations. In most cases, this isn't a problem, because physicists can function under one set of those limitations and get results that are incredibly accurate. But in cases where the physical situation is on the boundary of those limitations, the physics gets a little less clear, and neither theory by itself can give the complete picture. The goal of a theory of quantum gravity is to resolve the conflict between quantum theory and the theory of relativity, and hopefully provide answers in these situations.

In this chapter, we narrow in on ten tests that any theory of quantum gravity (whether it be string theory or some other proposal) would have to meet for physicists to really consider it a plausible theory of quantum gravity that's worth further study.

Reproduce Gravity

The theory of general relativity is our current physical theory for explaining gravity, and as we note throughout this book, it works exceptionally well at that — except in the quantum realm, where we really have no way of talking about gravity. A theory of quantum gravity would need to explain gravity at the quantum level, but it would *also* need to be able to replicate all the nice features of general relativity, which works extraordinarily well in describing the behavior of planets and stars.

Compute Quantum Corrections

If our new theory of quantum gravity reproduces general relativity and recognizes the quantum nature of the universe, we should be able to use it to compute some quantum effects due to gravity.

For instance, if we start from a case where general relativity works (like the description of the solar system), the quantum effects should be a tiny correction on top of what we know. For something like a dense star, they should be bigger. The new theory should automatically identify these different scales of magnitude for the quantum effects and give us a way to compute them.

Describe How Gravity and Matter Interact

In general relativity, matter interacts with the physical structure of space-time: Space-time literally bends in response to the presence of physical matter. And while this description works pretty well in predicting the physical behavior of systems of matter, the fact is, general relativity doesn't really explain *why* it works or the exact mechanism that causes physical matter to bend space-time.

A theory of quantum gravity, if it's truly going to reconcile the problems that exist in our current models, would need to include a clear description of how the interaction between gravity and matter takes place.

Explain Inflation

The theory of inflation explains how the universe quickly expanded from the size it was at the big bang, spreading random fluctuations of energy to large scales in a few highly energized moments of existence. Inflation is broadly accepted in the field of cosmology, and ideally, we'd be able to use a theory of quantum gravity to understand, explain, and study the mechanism by which space-time within the early universe could inflate at the rates needed to support the theory.

It's possible, of course, that a successful theory of quantum gravity can provide another explanation for the flatness problem in cosmology that accounts for the observed data in some way other than the current inflationary model. This, too, would be a success for a theory of quantum gravity, particularly if there was some experimental way to test between the inflation model and whatever alternative is presented by the proposed theory.

Explain What Happens When Someone Enters a Black Hole

Black holes are very strange objects, and they are extremely important in any theory of quantum gravity. In fact, they are one of the few objects where gravitational interactions are so significant that quantum effects cannot be overlooked.

Black holes are characterized by an event horizon: a surface of no return, from which we cannot escape. However, general relativity predicts in no uncertain terms that someone falling into a black hole wouldn't even notice the existence of the horizon. Many physicists believe this is at odds with quantum mechanics, and it's one of the incarnations of the famous black hole information paradox.

Any theory of quantum gravity should be able to compute what happens to someone who falls into a black hole and settle this decades-old debate.

Explain Whether Singularities Are Allowed

General relativity predicts that space can become infinitely warped, though this can happen only in very special circumstances. One such case is the singularity at the center of a black hole. Another possible singularity is the big bang itself.

Physicists are uneasy about infinite quantities, and they tend to view them as mathematical abstractions. After all, how can we ever measure infinity? Many physicists believe that all these infinite-curvature singularities are artifacts of general relativity and that they shouldn't be present in quantum gravity. Any theory of quantum gravity should make sense of singularities and explain whether they can actually exist in nature.

Explain the Birth and Death of Black Holes

We did say that black holes are important, didn't we? Physicists believe that black holes can form when certain stars collapse, and we're pretty sure they exist even in our own galaxy. Thanks to Stephen Hawking, we believe that over an incredibly long period of time, they will all evaporate out of existence.

Hawking came to this conclusion by trying to estimate quantum corrections of classical black holes without having a theory of quantum gravity at his disposal. His work is, however, an approximation.

A true theory of quantum gravity should allow us to follow the birth and death of a black hole. This is important because physicists cannot agree on precisely how this would happen and how it would be compatible with the rules of quantum physics. (Again, this is the black hole information paradox.)

Explain the Holographic Principle

Over the last couple of decades, substantial work in string theory has been devoted to testing different applications of the holographic principle, showing that it provides useful insights for comparing different theoretical frameworks.

The most concrete examples of holography come from string theory, but they look far removed from the real world: They occur in a strange number of dimensions, involve lots of supersymmetry, and make use of the anti-de Sitter geometry (see Chapter 13), which doesn't really show up in our universe.

Yet many physicists believe that holography is more general than that. Therefore, a holographic equivalent of our universe may exist! A complete theory of quantum gravity should explain what that is.

Provide Testable Predictions

At this point, it's (hopefully) pretty clear that one of the things we'd definitely like from any theory of quantum gravity is to be able to test it scientifically — not just with mathematical models, but with actual physical experiments.

For that to happen in a way that would be most satisfying to physicists, the theory has to be able to make clear, well-determined, testable predictions that we would expect to have different results if the theory were untrue. That would be really great.

Describe Its Own Limitations

Every physical theory contains certain limitations on how it can be applied. Consider the two big theories that exist in physics: general relativity and quantum theory.

The theory of general relativity describes the behavior of very large systems, like solar systems, extremely well, but by its nature, it's limited to describing the *gravitational* behavior of those systems. That's what, by definition, the theory of general relativity does.

Quantum theory, on the other hand, describes the behavior of much smaller systems, and over the last century, it has been refined to cover the strong nuclear force, weak nuclear force, and electromagnetic force. But it says nothing about gravity.

And more important, these theories make it clear to which domains they do and do not apply. If you understand the theories, then you understand their limitations.

A theory of quantum gravity would have different limitations, of course, but it would still have limitations. A successful theory should describe those limitations, so physicists aren't left with confusion over when the theory of quantum gravity applies or which types of approaches are outside the theory.

Index

C

D

E

F

G

H

I

J

K

L

M

N

O

P

Q

R

S

T

U

V

W

Y

Z

About the Authors

Andrew Zimmerman Jones is a science communicator and educator. He holds a bachelor's degree in physics from Wabash College, where he also studied mathematics and philosophy, and a master's degree in mathematical education from Purdue University. Andrew has communicated science and mathematics concepts across many publications and platforms, including About.com Physics, Discovery Education, McGraw Hill Education, and PBS's *NOVA*. He has also written science fiction and fantasy short stories, as well as philosophy essays that bridge the realms of science and philosophy in publications such as *The Big Bang Theory and Philosophy* and *The Avengers and Philosophy*. He currently spends his days working in the Office of Student Assessment at the Indiana Department of Education. He has been a member of Mensa since the eighth grade and has been intensely interested in both science and science fiction since even earlier. Along the way, he's also become a member of the National Association of Science Writers, an Eagle Scout, a Master Mason in the Freemasons, and won the Harold Q. Fuller Prize in Physics at Wabash College. His plan for world domination nears completion with the publication of this book's second edition.

Andrew lives in central Indiana with his beautiful wife, Amber, and sons, Elijah and Gideon. He is currently working on the design of a science card game. More about his efforts can be found at `www.azjones.info`.

Alessandro Sfondrini is a theoretical physicist. He obtained his BSc and MSc in physics with honors and the Galilean School of Higher Education degree with honors from the University of Padua. As a student, he also spent time at the Max Planck Institute for Quantum Optics in Garching (Germany), where he worked with Ignacio Cirac and Norbert Schuch, as well as at the Perimeter Institute for Theoretical Physics in Waterloo (Ontario) with Razvan Gurau and Roberto Percacci.

Alessandro obtained his PhD *cum laude* from Utrecht University (Netherlands) under the supervision of Gleb Arutyunov, with a thesis on the AdS/CFT correspondence in string theory. After that, he was employed as a Marie Curie Fellow at Humboldt University in Berlin (Germany) in the group of Matthias Staudacher, and subsequently as an *assistent* and *oberassistent* at the Swiss Federal Institute of Technology (ETH Zurich, Switzerland) in the group of Matthias Gaberdiel.

Alessandro is currently a Rita Levi-Montalcini Fellow and assistant professor in theoretical physics and mathematical methods and models at the University of Padua (Italy), as well as a member and IBM Einstein Fellow at the Institute for Advanced Study in Princeton (New Jersey).

Alessandro's passions include good books, good food, and good wine as well as, of course, studying and teaching mathematics and physics. He has collaborated on several outreach projects involving the Swiss National Science Foundation, ETH Zurich, the University of Padua, and YouTube channels *CGP Grey* and *Kurzgesagt — In a Nutshell*.

Dedication

To our families.

Authors' Acknowledgments

Andrew must first profoundly thank his former agent, Barb Doyen, for initially approaching him with this project.

He would also like to give thanks to his wife, Amber, and sons, Elijah and Gideon, for putting up with him, even when he was driven frantic by tight deadlines and medical challenges.

Alessandro would like to thank Kelsey Baird for providing him the exciting opportunity to contribute to the revision of this book, as well as the School of Natural Sciences at the Institute for Advanced Study, where part of his work on this book was completed. He also would like to thank his colleagues at the Institute and at the University of Padua for many helpful discussions on how to best present some of the material in this book. Finally, he thanks all those colleagues who, through lectures, collaborations, and discussions, have taught him what he knows about mathematics and physics.

Our deepest thanks and appreciation go out to the wonderful editorial staff at Wiley for their work on both the first edition and this new second edition. We also appreciate the constructive, and at times critical, input of Lorenz Eberhardt at the Institute for Advanced Study in Princeton, who provided technical editing on the book.

Publisher's Acknowledgments

Acquisitions Editor: Kelsey Baird

Project Manager and Development Editor: Thomas Hill

Managing Editor: Kristie Pyles

Production Editor: Tamilmani Varadharaj

Cover Image: © vchal/Shutterstock